Concise Metals Engineering Data Book

Editor
Joseph R. Davis

ASM INTERNATIONAL

The Materials Information Society

Copyright © 1997
by
ASM International®
All rights reserved

No part of this book may be reproduced, stored in a retrieval system, or transmitted, in any form or by any means, electronic, mechanical, photocopying, recording, or otherwise, without the written permission of the copyright owner.

First printing, December 1997
Second printing, August 2007
Third printing, April 2008
Fourth printing, March 2009
Fifth printing, March 2010

Great care is taken in the compilation and production of this book, but it should be made clear that NO WARRANTIES, EXPRESS OR IMPLIED, INCLUDING, WITHOUT LIMITATION, WARRANTIES OF MERCHANTABILITY OR FITNESS FOR A PARTICULAR PURPOSE, ARE GIVEN IN CONNECTION WITH THIS PUBLICATION. Although this information is believed to be accurate by ASM, ASM cannot guarantee that favorable results will be obtained from the use of this publication alone. This publication is intended for use by persons having technical skill, at their sole discretion and risk. Since the conditions of product or material use are outside of ASM's control, ASM assumes no liability or obligation in connection with any use of this information. No claim of any kind, whether as to products or information in this publication, and whether or not based on negligence, shall be greater in amount than the purchase price of this product or publication in respect of which damages are claimed. THE REMEDY HEREBY PROVIDED SHALL BE THE EXCLUSIVE AND SOLE REMEDY OF BUYER, AND IN NO EVENT SHALL EITHER PARTY BE LIABLE FOR SPECIAL, INDIRECT OR CONSEQUENTIAL DAMAGES WHETHER OR NOT CAUSED BY OR RESULTING FROM THE NEGLIGENCE OF SUCH PARTY. As with any material, evaluation of the material under end-use conditions prior to specification is essential. Therefore, specific testing under actual conditions is recommended.

Nothing contained in this book shall be construed as a grant of any right of manufacture, sale, use, or reproduction, in connection with any method, process, apparatus, product, composition, or system, whether or not covered by letters patent, copyright, or trademark, and nothing contained in this book shall be construed as a defense against any alleged infringement of letters patent, copyright, or trademark, or as a defense against liability for such infringement.

Comments, criticisms, and suggestions are invited, and should be forwarded to ASM International.

Library of Congress Catalog Card Number: 97-75187

ISBN-13: 978-0-87170-606-5   ISBN-10: 0-87170-606-7
SAN: 204-7586

ASM International®
Materials Park, OH 44073-0002
www.asminternational.org

Printed in the United States of America

Foreword

Established in 1952 through a grant from ASM International (then known as the American Society for Metals), the ASM Foundation for Education and Research provides for the advancement of scientific and engineering knowledge through its support of education and research. Foundation activities, which are supported by individual contributions, encourage the development of capable, well-educated materials specialists.

In 1995, the Foundation Board approved sponsorship of a metals reference book to provide a quick look-up resource for college students and to introduce students to ASM as an important source of materials information. This concept has evolved into the present volume—a valuable resource for students as well as engineering professionals and others who need reliable, practical information about the properties and applications of metals. By bringing ideas and people together through educational programs and ASM products, the Foundation continues to encourage education in the materials field.

William E. Quist
President 1997
ASM Foundation for Education and Research

Preface

The purpose of the *Concise Metals Engineering Data Book* is to provide students, salespeople, purchasing agents, and metallurgists with the type of *practical* information that they require on a daily basis. It was designed with an easy-to-use format to allow quick retrieval of data from a wide range of metals-related subjects. Chemical compositions, physical properties, and mechanical properties are listed from hundreds of metals and alloys (see, for example, Chapters 7, 8, and 9). Included are useful tables that compare and rank the density, melting point, and elastic modulus of the various metallic elements used in structural alloys.

Some chapters serve as introductions to important metallurgical subjects. For example, what are the three crystal structures associated with almost all metals, and how do crystal defects influence plastic deformation? What is hardenability? How does it differ from hardness, and how is hardenability determined or measured? The answers to these questions can be found in Chapters 5 and 11, respectively.

Still other chapters serve as a guide for further study or information gathering. The last chapter provides an extensive bibliography divided into key subject areas: materials properties and selection, failure analysis, corrosion, mechanical testing, etc.

I wish to acknowledge the following individuals who assisted me in developing the outline for the book: Sunneva Collins, Napro Company; Monte Pool, University of Cincinnati; Dick Connell, University of Florida; and William W. Scott, Jr., ASM International. Some of these individuals were involved in a Student/Faculty Focus Group organized by ASM. I would also like to express my thanks to Veronica Flint of the ASM editorial staff for her support and encouragement during the project.

Joseph R. Davis

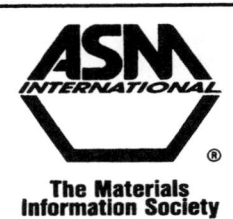

ASM International® is a Society whose mission is to gather, process and disseminate technical information. ASM fosters the understanding and application of engineered materials and their research, design, reliable manufacture, use and economic and social benefits. This is accomplished via a unique global information-sharing network of interaction among members in forums and meetings, education programs, and through publications and electronic media.

Contents

Forward iii
Preface v
1 Symbols for the Elements and the Periodic Table 1
    Table 1.1   Symbols and atomic numbers for the chemical elements 1
    Table 1.2   Important metallic element groups 3
2 Physical Properties of the Elements 4
    Table 2.1   Density rankings (lightest to heaviest) of metallic elements that
                serve as the base element for structural alloys 4
    Table 2.2   Melting point rankings (highest to lowest) of metallic elements
                that serve as the base element for structural alloys 5
    Table 2.3   Elastic modulus rankings (highest to lowest) of metallic elements
                that serve as the base element for structural alloys 5
    Table 2.4   Physical properties of the elements 6
3 Vapor Pressures of the Elements 14
    Table 3.1   Vapor pressures of the elements up to 1 atm (760 mm Hg) 14
4 Physical Properties of Gases and Liquids 16
    Table 4.1   Physical properties of common gases and liquids 16
    Table 4.2   Physical properties of common inorganic and organic acids 17
5 Crystal Structures 18
6 Phase Diagrams 29
7 Chemical Compositions of Metals and Alloys 48
    Table 7.1   Guide to the Unified Numbering System (UNS) for metals and alloys ... 48
    Table 7.2   SAE-AISI system of designations for carbon and alloy steels 49
    Table 7.3   Carbon steel compositions 50
    Table 7.4   Carbon steel compositions 52
    Table 7.5   Free-cutting (resulfurized) carbon steel compositions 53
    Table 7.6   Free-cutting (rephosphorized and resulfurized) carbon
                steel compositions 53
    Table 7.7   High-manganese carbon steel compositions 54
    Table 7.8   High-manganese carbon steel compositions 54
    Table 7.9   Low-alloy steel compositions applicable to billets, blooms, slabs, and
                hot-rolled and cold-finished bars 55
    Table 7.10   Composition ranges and limits for AISI-SAE standard low-alloy steel
                plate applicable for structural applications 58
    Table 7.11   Composition of carbon and carbon-boron H-steels 59
    Table 7.12   Composition of standard alloy H-steels 59
    Table 7.13   Composition limits of principal types of tool steels 61
    Table 7.14   Composition of selected standard and special stainless steels 64

Table 7.15	Compositions of maraging and high fracture toughness steels	67
Table 7.16	Four-digit numerical system used to identify wrought aluminum and aluminum alloys	67
Table 7.17	Four-digit numerical system used to identify cast aluminum and aluminum alloys	68
Table 7.18	Designations and nominal compositions of common wrought aluminum and aluminum alloys	68
Table 7.19	Designations and nominal compositions of common aluminum alloys used for casting	70
Table 7.20	Generic classification of copper alloys	71
Table 7.21	Nominal compositions of wrought copper and copper alloys	71
Table 7.22	Nominal compositions of cast copper and copper alloys	75
Table 7.23	Nominal compositions of common zinc alloy die castings and zinc alloy ingot for die casting	77
Table 7.24	Nominal compositions of zinc-aluminum foundry and die casting alloys directly poured to produce castings and in ingot form for remelting to produce castings	78
Table 7.25	Nominal compositions of zinc casting alloys used for sheet metal forming dies and for slush casting alloys in ingot form	78
Table 7.26	Nominal compositions of rolled zinc alloys per ASTM B 69	79
Table 7.27	Nominal compositions of magnesium casting alloys	79
Table 7.28	Nominal compositions of wrought magnesium alloys	80
Table 7.29	UNS categories and nominal compositions of various lead grades and lead-base alloys	80
Table 7.30	Chemical compositions of common titanium and titanium alloys	83
Table 7.31	Compositions of selected nickel and nickel-base alloys	84
Table 7.32	Nominal compositions of various cobalt-base alloys	86

8 Physical Properties of Metals and Alloys ... 87
Table 8.1	Density of metals and alloys	87
Table 8.2	Linear thermal expansion of metals and alloys	92
Table 8.3	Thermal conductivity of metals and alloys	95
Table 8.4	Electrical conductivity and resistivity of metals and alloys	97
Table 8.5	Approximate melting temperatures of metals and alloys	101

9 Mechanical Properties of Metals and Alloys ... 104
Table 9.1	Mechanical properties of selected carbon and alloy steels in the hot-rolled, normalized, and annealed condition	104
Table 9.2	Mechanical properties of selected carbon and alloy steels in the quenched-and-tempered condition	106
Table 9.3	Mechanical property data for stainless steels	109
Table 9.4	Typical mechanical properties of commonly used wrought aluminum alloys	112
Table 9.5	Typical tensile properties for separately cast test bars of common aluminum casting alloys	113
Table 9.6	Mechanical properties of wrought copper and copper alloys	114
Table 9.7	Mechanical properties of cast copper and copper alloys	117
Table 9.8	Typcial mechanical properties of zinc alloy die castings	120
Table 9.9	Minimum mechanical properties for magnesium alloys	120
Table 9.10	Tensile properties of common titanium and titanium alloys	121
Table 9.11	Mechanical properties of selected nickel-base alloys	122
Table 9.12	Mechanical properties of selected cobalt-base alloys	124

Table 9.13	Elastic constants for polycrystalline metals at 20 °C	124
10	**Mechanical Properties Charts for Steels**	**125**
11	**Hardenability Data for Steels**	**134**
12	**Hardness Conversion Tables**	**143**
Table 12.1	Approximate equivalent hardness numbers for nonaustenitic steels (Rockwell C hardness range)	144
Table 12.2	Approximate equivalent hardness numbers for nonaustenitic steels (Rockwell B hardness range)	145
Table 12.3	Approximate equivalent hardness numbers for austenitic stainless steel sheet (Rockwell C hardness range)	146
Table 12.4	Approximate equivalent hardness numbers for austenitic stainless steel sheet (Rockwell B hardness range)	147
Table 12.5	Approximate Brinell-Rockwell B hardness numbers for equivalent austenitic stainless steel plate in the annealed condition	148
Table 12.6	Approximate equivalent hardness numbers of alloyed white irons	149
Table 12.7	Approximate equivalent hardness numbers for wrought aluminum products	149
Table 12.8	Approximate equivalent hardness numbers for wrought coppers (>99% Cu, alloys C10200 through C14200)	150
Table 12.9	Approximate equivalent hardness numbers for cartridge brass (70% Cu, 30% Zn)	152
Table 12.10	Approximate equivalent hardness numbers for nickel and high-nickel alloy	154
13	**Corrosion Data**	**156**
Table 13.1	Relationships among some of the units commonly used for corrosion rates	156
Table 13.2	Corrosion rate calculation (from mass loss)	156
Table 13.3	Reference potentials and conversion factors	158
Table 13.4	Electromotive force (emf) series	158
Table 13.5	Tabular version (no specific potential values given) of the galvanic series in seawater at 25 °C (77 °F)	160
Table 13.6	Chemical resistance of cast iron to various environments	160
Table 13.7	Corrosion resistance of carbon steel to various environments	164
Table 13.8	Corrosion of structural steels in various atmospheric environments	166
Table 13.9	Relative corrosion resistance of standard (AISI) stainless steels	167
Table 13.10	Relative corrosion resistance of standard stainless steel grades for different environments	168
Table 13.11	Relative ratings of resistance to general corrosion and to SCC of wrought aluminum alloys	169
Table 13.12	Relative ratings of resistance to general corrosion and to SCC of cast aluminum alloys	170
Table 13.13	Relative SCC ratings for wrought products of high-strength aluminum alloys	171
Table 13.14	Weathering data for 0.89mm (0.035 in.) thick aluminum alloy sheet after 20-year exposure	173
Table 13.15	Corrosion ratings of wrought copper alloys in various corrosive media	175
Table 13.16	Corrosion rating of coast copper alloys in various media	183
Table 13.17	Atmospheric corrosion of selected copper alloys	188

14 Coefficients of Friction 189
Table 14.1 Friction coefficient data for metals sliding on metals 190
Table 14.2 Friction coefficient data for ceramics sliding on various materials 192
Table 14.3 Friction coefficient data for polymers sliding on various materials 193
Table 14.4 Friction coefficient data for coatings sliding on various materials 195
Table 14.5 Friction coefficient data for miscellaneous materials 197

15 Engineering/Scientific Constants 200
Table 15.1 Fundamental physical constants 200

16 Metric Practice Guide 201
Table 16.1 SI prefixes—names and symbols 201
Table 16.2 Base, supplementary, and derived SI units 202
Table 16.3 Conversion factors classified according to the quantity/property of interest 204
Table 16.4 Alphabetical listing of common conversion factors 205

17 Sheet Metal and Wire Gages 211
Table 17.1 Sheet metal gage thickness conversions 211
Table 17.2 Wire gage diameter conversions 211

18 Pipe Dimensions 213
Table 18.1 Dimensions of welded and seamless pipe manufactured in the United States 213
Table 18.2 Dimensions of welded and seamless pipe manufactured in the United States 214

19 Glossary of Abbreviations, Acronyms, and Symbols 215
Table 19.1 Common abbreviations, acronyms, and symbols found in the materials science literature 215
Table 19.2 Mathematical signs and symbols 228
Table 19.3 Greek alphabet 228

20 Directory of Standards Organizations and Technical Associations 229
Table 20.1 Technical associations and standards organizations located in North America 229
Table 20.2 Selected international standards organizations arranged according to country/region of origin 232

21 Bibliography of Selected References 238

1 Symbols for the Elements and the Periodic Table

Table 1.1 Symbols and atomic numbers for the chemical elements

Name	Symbol	Atomic No.	Name	Symbol	Atomic No.	Name	Symbol	Atomic No.
Actinium	Ac	89	Hafnium	Hf	72	Praseodymium	Pr	59
Aluminum	Al	13	Hahnium	Ha	105	Promethium	Pm	61
Americium	Am	95	Helium	He	2	Protactinium	Pa	91
Antimony	Sb	51	Holmium	Ho	67	Radium	Ra	88
Argon	Ar	18	Hydrogen	H	1	Radon	Rn	86
Arsenic	As	33	Indium	In	49	Rhenium	Re	75
Astatine	At	85	Iodine	I	53	Rhodium	Rh	45
Barium	Ba	56	Iridium	Ir	77	Rubidium	Rb	37
Berkelium	Bk	97	Iron	Fe	26	Ruthenium	Ru	44
Beryllium	Be	4	Krypton	Kr	36	Rutherfordium	Rf	104
Bismuth	Bi	83	Lanthanum	La	57	Samarium	Sm	62
Boron	B	5	Lawrencium	Lr (Lw)	103	Scandium	Sc	21
Bromine	Br	35	Lead	Pb	82	Selenium	Se	34
Cadmium	Cd	48	Lithium	Li	3	Silicon	Si	14
Calcium	Ca	20	Lutetium	Lu	71	Silver	Ag	47
Californium	Cf	98	Magnesium	Mg	12	Sodium	Na	11
Carbon	C	6	Manganese	Mn	25	Strontium	Sr	38
Cerium	Ce	58	Mendelevium	Md	101	Sulfur	S	16
Cesium	Cs	55	Mercury	Hg	80	Tantalum	Ta	73
Chlorine	Cl	17	Molybdenum	Mo	42	Technetium	Tc	43
Chromium	Cr	24	Neodymium	Nd	60	Tellurium	Te	52
Cobalt	Co	27	Neon	Ne	10	Terbium	Tb	65
Copper	Cu	29	Neptunium	Np	93	Thallium	Tl	81
Curium	Cm	96	Nickel	Ni	28	Thorium	Th	90
Dysprosium	Dy	66	Niobium	Nb	41	Thulium	Tm	69
Einsteinium	Es	99	Nitrogen	N	7	Tin	Sn	50
Erbium	Er	68	Nobelium	No	102	Titanium	Ti	22
Europium	Eu	63	Osmium	Os	76	Tungsten	W	74
Fermium	Fm	100	Oxygen	O	8	Uranium	U	92
Fluorine	F	9	Palladium	Pd	46	Vanadium	V	23
Francium	Fr	87	Phosphorus	P	15	Xenon	Xe	54
Gadolinium	Gd	64	Platinum	Pt	78	Ytterbium	Yb	70
Gallium	Ga	31	Plutonium	Pu	94	Yttrium	Y	39
Germanium	Ge	32	Polonium	Po	84	Zinc	Zn	30
Gold	Au	79	Potassium	K	19	Zirconium	Zr	40

Note: Elements 106 and 107 have been reported, but no official names or symbols have yet been assigned.

2 Concise Metals Engineering Data Book

Fig. 1.1 Periodic table of the elements. Numbers in parentheses are mass numbers of the most stable isotope of that element. (continued)

Symbols for the Elements 3

	58 +3 Ce +4	59 +3 Pr	60 +3 Nd	61 +3 Pm	62 +2 Sm +3	63 +2 Eu +3	64 +3 Gd	65 +3 Tb	66 +3 Dy	67 +3 Ho	68 +3 Er	69 +3 Tm	70 +2 Yb +3	71 +3 Lu	
*Lanthanides	140.12 -20-8-2	140.9077 -21-8-2	144.24 -22-8-2	147 -23-8-2	150.4 -24-8-2	151.96 -25-8-2	157.25 -25-9-2	158.925 -27-8-2	162.50 -28-8-2	164.9304 -29-8-2	167.26 -30-8-2	168.9342 -31-8-2	173.04 -32-8-2	174.967 -32-9-2	-N-O-P
	90 +4 Th	91 +5 Pa +4	92 +3 U +4 +5 +6	93 +3 Np +4 +5 +6	94 +3 Pu +4 +5 +6	95 +3 Am +4 +5 +6	96 +3 Cm	97 +3 Bk +4	98 +3 Cf	99 +3 Es	100 +3 Fm	101 +2 Md +3	102 +2 No +3	103 +3 Lr	
**Actinides	232.038 -18-10-2	231.0359 -20-9-2	238.029 21-9-2	237.0482 -22-9-2	239.052 -24-8-2	(243) -25-8-2	(247) -25-9-2	(247) -27-8-2	(251) -28-8-2	(254) -29-8-2	(257) -30-8-2	(258) -31-8-2	(259) -32-8-2	(260) -32-9-2	-O-P-Q

Numbers in parentheses are mass numbers of most stable isotope of that element.

Fig. 1.1 (continued) Numbers in parentheses are mass numbers of the most stable isotope of that element.

Table 1.2 Important metallic element groups

Group	Definition
Rare earth metals	The rare earth metals include Group IIIA elements scandium, yttrium, and the lanthanide elements (lanthanum, cerium, praseodymium, neodymium, promethium, samarium, europium, gadolinium, terbium, dysprosium, holmium, erbium, thulium, ytterbium, and lutetium) in the periodic table of the elements. A mixture of rare earth elements is referred to as mischmetal, which typically contains 50% cerium with the remainder being principally lanthanum and neodymium.
Refractory metals	The refractory metals group includes niobium (previously known as columbium), tantalum, molybdenum, tungsten, and rhenium. The name of this group arises from their high melting temperature, which range from 2468 to 3410 °C (4474 to 6170 °F).
Precious metals	The eight precious metals, listed in order of their atomic number as found in periods 5 and 6 (Groups VIII and Ib) of the periodic table of the elements, are ruthenium, rhodium, palladium, silver, osmium, iridium, platinum, and gold.
Platinum-group metals	The platinum-group metals include the precious metals ruthenium, rhodium, palladium, osmium, iridium, and platinum. They are so named because they are closely related (in terms of properties) and commonly occur together in nature.

2 Physical Properties of the Elements

Table 2.1 Density rankings (lightest to heaviest) of metallic elements that serve as the base element for structural alloys

See Table 2.4 for density values for these and other elements.

Element	Room-temperature density, g/cm^3
Magnesium (Mg)	1.74
Aluminum (Al)	2.70
Titanium (Ti)	4.51
Zirconium (Zr)	6.51
Zinc (Zn)	7.13
Tin (Sn)	7.30
Iron (Fe)	7.87
Niobium (Nb)	8.57
Cobalt (Co)	8.83
Nickel (Ni)	8.90
Copper (Cu)	8.93
Molybdenum (Mo)	10.22
Lead (Pb)	11.34
Tantalum (Ta)	16.60
Tungsten (W)	19.25

Physical Properties of the Elements 5

Table 2.2 Melting point rankings (highest to lowest) of metallic elements that serve as the base element for structural alloys

See Table 2.4 for melting point values for these and other elements.

Element	Melting point °C	Melting point °F
Tungsten (W)	3422	6192
Tantalum (Ta)	3020	5468
Molybdenum (Mo)	2623	4753
Niobium (Nb)	2469	4476
Zirconium (Zr)	1855	3371
Titanium (Ti)	1670	3038
Iron (Fe)	1538	2800
Cobalt (Co)	1495	2723
Nickel (Ni)	1455	2651
Copper (Cu)	1085	1985
Aluminum (Al)	660	1220
Magnesium (Mg)	650	1202
Zinc (Zn)	420	788
Lead (Pb)	327.5	621.5
Tin (Sn)	232	450

Source: *Alloy Phase Diagrams*, Vol 3, *ASM Handbook*, ASM International, 1992, p 4–5 to 4–6

Table 2.3 Elastic modulus rankings (highest to lowest) of metallic elements that serve as the base element for structural alloys

See Table 4.4 for elastic modulus values for these and other elements.

Element	Modulus of elasticity in tension GPa	Modulus of elasticity in tension 10^6 psi
Tungsten (W)	345	50
Molybdenum (Mo)	324	47
Cobalt (Co)	211	30.6
Iron (Fe)	208.2	30.2
Nickel (Ni)	207	30
Tantalum (Ta)	186	26.9
Copper (Cu)	128	18.6
Titanium (Ti)	115.8	16.8
Niobium (Nb)	103	14.9
Zirconium (Zr)	99.3	14.4
Zinc (Zn)	(a)	(a)
Aluminum (Al)	62	8.99
Magnesium (Mg)	44	6.38
Tin (Sn)	42.9	6.3
Lead (Pb)	13.8	2

(a) Pure zinc has no clearly defined modulus of elasticity. Values range from about 69 to 138 GPa (10 to 20 psi × 10^6).

6 Concise Metals Engineering Data Book

Table 2.4 Physical properties of the elements

Element	Atomic No.	Atomic weight	Density(a), g/cm³ (lb/in.³)	Melting point, °C (°F)	Boiling point, °C (°F)	Specific heat(b), cal/g · °C (J/kg · K)	Heat of fusion, cal/g (Btu/lb)
Actinium (Ac)	89	227	...	1050 ± 50 (1920 ± 90)	...	...	...
Aluminum (Al)	13	26.98	2.70 (0.0974)	660 (1220)	2450 (4442)	0.215 (900)	94.5 (170)
Americium (Am)	95	243	11.87 (0.4285)	...	...	...	...
Antimony (Sb)	51	121.76	6.65 (0.240)	630.5 ± 0.1 (1166.9 ± 0.2)	1380 (2516)	0.049 (205)	38.3 (68.9)
Argon (A)	18	39.99	1.784 (0.06440)(g)	−189.4 ± 0.2 (−308.9 ± 0.4)	−185.8 (−302.4)	0.125 (523)	6.7 (12)
Arsenic (As)	33	74.91	5.72 (0.206)	817 (1503)(j)	613 (1135)(k)	0.082 (343)	88.5 (159.3)
Astatine (At)	85	211	...	302 (576)(m)	...	...	...
Barium (Ba)	56	137.36	3.6 (0.13)	714 (1317)	1640 (2980)	0.068 (285)	...
Berkelium (Bk)	97	247	...	...	...	...	...
Beryllium (Be)	4	9.01	1.85 (0.0668)	1277 (2332)	2770 (5020)	0.45 (190)	260 (470)
Bismuth (Bi)	83	209.00	9.80 (0.354)	271.3 (520.3)	1560 (2840)	0.0294 (123)	12.5 (22.5)
Boron (B)	5	10.82	2.45 (0.0884)	2030 (3690)(q)	...	0.309 (1290)	...
Bromine (Br)	35	79.92	3.12 (0.113)	−7.2 ± 0.2 (19.0 ± 0.4)	58 (136)	0.070 (290)	16.2 (29.2)
Cadmium (Cd)	48	112.41	8.65 (0.312)	320.9 (609.6)	765 (1409)	0.055 (230)	13.2 (23.8)
Calcium (Ca)	20	40.08	1.55 (0.0560)	838 (1540)	1440 (2625)	0.149 (624)(u)	52 (93.6)
Californium (Cf)	98	251	...	...	...	...	...
Carbon, graphite (C)	6	12.01	2.25 (0.0812)	3727 (6740)(k)	4830 (8730)	0.165 (691)	...
Cerium (Ce)	58	140.13	6.77 (0.244)	804 (1479)	3470 (6280)	0.045 (190)	8.5 (15.9)
Cesium (Cs)	55	132.91	1.87 (0.0675)	28.7 (83.6)	690 (1273)	0.04817 (201.7)	3.8 (6.8)
Chlorine (Cl)	17	35.46	3.214 (0.1160)(g)	−100.99 (−149.78)	−34.7 (−30.5)	0.116 (486)	21.6 (38.9)
Chromium (Cr)	24	52.01	7.19 (0.260)	1875 (3407)	2665 (4829)	0.11 (460)	96 (173)
Cobalt (Co)	27	58.94	8.85 (0.319)	1495 ± 1 (2723 ± 1.8)	2900 (5250)	0.099 (410)	58.4 (105)
Copper (Cu)	29	63.54	8.96 (0.323)	1083.0 ± 0.1 (1981.4 ± 0.18)	2595 (4703)	0.092 (380)	50.6 (91.1)
Curium (Cm)	96	247	7 (0.3)	...	...	...	...
Dysprosium (Dy)	66	162.51	8.55 (0.309)	1407 (2565)	2330 (4230)	0.041 (170)	25.2 (45.4)
Einsteinium (E)	99	254	...	...	...	...	...
Erbium (Er)	68	167.27	9.15 (0.330)	1497 (2727)	2630 (4770)	0.040 (170)	24.5 (44.1)
Europium (Eu)	63	152.0	5.24 (0.189)	826 (1519)	1490 (2710)	0.039 (160)	16.5 (29.6)

(continued)

(a) Density may depend considerably on previous treatment. (b) At 20 °C (68 °F). (g) Gas, grams per liter at 20 °C (68 °F) and 760 mm (30 in.). (j) 28 atm. (k) Sublimes. (m) Estimated. (q) Approximate. (u) From 0 to 100 °C (32 to 212 °F).

Physical Properties of the Elements

Table 2.4 (continued)

Symbol	Coefficient of linear thermal expansion(c), μin./in. °C (μin./in. °F)	Thermal conductivity(c), cal/cm²/cm/s/°C	Electrical resistivity, μΩ·cm	Modulus of elasticity in tension, 10⁶ psi	Lattice parameters(b), Å a	b	c (or axial angle)	Closest approach of atoms
Ac	...	...	...	...	...	...	...	...
Al	23.6 (13.1)(d)	0.53	2.6548(b)	9	4.0491	...	...	2.862
Am	...	...	...	...	...	...	...	...
Sb	8.5–10.8 (4.7–6)(e)	0.045	39.0(f)	11.3	4.5065	...	57° 6.5'	2.904
A	...	0.406 × 10⁻⁴	...	...	5.43(h)	...	...	3.84
As	4.7 (2.6)	...	33.3(b)	...	4.159	...	53° 49'	...
At	...	...	...	...	...	...	...	...
Ba	...	...	...	...	5.025	...	...	4.348
Bk	...	...	...	...	...	...	...	...
Be	11.6 (6.4)(n)	0.35	4(b)(p)	40–44	2.2858	...	3.5842	...
Bi	13.3 (7.4)	0.020	106.8(f)	4.6	4.7457	...	57° 14.2'	3.111
B	8.3 (4.6)(r)	...	1.8 × 10¹²(f)	...	17.89	8.95	10.15	...
Br	...	...	...	...	4.49(s)	6.68(s)	8.74(s)	2.27
Cd	29.8 (16.55)	0.22	6.83(f)	8(t)	2.9787	...	5.617	...
Ca	22.3 (12.4)(y)	0.3	3.91(f)	3.2–3.8(w)	5.582	...	...	...
Cf	...	...	...	...	...	...	...	...
C	0.6–4.3 (0.3–2.4)(d)	0.057	1375(f)	0.7	2.4614	...	6.7041	1.42
Ce	8 (4.44)	0.026(x)	75(y)	6(z)	5.16	...	...	...
Cs	97 (54)(aa)	...	20(b)	...	6.13(bb)	...	...	...
Cl	...	0.172 × 10⁻⁴	...	...	8.58(cc)	...	6.13(cc)	1.81
Cr	6.2 (3.4)	0.16	12.9(f)	36	2.884	...	...	2.498
Co	13.8 (7.66)	0.165	6.24(b)	30	2.5071	...	4.0686	2.4967
Cu	16.5 (9.2)	0.941 ± 0.005	1.6730(b)	16	3.6153	...	...	2.556
Cm	...	...	...	...	...	...	...	...
Dy	9 (5)	0.024(x)	57(y)	10–14(z)	3.59	...	5.65	...
E	...	...	...	...	...	...	...	...
Er	9 (5)	0.023(x)	107(y)	16(z)	3.65	...	5.58	...
Eu	26 (14.44)	...	90(y)	...	4.58	...	...	...

(continued)

(b) At 20 °C (68 °F). (c) Near 20 °C (68 °F). (d) From 20 to 100 °C (68 to 212 °F). (e) From 20 to 60 °C (68 to 140 °F). (f) At 0 °C (32 °F). (h) At −233 °C (−387 °F). (n) From 25 to 100 °C (77 to 212 °F). (p) Annealed, commercial purity. (r) From 20 to 750 °C (68 to 1380 °F). (s) At −150 °C (−238 °F). (t) Sand cast. (w) Annealed. F). (x) At 28 °C (82 °F). (y) At 25 °C (77 °F. (z) Measured from stress-strain relationship on as-cast metal. (aa) From 0 to 26 °C (32 to 70 °F). (bb) At −10 °C (14 °F). (cc) At −185 °C (−300 °F)

Table 2.4 (continued)

Element	Atomic No.	Atomic weight	Density(a), g/cm³ (lb/in.³)	Melting point, °C (°F)	Boiling point, °C (°F)	Specific heat(b), cal/g · °C (J/kg · K)	Heat of fusion, cal/g (Btu/lb)
Fermium (Fm)	100	253	...	...	...	...	...
Fluorine (F)	9	19.00	1.696 (0.06123)(g)	−219.6 (−363.3)	−188.2 (−306.8)	0.18 (750)	10.1 (18.2)
Francium (Fr)	87	223	...	27 (81)(m)	...	...	...
Gadolinium (Gd)	64	157.26	7.86 (0.284)	1312 (2394)	2730 (4950)	0.071 (300)	23.5 (42.4)
Gallium (Ga)	31	69.72	5.91 (0.213)	29.78 (85.60)	2237 (4059)	0.079 (330)	19.16 (34.49)
Germanium (Ge)	32	72.60	5.32 (0.192)	937.4 ± 1.5 (1719.3 ± 2.7)	2830 (5125)	0.073 (310)	...
Gold (Au)	79	197.0	19.3 (0.697)	1063.0 ± 0.0 (1945.4 ± 0.0)	2970 (5380)	0.0312 (131)(jj)	16.1 (29.0)
Hafnium (Hf)	72	178.58	13.1 (0.473)	2222 ± 30 (4032 ± 54)	5400 (9750)	0.0351 (147)	...
Helium (He)	2	4.00	0.1785 (0.006444)(g)	−269.7 (−453.5)	−268.9 (−452.0)	1.25 (5230)	24.9 (44.7)
Holmium (Ho)	67	164.94	6.79 (0.245)	1461 (2662)	2330 (4230)	0.039 (160)	15.0 (27.0)
Hydrogen (H)	1	1.008	0.0899 (0.00325)(g)	−259.19 (−434.54)	−252.7 (−422.9)	3.45 (14.400)	6.8 (12.2)
Indium (In)	49	114.82	7.31 (0.264)	156.2 (313.1)	2000 (3632)	0.057 (240)	14.2 (25.6)
Iodine (I)	53	126.91	4.94 (0.178)	113.7 (236.7)	183 (361)	0.052 (220)	...
Iridium (Ir)	77	192.2	22.65 (0.8177)	2454 ± 3 (4449 ± 5)	5300 (9570)	0.0307 (129)	65.5
Iron (Fe)	26	55.85	7.87 (0.284)	1536.5 ± 1 (2797.7 ± 1.8)	3000 ± 150 (5430 ± 270)	0.11	...
Krypton (Kr)	36	83.8	3.743 (0.1351)(g)	−157.3 (−251.1)	−152 (−242)	...	...
Lanthanum (La)	57	138.92	6.15 (0.222)	920 (1688)	3470 (6280)	0.048 (200)	17.3 (31.1)
Lawrencium (Lw)	103	257	...	...	...	...	...
Lead (Pb)	82	207.21	11.34 (0.4094)	327.4258 (621.3664)	1725 (3137)	0.0309 (129)(f)	6.26 (11.27)
Lithium (Li)	3	6.94	0.534 (0.193)	180.54 (356.97)	1330 (2426)	0.79 (3300)	104.2 (187.6)
Lutetium (Lu)	71	174.99	9.85 (0.356)	1652 (3006)(uu)	1930 (3510)	0.037 (150)	26.29 (47.32)
Magnesium (Mg)	12	24.32	1.74 (0.0628)	650 ± 2 (1202 ± 4)	1107 ± 10 (2025 ± 20)	0.245 (1030)	88 ± 2 (158 ± 4)
Manganese (Mn)	25	54.94	7.43 (0.268)	1245 (2273)	2150 (3900)	0.115 (481)(xx)	63.7 (114.7)
Mendelevium (Mv)	101	256	...	...	...	...	...
Mercury (Hg)	80	200.61	13.55 (0.4892)	−38.36 (−37.05)	357 (675)	0.033 (140)	2.8 (5.0)
Molybdenum (Mo)	42	95.95	10.2 (0.368)	2610 (4730)	5560 (10,040)	0.066 (280)	69.8 (125.6)(m)
Neodymium (Nd)	60	144.27	7.00 (0.253)	1019 (1866)	3180 (5756)	0.045 (190)	11.78 (21.20)
Neon (Ne)	10	20.18	0.8999 (0.03249)(g)	−248.6 ± 0.3 (−415.5 ± 0.5)	−246.0 (−410.8)	...	...
Neptunium (Np)	93	237	20.5 (0.740)	637 ± 2 (1179 ± 4)	...	...	...

(continued)

(a) Density may depend considerably on previous treatment. (b) At 20 °C (68 °F). (f) At 0 °C (32 °F). (g) Gas, grams per liter at 20 °C (68 °F) and 760 mm (30 in.). (m) Estimated. (jj) At 18 °C (64 °F). (uu) Distilled metal. (xx) For alpha; gamma is 0.120; both at 25.2 °C (77.3 °F).

Table 2.4 (continued)

Symbol	Coefficient of linear thermal expansion(c), µin./in. °C (µin./in. °F)	Thermal conductivity(c), cal/cm²/cm/s/°C	Electrical resistivity, µΩ·cm	Modulus of elasticity in tension, 10⁶ psi	Lattice parameters(b), Å a	b	c (or axial angle)	Closest approach of atoms
Fm	...	...	...	...	...	...	...	...
F	...	...	...	...	...	...	...	...
Fr	...	...	...	...	...	...	...	...
Gd	4 (2.22)(dd)	0.021(x)	140.5(y)	8–14(z)	3.64	...	5.78	...
Ga	18 (10)(ee)	0.07–0.09(ff)	17.4(gg)	...	4.524(y)	4.523(y)	7.661(y)	2.437
Ge	5.75 (3.19)	0.14	46(hh)	...	5.658	...	...	2.449
Au	14.2 (7.9)	0.71	2.35(b)	11.6	4.078	...	...	2.882
Hf	5.19 (288)(kk)	0.223(mm)	35.1(y)	...	3.1883	...	5.0422	...
He	...	3.32 × 10⁻⁴	...	...	3.58(nn)	...	5.84(nn)	3.58
Ho	...	...	87(y)	11(z)	3.58	...	5.62	...
H	...	4.06 × 10⁻⁴	...	...	3.76(pp)	...	6.13(pp)	...
In	33 (18)	0.057	8.37(b)	1.57	4.594	...	4.951	3.25
I	93 (52)	10.4 × 10⁻⁴	1.3 × 10¹⁵ (b)	...	4.787	7.266	9.793	2.71
Ir	6.8 (3.8)	0.14	5.3(b)	76	3.8389	...	...	2.714
Fe	11.76 (6.53)(qq)	0.18(rr)	9.71(b)	28.5 ± 0.5	2.8664(y)	...	...	2.4824
Kr	...	0.21 × 10⁻⁴	...	...	5.69(ss)	...	...	4.03
La	5 (2.77)	0.033(x)	57(y)	10–11(z)	3.77	...	12.16	...
Lw	...	...	...	...	...	...	...	...
Pb	29.3 (16.3)(tt)	0.083(f)	20.648(b)	2	4.9489	...	...	3.499
Li	56 (31)	0.17	8.55(f)	...	3.5089	...	...	3.0387
Lu	...	...	79(y)	...	3.50	...	5.50	...
Mg	27.1 (15.05)(vv)	0.367	4.45(b)	6.35(ww)	3.2088(y)	...	5.2095(y)	3.196
Mn	22 (12.22)(yy)	...	185(zz)	23	8.912	...	...	...
Mv	...	...	...	...	...	...	...	...
Hg	...	0.0196(f)	98.4(aaa)	...	3.005(bbb)	...	70° 31.7' (bbb)	3.005
Mo	4.9 (2.7)(d)	0.34	5.2(f)	47	3.1468(y)	...	...	2.725
Nd	6 (3.33)	0.031(ccc)	64(y)	...	3.66	...	11.80	...
Ne	...	0.00011	...	...	4.53(ddd)	...	...	3.21
Np	...	...	...	...	...	...	...	...

(continued)

(b) At 20 °C (68 °F). (c) Near 20 °C (68 °F). (d) From 20 to 100 °C (68 to 212 °F). (f) At 0 °C (32 °F). (u) From 0 to 100 °C (32 to 212 °F). (x) At 28 °C (82 °F). (y) At 25 °C (77 °F). (z) Measured from stress-strain relationship on as-cast metal. (dd) Near 40 °C (105 °F); the coefficient of expansion of gadolinium changes rapidly between −100 and +100 °C (−150 and +212 °F). (ee) From 0 to 30 °C (32 to 86 °F). (ff) At melting point. (gg) For a-axis; 8.1 for b-axis and 54.3 for c-axis. (hh) Ohm·cm of intrinsic germanium at 300 K. (jj) At 18 °C (64 °F). (kk) From 20 to 200 °C (68 to 390 °F). (mm) W/cm/°C at 50 °C (120 °F). (nn) At −271 °C (−456.7 °F). (pp) At −271 °C (−455.8 °F). (qq) At 25 °C (77 °F) for high-purity k iron (rr) For ingot iron at 0 °C (32 °F). (ss) At −191 °C (−311.8 °F). (tt) From 17 to 100 °C (63 to 212 °F). (vv) Along a-axis; 24.3 along c-axis. (ww) Dynamic; static, 5.77; both for 99.98% magnesium. (yy) Alpha; gamma, 14; both from 0 to 100 °C (32 to 212 °F). (zz) Alpha at 20 °C (68 °F). (aaa) At 50 °C (122 °F). (bbb) At −50 °C (−58 °F). (ccc) At −2.22 °C (28 °F). (ddd) At −268 °C (−450.4 °F). (eee) At 0 °C (32 °F), unmagnetized. (fff) At −234 °C (−389 °F). (ggg) At 50 °C (122 °F), parallel to a-axis, mean value; parallel to c-axis at 50 °C (122 °F), 5.8. (hhh) At 26 °C (78.8 °F). (jjj) At −225 °C (−373 °F).

10 Concise Metals Engineering Data Book

Table 2.4 (continued)

Element	Atomic No.	Atomic weight	Density(a), g/cm³ (lb/in.³)	Melting point, °C (°F)	Boiling point, °C (°F)	Specific heat(b), cal/g · °C (J/kg · K)	Heat of fusion, cal/g (Btu/lb)
Nickel (Ni)	28	58.71	8.9 (0.32)	1453 (2647)	2730 (4950)	0.105 (440)	73.8 (132.8)
Niobium (Nb)	41	92.91	8.57 (0.309)	2468 ± 10 (4474 ± 18)	4927 (8901)	0.065 (270)(f)	69 (124.2)
Nitrogen (N)	7	14.01	1.250 (0.04513)(g)	−209.97 (−345.95)	−195.8 (−320.4)	0.247 (1030)	6.2 (11.2)
Nobelium (No)	102	247	...	...	...	...	...
Osmium (Os)	76	190.2	22.61 (0.8162)	2700 ± 200 (4900 ± 350)(m)	5500 (9950)	0.031 (130)	...
Oxygen (O)	8	16.00	1.429 (0.05159)(g)	−218.83 (−361.89)	−183.0 (−297.4)	0.218 (913)	3.3 (5.9)
Palladium (Pd)	46	106.4	12.02 (0.4339)	1552 (2826)	3980 (7200)	0.0584 (245)(f)	34.2 (61.6)
Phosphorus, white (P)	15	30.98	1.83 (0.0661)	44.25 (111.65)	280 (536)	0.177 (741)	5.0 (9.0)
Platinum (Pt)	78	195.09	21.45 (0.7743)	1769 (3217)	4530 (8185)	0.0314 (131)(f)	26.9 (48.4)
Plutonium (Pu)	94	242	19.4 (0.700)	640 (1184)	3235 (6000)	0.033 (140)(qqq)	...
Polonium (Po)	84	210	9.40 (0.339)	254 ± 10 (489 ± 18)	...	...	...
Potassium (K)	19	39.10	0.86 (0.031)	63.7 (146.7)	760 (1400)	0.177 (741)	14.6 (26.3)
Praseodymium (Pr)	59	140.92	6.77 (0.244)	919 (1686)	3020 (5468)	0.045 (188)	11.71 (21.08)
Promethium (Pm)	61	145	...	1027 (1880)(m)	...	...	...
Proactinium (Pa)	91	231.1	15.4 (0.556)	1230 (2246)(m)	...	...	...
Radium (Ra)	88	226.05	5.0 (0.18)	700 (1292)	...	...	...
Radon (Rn)	86	222	9.960 (0.3596)(g)	−71 (−96)(m)	−61.8 (−79.2)	0.033 (140)	...
Rhenium (Re)	75	186.22	21.0 (0.76)	3180 ± 20 (5755 ± 35)	5900 (10,650)	0.059 (250)(f)	...
Rhodium (Rh)	45	102.91	12.41 (0.4480)	1966 ± 3 (3571 ± 5)	4500 (8130)	0.080 (330)	6.5 (11.79)
Rubidium (Rb)	37	85.48	1.53 (0.0552)	38.9 (102)	688 (1270)	0.057 (240)(f)	...
Ruthenium (Ru)	44	101.07	12.45 (0.4494)	2500 ± 100 (4530 ± 180)	4900 (8850)	0.042 (180)(xxx)	17.29 (31.12)
Samarium (Sm)	62	150.35	7.49 (0.270)	1072 (1962)	1630 (2966)	0.134 (561)	84.52 (152.14)
Scandium (Sc)	21	44.96	2.9 (0.10)	1539 (2802)	2730 (4946)	0.084 (350)(x)	16.4 (29.5)
Selenium (Se)	34	78.96	4.8 (0.17)	217 (423)	685 ± 1 (1265 ± 2)	0.162 (678)(f)	432 (778)
Silicon (Si)	14	28.09	2.33 (0.0841)	1410 (2570)	2680 (4860)	0.0559 (234)(f)	25 (45)
Silver (Ag)	47	107.88	10.49 (0.3787)	960.80 (1761.44)	2210 (4010)	0.295 (1240)	27.5 (49.5)
Sodium (Na)	11	22.99	0.9712 (0.03506)	97.82 (208.08)	892 (1638)	0.176 (737)	25 (45)
Strontium (Sr)	38	87.63	2.60 (0.0939)	768 (1414)	1380 (2520)	0.175 (733)	9.3 (16.7)
Sulfur, yellow (S)	16	32.07	2.07 (0.0747)	119.0 ± 0.5 (246.2 ± 0.9)	444.6 (832.3)		

(continued)

(a) Density may depend considerably on previous treatment. (b) At 20 °C (68 °F). (f) At 0 °C (32 °F). (g) Gas, grams per liter at 20 °C (68 °F) and 760 mm (30 in.). (m) Estimated. (x) At 28 °C (82 °F). (y) At 25 °C (77 °F). (uu) Distilled metal. (qqq) For alpha at 25 °C (77 °F). (www) At −173 °C (−279 °F). (xxx) Calculated.

Table 2.4 (continued)

Symbol	Coefficient of linear thermal expansion(c), μin./in. °C (μin./in. °F)	Thermal conductivity(c), cal/cm²/cm/s/°C	Electrical resistivity, μΩ·cm	Modulus of elasticity in tension, 10⁶ psi	Lattice parameters(b), Å a	b	c (or axial angle)	Closest approach of atoms
Ni	13.3 (7.39)(u)	0.22(y)	6.84(b)	30(eee)	3.5238	...	...	2.491
Nb	7.31 (4.06)	0.125(f)	12.5(f)	...	3.301	...	...	2.859
N	...	0.000060	...	...	4.04(fff)	...	6.60(fff)	...
No.	...	...	...	...	...	...	...	...
Os	4.6 (2.6)(ggg)	...	9.5(b)	81	2.7341(hhh)	...	4.3197(hhh)	2.750
O	...	0.000059	...	...	6.84(jjj)	...	...	...
Pd	11.76 (6.53)	1.68(jj)	10.8(b)	16.3	3.8902	...	...	2.750
P	125 (70)	...	1 × 10¹⁷(kkk)	...	7.18(mmm)	...	...	...
Pt	8.9 (4.9)	0.165(nnn)	10.6(b)	21.3(ppp)	3.9310(y)	...	...	2.775
Pu	55 (30.55)(rrr)	0.020(y)	141.4(sss)	14(ttt)	6.182(y)	4.826(y)	10.956(y)	...
Po	...	...	...	...	7.43	4.30	14.13	3.4
K	83 (46)	0.24	6.15(f)	...	5.334	...	...	4.624
Pr	4 (2.22)	0.028(ccc)	68(y)	7–14(z)	3.67	...	11.84	...
Pm	...	...	...	...	...	...	...	...
Pa	...	...	...	...	...	...	...	...
Ra	...	...	...	...	...	...	...	...
Rn	...	...	...	...	...	...	...	...
Re	6.7 (3.7)(uuu)	0.17	19.3(b)	66.7(b)	2.760	...	4.458	2.74
Rh	8.3 (4.6)	0.21(nnn)	4.51(b)	42.5(vvv)	3.804	...	...	2.689
Rb	90 (50)	...	12.5(b)	...	5.63(www)	...	...	4.88
Ru	9.1 (5.1)	...	7.6(f)	60(q)	2.7041	...	4.2814	...
Sm	...	...	88(y)	8(z)	8.99	...	23° 13'	...
Sc	...	...	61(yyy)	...	3.31	...	5.27	...
Se	37 (21)	7–18.3 × 10⁻⁴	12(f)	8.4	4.346	...	4.954	...
Si	2.8–7.3 (1.6–4.1)	0.20	10(f)	16.35(zzz)	5.428	...	...	2.351
Ag	19.68 (10.9)(u)	1.0(f)	1.59(b)	11	4.086	...	...	2.888
Na	71 (39)	0.32	4.2(f)	...	4.289	...	...	3.714
Sr	...	...	23(b)	...	6.087	...	...	4.31
S	64 (36)	6.31 × 10⁻⁴	2 × 10²³(b)	...	10.50	12.95	24.60	2.12

(continued)

(b) At 20 °C (68 °F). (c) Near 20 °C (68 °F). (e) From 20 to 60 °C (68 to 140 °F). (f) At 0 °C (32 °F). (q) Approximate. (u) From 0 to 100 °C (32 to 212 °F). (y) At 25 °C (77 °F). (z) Measured from stress-strain relationship on as-cast metal. (jj) At 18 °C (64 °F). (ccc) At –2.22 °C (28 °F). (eee) At 0 °C (32 °F), unmagnetized. (fff) At –234 °C (–389 °F). (ggg) At 50 °C (122 °F), parallel to a-axis, mean value; parallel to c-axis at 50 °C (122 °F), 5.8. (hhh) At 26 °C (78.8 °F). (jjj) At –225 °C (–373 °F). (kkk) At 11 °C (51.8 °F). (mmm) At –35 °C (–31 °F). (nnn) At 17 °C (63 °F). (ppp) For small cyclic strains. (rrr) From 21 to 104 °C (70 to 219 °F). (sss) At 107 °C (224.6 °F). (ttt) At 25 °C (77 °F), for cast metal. (uuu) From 20 to 500 °C (68 to 930 °F). (vvv) For hard wire. (www) At –173 °C (–279 °F). (yyy) Average value at 22 °C (72 °F). (zzz) Chill cast specimen 90.2 by 24.6 by 24.6 mm (3.55 by 0.97 by 0.97 in.). (aaaa) At 23 °C (73 °F). (bbbb) From 25 to 1000 °C (77 to 1830 °F), for iodide thorium. (cccc) At 100 °C (212 °F). (dddd) From 0 to 100 °C (32 to 212 °F), for polycrystalline metal. (eeee) At 0 °C (32 °F), for white tin. (ffff) Cast tin. (gggg) Btu · ft/h · ft² · °F at –400 °F. (hhhh) At 27 °C (80.6 °F). (kkkk) Rolled rods. (mmmm) At 70 °C (158 °F). (nnnn) Crystallographic average.

Table 2.4 (continued)

Element	Atomic No.	Atomic weight	Density(a), g/cm³ (lb/in.³)	Melting point, °C (°F)	Boiling point, °C (°F)	Specific heat(b), cal/g · °C (J/kg · K)	Heat of fusion, cal/g (Btu/lb)
Tantalum (Ta)	73	180.95	16.6 (0.599)	2996 ± 50 (5425 ± 90)	5425 ± 100 (9800 ± 200)	0.034 (140)(y)	38 (68)
Technetium (Tc)	43	98	11.5 (0.415)	2130 (3870)(m)	...	...	...
Tellurium (Te)	52	127.61	6.24 (0.225)	449.5 ± 0.3 (841.1 ± 0.5)	989.8 ± 3.8 (1813.6 ± 6.8)	0.047 (200)	32 (58)
Terbium (Tb)	65	158.93	8.25 (0.298)	1356 (2472)(uu)	2530 (4586)	0.044 (180)	24.54 (44.17)
Thallium (Tl)	81	204.39	11.85 (0.4278)	303 (577)	1457 (2655)	0.031 (130)	5.04 (9.07)
Thorium (Th)	90	232.05	11.5 (0.415)	1750 (3182)	3850 ± 350 (7000 ± 600)	0.034 (140)	<19.82 (<35.68)
Thulium (Tm)	69	168.94	9.31 (0.336)	1545 (2813)	1720 (3130)(www)	0.038 (160)	26.04 (46.87)
Tin (Sn)	50	118.70	7.30 (0.264)	231.912 ± 0.000 (449.442 ± 0.000)	2270 (4120)	0.054 (230)	14.5 (26.1)
Titanium (Ti)	22	47.90	4.51 (0.163)	1668 ± 10 (3035 ± 18)	3260 (5900)	0.124 (519)	104 (188)(m)
Tungsten (W)	74	183.86	19.3 (0.697)	3410 (6170)	5930 (10,706)	0.033 (140)	44 (70)
Uranium (U)	92	238.07	19.07 (0.6884)	1132.3 ± 0.8 (2070.4 ± 1.5)	3818 (6904)	0.02709 (113.4)(jjjj)	...
Vanadium (V)	23	50.95	6.11 (0.221)	1900 ± 25 (3450 ± 50)	3400 (6150)	0.119 (498)(t)	...
Xenon (Xe)	54	131.30	5.896 (0.2128)(g)	−111.9 (−169.4)	−108.0 (−162.4)	...	...
Ytterbium (Yb)	70	173.04	6.96 (0.251)	824 (1515)	1530 (2786)	0.035 (150)	12.71 (22.88)
Yttrium (Y)	39	88.92	4.47 (0.161)	1509 (2748)(uu)	3030 (5490)	0.071 (300)	46 (83)
Zinc (Zn)	30	65.38	7.13 (0.257)	419.5050 (787.1090)	906 (1663)	0.0915 (383)	24.09 (43.36)
Zirconium (Zr)	40	91.22	6.49 (0.234)	1852 (3366)	3580 (6470)	0.067 ± 0.001 (280 ± 4)	60 (110)(m)

(a) Density may depend considerably on previous treatment. (b) At 20 °C (68 °F). (g) Gas, grams per liter at 20 °C (68 °F) and 760 mm (30 in.). (m) Estimated. (t) Sand cast. (y) At 25 °C (77 °F). (uu) Distilled metal. (www) At −173 °C (−279 °F). (jjjj) At 27 °C (80 °F).

Table 2.4 (continued)

Symbol	Coefficient of linear thermal expansion(c), μin./in. °C (μin./in. °F)	Thermal conductivity(c), cal/cm²/cm/s/°C	Electrical resistivity, μΩ·cm	Modulus of elasticity in tension, 10⁶ psi	Lattice parameters(b), Å a	b	c (or axial angle)	Closest approach of atoms
Ta	6.5 (3.6)	0.130	12.45(y)	27(b)	3.303	...	...	2.859
Tc	...	...	...	...	...	...	...	...
Te	16.75 (9.3)	0.014	436,000(aaaa)	6	4.4570	...	5.9290	2.571
Tb	7 (3.88)	...	...	...	3.60	...	5.69	...
Tl	28 (16)	0.093	18(f)	...	3.457	...	5.525	3.408
Th	12.5 (6.9)(bbbb)	0.090(cccc)	13(f)	...	5.09	...	...	3.60
Tm	...	...	79(y)	...	3.53	...	5.55	...
Sn	23 (13)(dddd)	1.50(e)	11(eeee)	6–6.5(ffff)	5.8314	...	3.1815	...
Ti	8.41 (4.67)	6.6(gggg)	42(b)	16.8	2.95030	...	4.68312	...
W	4.6 (2.55)	0.397(e)	5.65(hhhh)	50	3.158	...	...	2.734
U	6.8–14.1 (3.8–7.8)(kkkk)	0.07(mmmm)	30(nnnn)	24	2.8545(y)	5.8681(y)	4.9566(y)	...
V	8.3 (4.6)(pppp)	0.074(cccc)	24.8–26.0(b)	18–20	3.039	...	...	2.632
Xe	...	1.24 × 10⁻⁴	...	...	6.25(rrrr)	...	...	4.42
Yb	25 (13.9)	...	29(y)	...	5.49	...	...	...
Y	...	0.035(ccc)	57(ssss)	17(z)	3.65	...	5.73	...
Zn	39.7 (22.0)(ssss)	0.27(y)	5.916(b)	(tttt)	2.6649	...	4.9470	2.6648
Zr	5.85 (3.2)(uuuu)	0.211(vvvv)	40	13.7	3.2312(y)	...	5.1477(y)	3.17

(b) At 20 °C (68 °F). (c) Near 20 °C (68 °F). (e) From 20 to 60 °C (68 to 140 °F). (f) At 0 °C (32 °F). (y) At 25 °C (77 °F). (z) Measured from stress-strain relationship on as-cast metal. (ccc) At −2.22 °C (28 °F). (aaaa) At 23 °C (73 °F). (bbbb) From 25 to 1000 °C (77 to 1830 °F), for iodide thorium. (cccc) At 100 °C (212 °F). (dddd) From 0 to 100 °C (32 to 212 °F), for polycrystalline metal. (eeee) At 0 °C (32 °F), for white tin. (ffff) Cast tin. (gggg) Btu · ft/h · ft² · °F at −400 °F. (hhhh) At 27 °C (80.6 °F). (kkkk) Rolled rods. (mmmm) At 70 °C (158 °F). (nnnn) Crystallographic average. (pppp) From 23 to 100 °C (73 to 212 °F). (rrrr) Polycrystalline; c-axis, 135; basal plane, 72. (ssss) From 20 to 250 °C (68 to 480 °F), for polycrystalline metal. (tttt) Pure zinc has no clearly defined modulus of elasticity. (uuuu) Alpha, polycrystalline. (vvvv) W/cm/°C at 27 °C (80.6 °F)

3 Vapor Pressures of the Elements

Table 3.1 Vapor pressures of the elements up to 1 atm (760 mm Hg)

	\multicolumn{6}{c}{Pressure, atm}
	0.0001		0.001		0.01	
Element	°C	°F	°C	°F	°C	°F
Aluminum	1110	2030	1263	2305	1461	2662
Antimony	759	1398	872	1602	1013	1855
Arsenic	308	586	363	685	428	802
Bismuth	914	1677	1008	1846	1121	2050
Cadmium	307(a)	585(a)	384(b)	723(b)	471	880
Calcium	688	1270	802(c)	1476(c)	958(b)	1756(b)
Carbon	3257	5895	3547	6417	3897	7047
Chromium	1420(a)	2588(a)	1594(b)	2901(b)	1813	3295
Copper	1412	2574	1602	2916	1844	3351
Gallium	1178	2152	1329	2424	1515	2759
Gold	1623	2953	1839	3342	2115	3839
Iron	1564	2847	1760	3200	2004	3639
Lead	815	1499	953	1747	1135	2075
Lithium	592	1098	707	1305	858	1576
Magnesium	516	961	608(a)	1126(a)	725(b)	1337(b)
Manganese	1115(d)	2039(d)	1269(b)	2316(b)	1476	2889
Mercury	77.9(b)	172.2(b)	120.8	249.4	176.1	349.0
Molybdenum	2727	4941	3057	5535	3477	6291
Nickel	1586	2887	1782	3240	2025	3677
Platinum	2367	4293	2687	4869	3087	5589
Potassium	261	502	332	630	429	804
Rubidium	223	433	288	550	377	711
Selenium	282	540	347	657	430	806
Silicon	1572	2862	1707	3105	1867	3393
Silver	1169	2136	1334	2433	1543	2809
Sodium	349	660	429	804	534	993
Strontium	...	...	(a)	(a)	877(b)	1629(b)
Tellurium	(a)	(a)	509(b)	948(b)	632	1170
Thallium	692	1277	809	1488	962	1764
Tin	...	...	...	...	...	...
Tungsten	3547	6417	3937	7119	4437	8019
Zinc	399(a)	750(a)	477(b)	891(b)	579	1074

(a) In the solid state. (b) In the liquid state. (c) β. (d) γ. Source: K.K. Kelley, *Bur. Mines Bull.*, Vol 383, 1935

Vapor Pressures for the Elements

Table 3.1 (continued)

Element	Pressure, atm					
	0.1		0.5		1.0	
	°C	°F	°C	°F	°C	°F
Aluminum.	1713	3115	1940	3524	2056	3733
Antimony	1196	2185	1359	2478	1440	2624
Arsenic	499	930	578	1072	610	1130
Bismuth	1254	2289	1367	2493	1420	2588
Cadmium	594	1101	708	1306	765	1409
Calcium	1175	2147	1380	2516	1487	2709
Carbon.	4317	7803	4667	8433	4827	8721
Chromium.	2097	3807	2351	4264	2482	4500
Copper.	2162	3924	2450	4442	2595	4703
Gallium	1751	3184	1965	3569	2071	3760
Gold	2469	4476	2796	5065	2966	5371
Iron.	2316	4201	2595	4703	2735	4955
Lead	1384	2523	1622	2952	1744	3171
Lithium	1064	1947	1266	2311	1372	2502
Magnesium	886	1627	1030	1886	1107	2025
Manganese	1750	3182	2019	3666	2151	3904
Mercury	251.3	484.3	321.5	610.7	357	675
Molybdenum	4027	7281	4537	8199	4804	8679
Nickel	2321	4210	2593	4699	2732	4950
Platinum.	3637	6579	4147	7497	4407	7965
Potassium	565	1051	704	1299	774	1425
Rubidium	497	927	617	1143	679	1254
Selenium	540	1004	634	1173	680	1256
Silicon	2057	3735	2217	4023	2287	4149
Silver.	1825	3317	2081	3778	2212	4014
Sodium	679	1254	819	1506	892	1638
Strontium	1081	1978	1279	2334	1384	2523
Tellurium	810	1490	991	1816	1087	1989
Thallium.	1166	2131	1359	2478	1457	2655
Tin	1932(b)	3510(b)	2163	3925	2270	4118
Tungsten.	5077	9171	5647	10197	5927	10701
Zinc	717	1323	842	1548	907	1665

(a) In the solid state. (b) In the liquid state. (c) β. (d) γ. Source: K.K. Kelley, *Bur. Mines Bull.*, Vol 383, 1935

4 Physical Properties of Gases and Liquids

Table 4.1 Physical properties of common gases and liquids

Name	Formula	Molecular weight	Density(a), g/L	Melting point, °C	Boiling point, °C	Auto-ignition point, °C	Explosive limits, percent by volume air Lower	Upper
Acetylene	C_2H_2	26.04	1.173	−81	−83.6 subl.(c)	335	2.5	80.0
Air	...	28.97(b)	1.2929	...	...	...	...	...
Ammonia	NH_3	17.03	0.7710	−77.7	−33.4	780	16.0	27.0
Argon	Ar	39.94	1.784	−189.2	−185.7	...	...	...
Butane-*n*	C_4H_{10}	58.12	2.703	−138	−0.6	430	1.6	8.5
Butane-*i*	C_4H_{10}	58.12	2.637	−159	−11.7	...	...	...
Butylene-*n*.	C_4H_8	56.10	2.591	−185	−6.3	...	1.7	9.0
Carbon dioxide	CO_2	44.01	1.977	−57 (at 5 atm)−78.5 subl.(c)		...	...	...
Carbon monoxide.	CO	28.01	1.250	−207	−191	650	12.5	74.2
Chlorine	Cl_2	70.91	3.214	−101	−34	...	...	...
Ethane	C_2H_6	30.07	1.356	−172	−88.6	510	3.1	15.0
Ethylene	C_2H_4	28.05	1.261	−169	−103.7	543	3.0	34.0
Helium.	He	4.003	0.1785	−272	−268.9	...	...	...
Heptane-*n*	C_7H_{16}	100.20	0.684 g/cm^3	−90.6	98.4	233	1.0	6.0
Hexane-*n*	C_6H_{14}	86.17	0.6594 g/cm^3	−95.3	68.7	248	1.2	6.9
Hydrogen	H_2	2.016	0.0899	−259.2	−252.8	580	4.1	74.2
Hydrogen chloride	HCl	36.47	1.639	−112	−84	...	...	...
Hydrogen fluoride	HF	20.01	0.921	−92.3	19.5	...	...	...
Hydrogen sulfide	H_2S	34.08	1.539	−84	−62	...	4.3	45.5
Methane	CH_4	16.04	0.7168	−182.5	−161.5	538	5.3	13.9
Nitrogen	N_2	28.016	1.2506	−209.9	−195.8	...	...	...
Octane-*n*.	C_8H_{18}	114.23	0.7025 g/cm^3	−56.8	125.7	232	0.8	3.2
Oxygen	O_2	32.00	1.4290	−218.4	−183.0	...	...	...
Pentane-*n*	C_5H_{12}	72.15	0.016 g/cm^3	−131	36.2	310	1.4	8.0
Propane	C_3H_8	44.09	2.020	−189	−44.5	465	2.4	9.5
Propylene	C_3H_6	42.05	1.915	−184	−48	458	2.0	11.1
Sulfur dioxide.	SO_2	64.06	2.926	−75.7	−10.0	...	...	...

(a) Density of gases is given in g/L at 0 °C and 760 mm Hg (1 atm). Density of liquids is given in g/cm^3 at 20 °C. (b) Because air is a mixture, it does not have a true molecular weight. This is the average molecular weight of its constituents. (c) subl. indicates that the substance sublines at the temperature listed. Source: *Corrosion Tests and Standards: Application and Interpretation*, ASTM, 1995, p 27

Table 4.2 Physical properties of common inorganic and organic acids

Name	Formula	Molecular weight	Specific gravity(a)	Melting point, °C	Boiling point, °C
Inorganic acids					
Boric acid	H_3BO_3	61.84	1.435 (15 °C)	185 dec.(b)	...
Hydrochloric acid	HCl	36.47	1.268 (0 °C)	−111	−85
Hydrofluoric acid	HF	20.01	0.988 (13.6 °C)	−83	19.4
Nitric acid	HNO_3	63.02	1.502	−42	86
Phosphoric acid (ortho-)	H_3PO_4	98.00	1.834 (18.2 °C)	42.35	−½H₂O, 213(c)
Sulfamic acid	NH_2SO_3H	97.09	2.03 (12/4 °C)	205 dec.(b)	...
Sulfuric acid	H_2SO_4	98.08	1.834 (18/4 °C)	10.49	340 dec.(b)
Sulfurous acid	H_2SO_3	82.08	1.03	...	...
Organic acids					
Acetic acid	CH_3CO_2H	60.05	1.049 (20/4 °C)	16.7	118.1
Benzoic acid	$C_6H_5CO_2H$	122.12	1.266 (15/4 °C)	121.7	249.2
Formic acid	HCO_2H	46.03	1.220 (20/4 °C)	8.6	100.8
Malic acid (dl-)	$HO_2CCH_2CH(OH)CO_2H$	134.09	1.601 (20/4 °C)	128–129	150 dec.(b)
Oleic acid	$C_8H_{17}CH:CH(CH_2)_7CO_2H$	282.45	0.854 (78/4 °C)	14	285–286
Oxalic acid	$HO_2C \cdot CO_2H \cdot 2H_2O$	126.07	1.653 (19/4 °C)	101.5	...
Propionic acid	$CH_3CH_2CO_2H$	74.08	0.992 (20/4 °C)	−22	141.1

(a) Specific gravity values are given at room temperature (15 to 20 °C) unless indicated otherwise by the temperatures given in parenthesis that follow the value: thus 2.03 (12/4 °C) indicates a specific gravity of 2.03 for the acid at 18 °C, referred to water at 4 °C. (b) dec. indicates that the substance decomposes at the temperature listed. (c) −½H₂O, 213 indicates a loss of ½ mole of water per formula weight of the acid at a temperature of 213 °C. Source: *Perry's Chemical Engineers' Handbook*, 6th ed., McGraw-Hill, 1984

5 Crystal Structures

Crystal structure, as defined broadly, is the arrangement of atoms in the solid state. Crystal structure also involves consideration of defects, or abnormalities, in the idealized atomic arrangements. The collective arrangement of these atoms on a scale much greater than that of the individual atom is referred to as the microstructure of the material.

This chapter briefly reviews the basic concepts associated with metallic crystal structures and atomic coordination and describes the common crystal defects. More detailed information on the crystal structure of metals can be found in Volume 9, *Metallography and Microstructures*, of the *ASM Handbook* and in the Selected References listed at the conclusion of this chapter.

Basic Concepts of Crystal Structure and Atomic Coordination

The arrangement of atoms in most solid metals demonstrates a long-range pattern. That is, the atomic packing is repetitive over distances large in comparison to the atomic size. Such an arrangement is called *crystalline*, and the repetitive pattern can be described by a fundamental repeating unit or *unit cell*.

Almost all metals crystallize in one of three patterns: face-centered cubic (fcc), hexagonal close packed (hcp), or body-centered cubic (bcc). The atomic arrangements in these cells are depicted in Fig. 1. The positions of atom centers are noted in the left sides of each figure, and the atoms are represented by spheres (or partial spheres when an atom is shared by adjacent unit cells) in the right sides of the figures. All the arrangements are characterized by efficient atomic packing. Indeed, the fcc array (Fig. 1a) represents the most efficient possible atomic packing as is manifested by the high coordination number (CN = 12) of this structure. (The coordination number refers to the number of nearest neighbors in an atom in a solid.) A viewing of a face of an fcc cell shows that an atom in a face-center is coordinated by four other atoms at cell corners. The distance separating the atom centers is the atomic diameter (also equal to $a/(2)^{1/2}$ where a is the edge length or *lattice parameter* of the unit cell). However, the atom at the face-center is this same distance from four other atoms on the centers of the four adjoining cell faces (Fig. 2b). In addition, the reference atom is likewise coordinated to four atoms in the centers of adjacent faces in the unit cell directly in front of the unit cell of Fig. 2(b). Thus CN = 12 for the fcc structure.

An alternative view of fcc packing permits another way of seeing that it is efficiently packed. Figure 3 is a view of a close-packed plane in the fcc structure. A plane is defined by two nonparallel directions; in Fig. 3, these are taken as two face-diagonals. The atoms in this plane are arranged as billiard balls are in a cue rack. When these atomic planes are stacked vertically, and in a direction parallel to the cube diagonal, atoms of one plane lie in the vertices of atoms in the plane beneath (Fig. 3b). Such a stacking pattern generates a close-packed structure. In the fcc pattern, the positions of atom centers repeat every fourth of these planes. That is, atom centers in the fourth plane lie directly above atom centers in the first, atom centers in the fifth plane are directly above those in the second, and so forth. The stacking is thus described as ... ABCABC ....

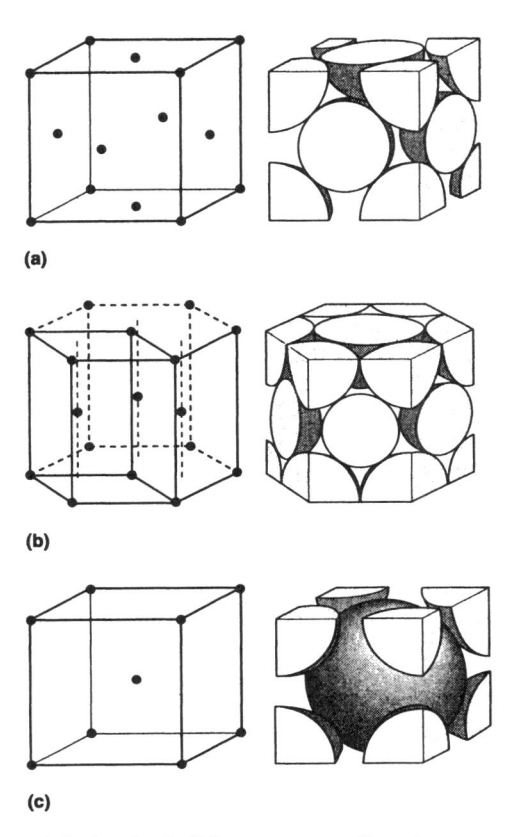

Fig. 1 Representation of several simple unit cells. Points represent positions of atom centers (left), and atoms are represented by spheres or portions of spheres (right). (a) Face-centered cubic unit cell. (b) Hexagonal close-packed unit cell. (c) Body-centered cubic unit cell

The ideal hcp structure (Fig. 1b and 2c) is packed as efficiently as the fcc structure. Atoms in the close-packed (basal) plane have an atomic arrangement identical to that in a close-packed fcc plane. However, in the hcp structure these planes repeat every other layer; that is, atom centers in the third layer lie directly above atom centers in the first, atom centers in the fourth layer are directly above atom centers in the second, and so forth. This stacking is therefore described as ..., ABAB ....

Two lattice parameters (c and a) are needed to define the hcp unit cell (Fig. 1b and 2c). An hcp cell has the maximum atomic-packing efficiency only when a definite relationship between c and a ($c/a = 1.63$) exists. Few hcp metals exhibit this ratio (most have $c/a < 1.63$). In these situations, the hcp structure can no longer be viewed as being as efficiently packed as the fcc structure.

The CN for the bcc structure (Fig. 1c and 2a) is 8. This can be deduced with reference to the atom in the center of the bcc unit cell; it is equidistant from 8 atoms at the cell corners (Fig. 2a). Because the atomic packing is less efficient in bcc, the closest-packed plane in this structure is also less densely packed than in the corresponding fcc plane. A view (Fig. 4) of the closest-packed bcc plane (which is defined by a cell edge and a face-diagonal) shows that atoms within

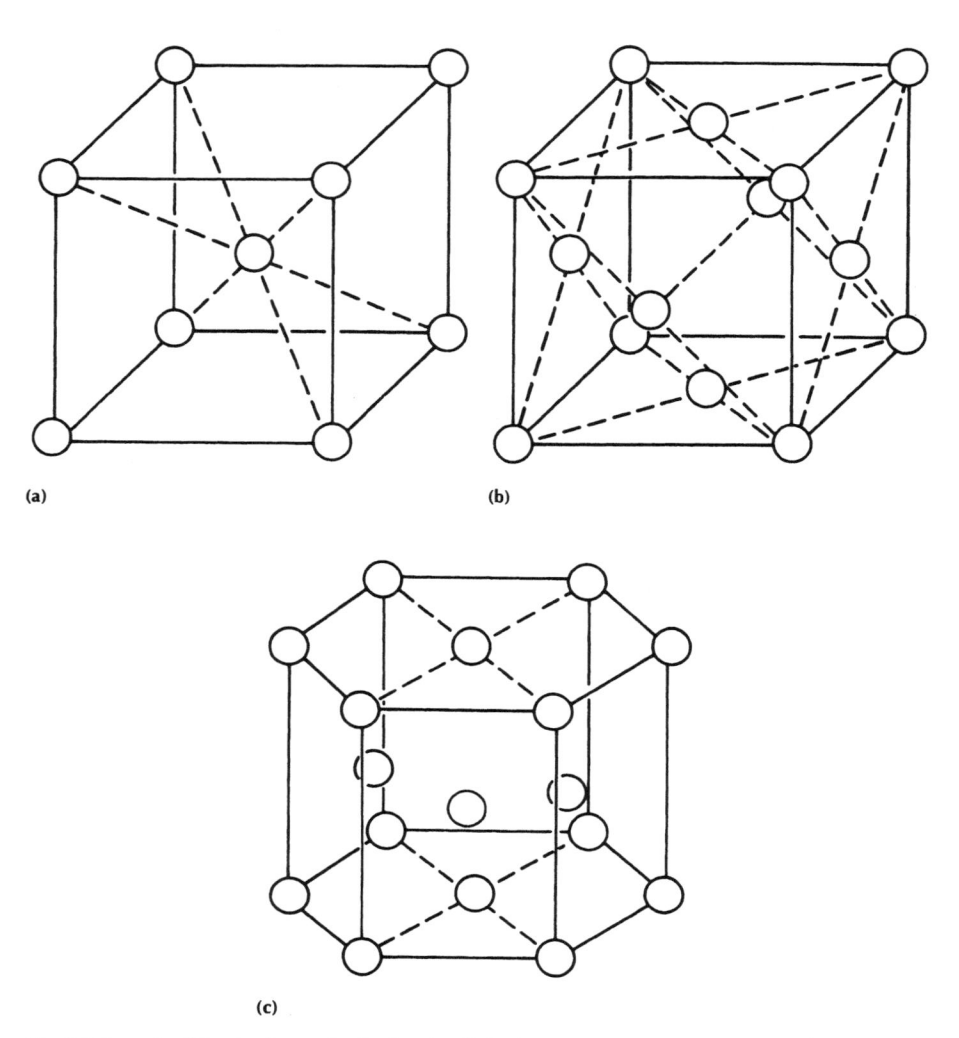

Fig. 2 Schematic of the atomic coordination in unit cells for the most common crystal structures found in metals and alloys. (a) bcc. (b) fcc. (c) hcp

this plane touch along the cube diagonals. There are two nonparallel *close-packed directions* of this plane; the CN for the fcc close-packed plane is 3.

Many metals exist in more than one crystalline form, depending on pressure and temperature. At one atmosphere, for example, iron is bcc at temperatures below 912 °C (1674 °F), fcc between 912 and 1394 °C (1674 and 2541 °F), and above 1394 °C (2541 °F) iron reverts to the bcc form until it melts at 1538 °C (2800 °F). Titanium, zirconium, and hafnium all exhibit a transition from a hcp structure to bcc on heating. Many other metals also exhibit such *allotropic* transformations.

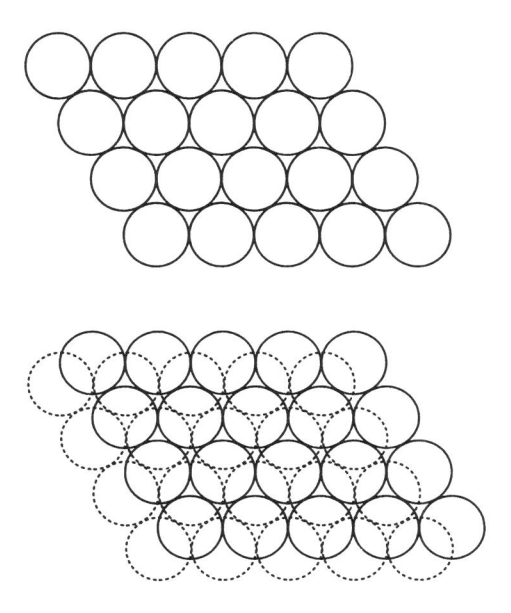

Fig. 3 (a) Plan view of a close-packed plane in the fcc structure. The directions along which atoms touch are face-diagonals. (b) Plan view of two close-packed planes of spheres, with spheres in the top plane (solid circles) situated in interstices in the bottom plane (broken circles)

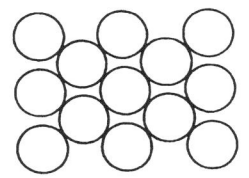

Fig. 4 Plan view of atomic packing in a close-packed bcc plane. Atoms touch along two nonparallel close-packed directions (the cube diagonals).

Crystal Structure and Plastic Deformation

Further details of atomic arrangements are described here, because they are relevant to plastic deformation of crystalline materials. As shown in Fig. 3, the close-packed plane in the fcc structure is defined by two face-diagonals and within such a plane there are three nonparallel close-packed directions. In addition, there are four *nonparallel* planes of this nature in the fcc crystal structure (Fig. 5). (To better illustrate the point, the planes in Fig. 5 are taken from adjacent cells.) There are thus 12 combinations of nonparallel planes and directions (four planes times three directions per plane) in the fcc lattice.

The above is germane for plastic deformation and takes place by slip (sliding) of close-packed planes over one another. A reason for this slip plane preference is that the separation between

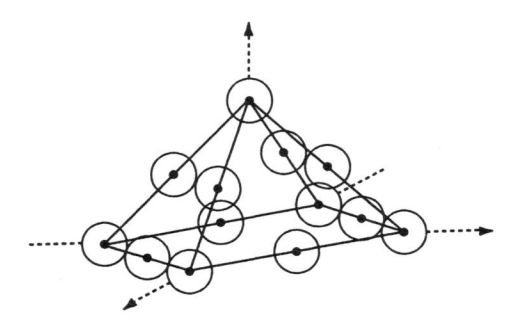

Fig. 5 Four nonparallel close-packed planes characterize the fcc structure. There are three nonparallel close-packed directions within each plane, giving rise to 12 slip systems.

close-packed planes is greater than for other crystal planes, and this makes their relative displacement easier. Furthermore, the slip transit direction (or *slip direction*) is a close-packed direction. The combination of planes and directions on which slip takes place (12 for the fcc structure) constitutes the slip systems in the material. In polycrystalline materials, which are defined below in the section "Crystalline Defects" in this chapter, a certain number of slip systems must be available in order for the material to be capable of plastic deformation. Other things being equal, the greater the number of slip systems the greater the capacity for this deformation. Face-centered cubic metals have a large number of slip systems and, indeed, all of them except iridium and rhodium are capable of moderate to extensive plastic deformation even at temperatures approaching 0 K.

Materials having the bcc structure also often display 12 slip systems, although this number comes about differently than it does for the fcc lattice. A closest-packed bcc plane is defined by a unit cell edge and face-diagonal (Fig. 3). There are only two close-packed directions (the cube diagonals) in the closest-packed bcc plane, but there are six nonparallel planes of this type. Over certain temperature ranges some bcc metals display slip on other than close-packed planes, although the slip direction remains a close-packed one. Thus bcc metals have the requisite number of slip systems to allow for their plastic deformation. Some of the bcc metals become "brittle" at low temperatures as a result of the strong temperature sensitivity of their yield strength that results in fracture rather than significant plastic deformation.

Depending on the c/a ratio, polycrystalline hcp metals may or may not have the necessary number of slip systems to allow for appreciable plastic deformation. The ideal hcp structure has only three slip systems as there is only one nonparallel close-packed plane in it (the basal plane, which contains three nonparallel close-packed directions). This number (three) of slip systems is insufficient to permit polycrystalline plastic deformation, and so hcp polycrystals for which slip is restricted to the basal plane are not malleable. When c/a is less than the ideal ratio, basal planes become less widely separated and other planes compete with them for slip activity. In these instances, the number of slip systems increases and material ductility is beneficially affected.

Crystalline Defects

Atomic arrangements in crystals deviate slightly from the ideal ones described above. Such deviations are called *crystalline defects* (or *imperfections*), although these "defects" often lead to improved material performance. Regardless of the term used to describe defects, they can be

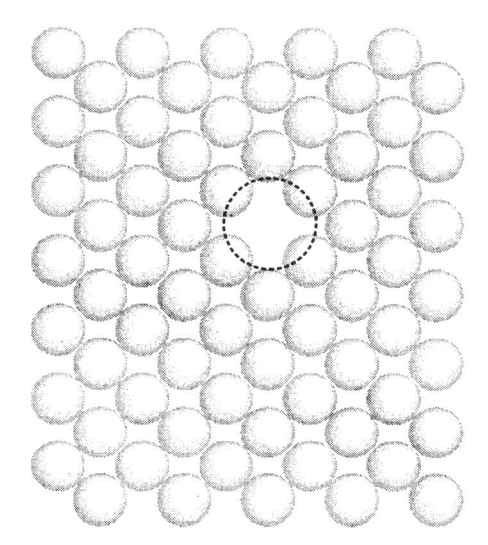

Fig. 6 Two-dimensional representation of a crystal illustrating a vacant lattice site

classified by their scale or size. The smallest sized deviation in the ideal crystal arrangement has a volume comparable to that of an atom; such a defect is termed a *point defect*.

Point defects are of two types—*impurity atoms* and *vacancies*. A vacancy is schematically illustrated in Fig. 6. Rather than having all lattice sites occupied, one site is vacant. Vacancies arise as a result of entropic effects, and the fraction of vacant lattice sites increases with temperature. This fraction is zero at 0 K and is on the order of 10^{-3} for many metals at or close to their melting point.

Vacancies alter properties. Density is (very slightly) decreased by them. Material strength is also slightly (and counterintuitively) increased by vacancies. Vacancies increase the electrical resistivities of metals. Vacancies also enhance atomic *diffusion*. Diffusion refers to the macroscopic atomic mixing that takes place as a result of the motion of many individual atoms. If a layer of copper is placed on one of nickel, for example, and then held at an elevated temperature for a long time, the resultant solid displays a uniform composition as a result of the interdiffusion of copper and nickel atoms.

Impurity atoms are also termed point defects. An fcc unit cell of an alloy of composition 75 at.% Cu-25 at.% Ni, for example, contains—on the average—three times as many copper atoms as nickel ones. The substituted nickel atoms are considered defects because their size differs from that of the host copper atom, and this causes a local distortion of the unit cell. Impurity atoms also affect properties. Electrical and thermal conductivities in metals are reduced by them. However, metallic strengths are increased by impurities. This *solid-solution hardening* is used to strengthen a number of metals. Adding zinc to copper, as in brasses, is a technologically important example.

Small impurity atoms do not substitute for the host atoms, but enter into *interstitial* spaces among them and are referred to as interstitials. Typical interstitials in metals are nitrogen, carbon, and oxygen. Interstitials generally strengthen a metal more than substitutional atoms do, because the interstitials cause more distortion. Carbon atoms in the bcc form of iron are particularly po-

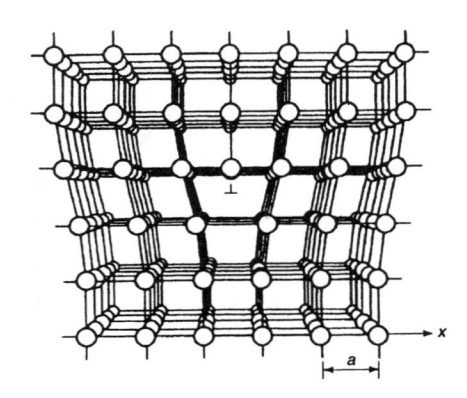

Fig. 7 A schematic of an edge dislocation, represented by a partial atomic plane, in a crystal. The "core" of the dislocation is localized at the partial plane termination. Atomic positions are distorted in region of this core, making slip easier in the vicinity of the dislocation.

tent hardeners in this respect. The effect is used beneficially in strengthening of quenched-and-tempered steels.

A *line defect* has two dimensions comparable to an atomic diameter and one much greater. An example of a particular line defect, an *edge dislocation,* is shown in Fig. 7. The upper half of the crystal shown contains one more atom column than the lower half of it. The resultant atomic disregistry is centered about a small region; as suggested by Fig. 7 the disregistry is accommodated in an approximately cylindrical volume having a radius comparable to that of an atom and extending along the termination of the atomic column for distances much greater than this. Dislocations are found in all crystalline solids, but the extent to which they exist varies among the material classes. The quantity of dislocations (the *dislocation density*) can be expressed in terms of their number per unit area. With reference to Fig. 7, for example, the dislocation density would be the number of dislocations emerging from a surface divided by the area of the depicted crystal plane. Dislocation densities in metals range from about 10^{10} to $10^{15}/m^2$.

Dislocations are important because their motion in response to an applied stress is responsible for plastic deformation in most crystalline solids. As mentioned, plastic deformation takes place by the relative displacement of atomic planes. This is easier to accomplish when dislocations are present. The atomic disruption in the dislocation vicinity is responsible for the easier slippage of planes on which dislocations are situated. In fact, the stress required to cause dislocations to move is orders of magnitude less than the stress needed to cause slip plane displacement in a "perfect" crystal.

The mechanism of plastic deformation (flow) by the slip process, which is actually produced by dislocation movement, is illustrated schematically in Fig. 8. If forces, as indicated by the arrows in Fig. 8, are applied to a crystal, such as the perfect crystal shown in Fig. 8(a), one part of the crystal will slip. The edge of the slipped region, shown as a dashed line in Fig. 8(b), is a dislocation. The portion of this line at the left near the front of the crystal and perpendicular to the arrows, in Fig. 8(b), is an edge dislocation, because the displacement involved is perpendicular to the dislocation.

The slip deformation in Fig. 8(b) has also formed another type of dislocation. The part of the slipped region near the right side, where the displacement is parallel to the dislocation, is termed a *screw dislocation.* In this part, the crystal no longer is made of parallel planes of atoms, but instead consists of a single plane in the form of a helical ramp (screw).

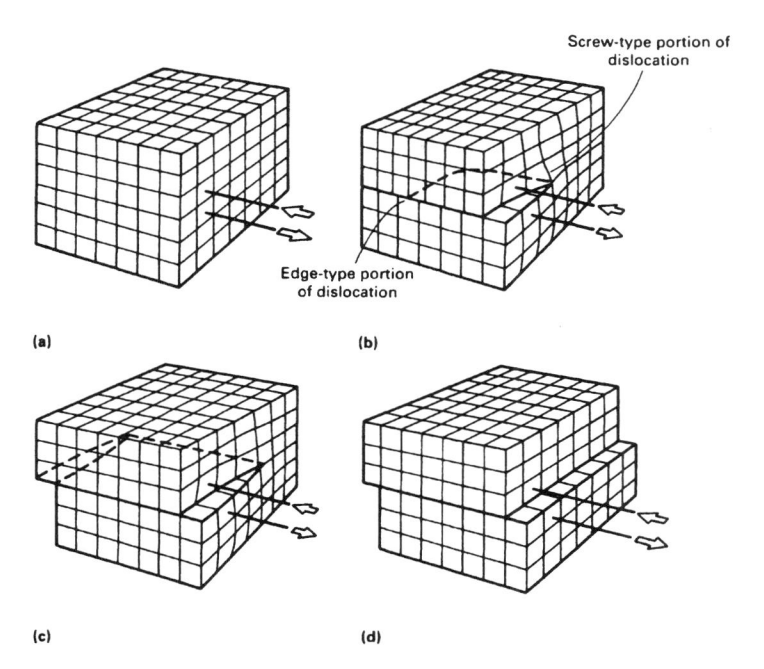

Fig. 8 Schematic representations of four stages of slip deformation by formation and movement of a dislocation (dashed line) through a crystal. (a) Crystal before displacement. (b) Crystal after some displacement. (c) Complete

Fig. 9 Schematic representation of the orientations of individual grains in a polycrystal. Within individual grains, a set of atomic planes has the same orientation in space. At a grain boundary, the orientation changes abruptly

As the slipped region spreads across the slip plane, the edge-type portion of the dislocation moved out of the crystal, leaving the screw-type portion still embedded, as shown in Fig. 8(c). When all of the dislocation finally emerged from the crystal, the crystal was again perfect but with the upper part displaced one unit from the lower part, as shown in Fig. 8(d).

The role of dislocations in plastic flow is verified by the exceptionally high strengths of metal crystals not containing (or containing very few) dislocations. It might be thought that the greater the dislocation density, the lesser the stress required for plastic deformation. This is true for mate-

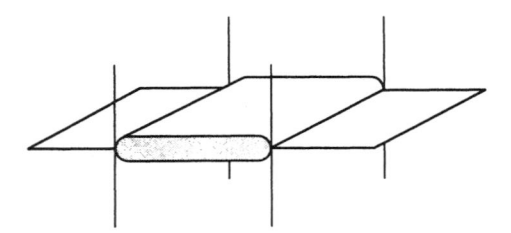

Fig. 10 A three-dimensional illustration of a stacking fault in a fcc crystal. The fault is a narrow ribbon of thickness several atomic diameters. It is bonded by partial dislocation (the lines AB and CD).

rials containing relatively few dislocations (e.g., less than approximately $10^8/m^2$). Paradoxically, though, when the dislocation density becomes high enough the stress required to cause plastic flow increases with dislocation density. This is so because dislocations mutually impede each other's motion. Dislocations in metals also multiply—sometimes substantially—when they are plastically deformed. This is accompanied by an increase in the stress required to continue deformation. This phenomenon of *work hardening* is used to manipulate strengths of a number of metallic materials, including conventional stainless steels and copper and its alloys.

Crystalline solids also contain internal surface defects. A surface defect has one dimension comparable to the atomic size, and two dimensions much larger. The most important surface defect is a *grain boundary*. As indicated in Fig. 9, such boundaries separate differently spatially oriented crystals, and the collective aggregate is termed a *polycrystal* (or polycrystalline solid). The average diameter of the individual grains within a polycrystal defines the material grain size. Grain sizes in engineered materials vary by quite a bit. They are usually less in nonmetals than in metals, and can be as fine as 0.1 μm in some ceramics. Metallic grain sizes typically range from several micrometers to, in the case of slowly cooled castings, several centimeters. Some recently developed processes—e.g., rapid solidification and mechanical alloys—produce materials having grain sizes on the order of nanometers. To put this in perspective, the diameter of a typical atom is about 0.25 nm. Thus, grains having a diameter of several nanometers are about ten atoms across.

Grain size affects mechanical properties. The yield strength increases with decreases in grain size, because the distance over which dislocations can move freely is limited to the grain diameter. (Dislocations are restricted from crossing grain boundaries.) Fracture resistance also generally improves with reductions in grain size. The reason for the improved fracture resistance is that cracks formed during deformation, and which are the precursors to those causing fracture, are limited in size to the grain diameter.

Stacking faults and *twin boundaries* are other internal surface defects. While found in all crystal structures, they are most easily described with reference to the fcc one. A stacking fault in a fcc lattice corresponds to a "mistake" made in the close-packed plane stacking sequence. Instead of the usual ... ABCABCABC ... sequence, an ... ABCABABCAB ... one is found. The placing of a plane in the A, rather than C, position results in a thin layer of hcp-like material (denoted by **ABAB**). The thickness of this defect is only several atomic diameters in the direction normal to the close-packed planes. Stacking faults in fcc materials generally occur as ribbons (Fig. 10). The fault extends normal to the plane of this figure over distances that are large compared to an atomic size. The ribbon width (the distance between points A and C or B and D in Fig. 10) is highly variable, ranging in size from the order of one to many atomic diameters. Generally, if the energy of the hcp and fcc allotropic forms of the solid are comparable, the width is large and vice-versa. The boundaries at the edges of the faults (lines AB and CD in Fig. 10) are defined by a

Fig. 11 The microstructure of annealed cartridge brass (70% Cu-30% Zn), illustrating both grain boundaries and annealing twins. The twins are the regions with parallel sides within the grains.

special type of dislocation that accommodates the disregistry between the hcp and fcc stacking at the boundaries. Stacking faults play an important role in the work-hardening behavior of some fcc metals and alloys. If their width is large, the material work hardens more than if it is small.

The stacking sequence across a twin boundary is ... ABCABACBA ...; the position of the boundary is denoted by **B**. Note that to either side of this boundary the stacking sequence is typical of fcc. (ACBACB ... represents the same stacking as does ABCABC ... in that close-packed layers repeat every fourth layer.) At the twin boundary a layer of ABA (hcp stacking) exists, so twin boundaries are somewhat akin to stacking faults. However, there are differences between these types of defects. The differences arise from the different positioning of the atoms in the atomic plane twice removed from the respective boundaries. Twins also typically have a width much greater than do stacking faults. Examples of twins in a copper alloy are shown in Fig. 11. These twins developed in response to heat treatment, and for this reason they are called *annealing* twins. Twins do not affect mechanical behavior to the same degree that stacking faults do (an important exception is low-temperature deformation of bcc metals). Thus, of the several surface defects discussed, grain boundaries play an important role in plastic deformation, stacking faults affect the work-hardening behavior of fcc metals, but twins generally only play a minor role in plastic flow.

Lastly, *volume defects*—pores and microcracks—are often present in engineering solids. Volume defects have all three of their dimensions much larger than the atomic size, although the characteristic dimension may still be small (e.g., on the order of 10^{-7} m). Volume defects almost invariably reduce strength and fracture resistance. (An exception is for spherical pores having a radius on the order of nanometers. Such voids are sometimes found in materials exposed to high energy radiation, and a modest increase in strength attends their presence.) The reductions in strength and fracture resistance can be quite substantial, even when the defects constitute only several percentage by volume of the material. In metals, pores are much more likely to be found in cast than in wrought products. The shrinkage accompanying solidification in almost all metals is manifested in microporosity; i.e., in pores having diameters on the order of micrometers. The extensive deformation accompanying the production of wrought metals is usually sufficient to "heal" or close this microporosity. Powdered metals frequently contain pores. Powder products are typically fabricated by a pressing operation followed by a high-temperature heat treatment (sintering) that results in material densification. Full density is difficult to achieve through a "press-and-sinter" cycle, and thus residual porosity is usually found in the sintered product. Full density is more likely to be obtained when a stress is applied during sintering (as in hot pressing in which a uniaxial compressive stress is applied, or hot isostatic pressing in which the stress state is hydrostatic compression). Pore removal is facilitated by pressure for much the same reason deformation processing removes pores in the original ingot structure in wrought products.

ACKNOWLEDGMENT

The information in this chapter was adapted from T.H. Courtney, Fundamental Structure-Property Relationships in Engineering Materials, to be published in Vol 20, *Materials Selection and Design,* of the *ASM Handbook,* ASM International, fall of 1997.

SELECTED REFERENCES

- C.S. Barrett and T.B. Massalski, *Structure of Metals,* 3rd ed., Pergamon Press, 1980
- M.J. Buerger, *Elementary Crystallography,* John Wiley & Sons, 1963
- T. Hahn, Ed., *International Tables for Crystallography,* Vol A, Space-Group Tables, Kluwer Academic Publishers, 1983
- W.B. Pearson, *A Handbook of Lattice Spacings and Structures of Metals and Alloys,* Pergamon Press, Vol 1, 1958; Vol 2, 1967
- G.H. Stout and L.J. Jensen, *X-Ray Structure Determination,* Macmillan, 1968

6 Phase Diagrams

Alloy phase diagrams are useful to metallurgists, materials engineers, and materials scientists in four major areas: (1) development of new alloys for specific applications, (2) fabrication of these alloys into useful configurations, (3) design and control of heat-treatment procedures for specific alloys that will produce the required mechanical, physical, and chemical properties, and (4) solving problems that arise with specific alloys in their performance in commercial applications, thus improving product predictability. In all these areas, the use of phase diagrams allows research, development, and production to be done more efficiently and cost effectively.

In the area of alloy development, phase diagrams have proved invaluable for tailoring existing alloys to avoid overdesign in current applications, designing improved alloys for existing and new applications, designing special alloys for special applications, and developing alternative alloys or alloys with substitute alloying elements to replace those containing scarce, expensive, hazardous, or "critical" alloying elements. Application of alloy phase diagrams in processing includes their use to select proper parameters for working ingots, blooms, and billets, finding causes and cures for microporosity and cracks in castings and welds, controlling solution heat treating to prevent damage caused by incipient melting, and developing new processing technology.

In the area of performance, phase diagrams give an indication of which phases are thermodynamically stable in an alloy and can be expected to be present over a long time when the part is subjected to a particular temperature (e.g., in an automotive exhaust system). Phase diagrams also are consulted when attacking service problems such as pitting and intergranular corrosion, hydrogen damage, and hot corrosion.

In a majority of the more widely used commercial alloys, the allowable composition range encompasses only a small portion of the relevant phase diagram. The nonequilibrium conditions that are usually encountered in practice, however, necessitate the knowledge of a much greater portion of the diagram. Therefore, a thorough understanding of alloy phase diagrams in general and their practical use will prove to be of great help to a metallurgist expected to solve problems in any of the areas mentioned above.

This chapter provides examples of binary and ternary phase diagrams that form the basis of the most important classes of structural alloys: steels, aluminum alloys, and copper alloys. Other examples of phase diagrams can be found in Volume 3 of the *ASM Handbook*. This Volume also contains an article entitled "Introduction to Phase Diagrams," which (1) outlines the basic features of phase diagrams, (2) describes the thermodynamic principles associated with the phases formed in an alloy system, and (3) discusses practical applications of phase diagrams.

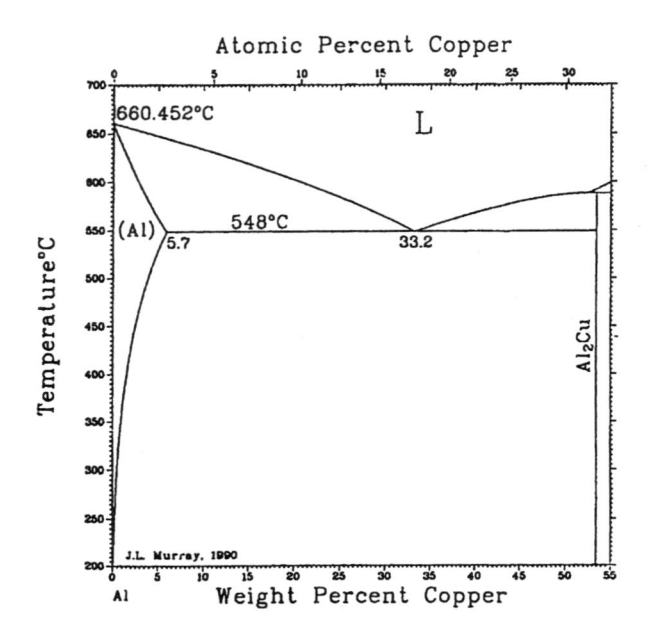

Fig. 6.1 The aluminum-rich portion of the aluminum-copper phase diagram. This system is the basis for the wrought 2xxx and cast 2xx.x aluminum alloys. See also Fig. 6.6.

Phase Diagrams 31

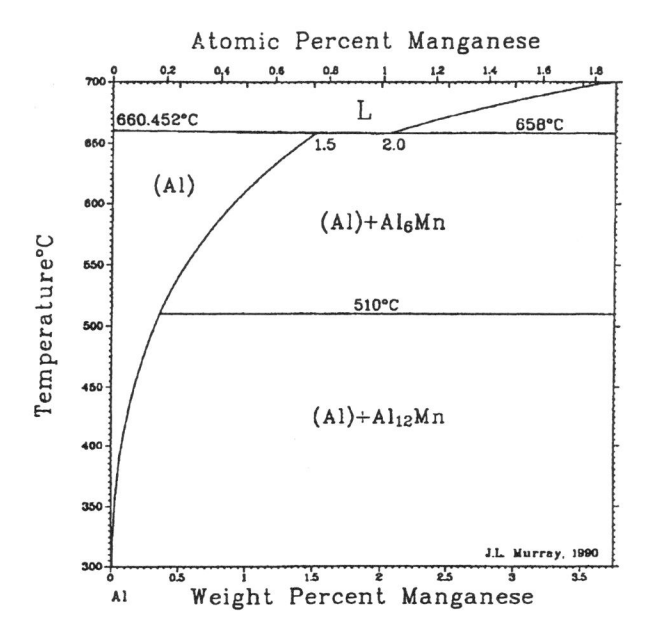

Fig. 6.2 The aluminum-rich portion of the aluminum-manganese phase diagram. This system is the basis for the wrought 3xxx aluminum alloys.

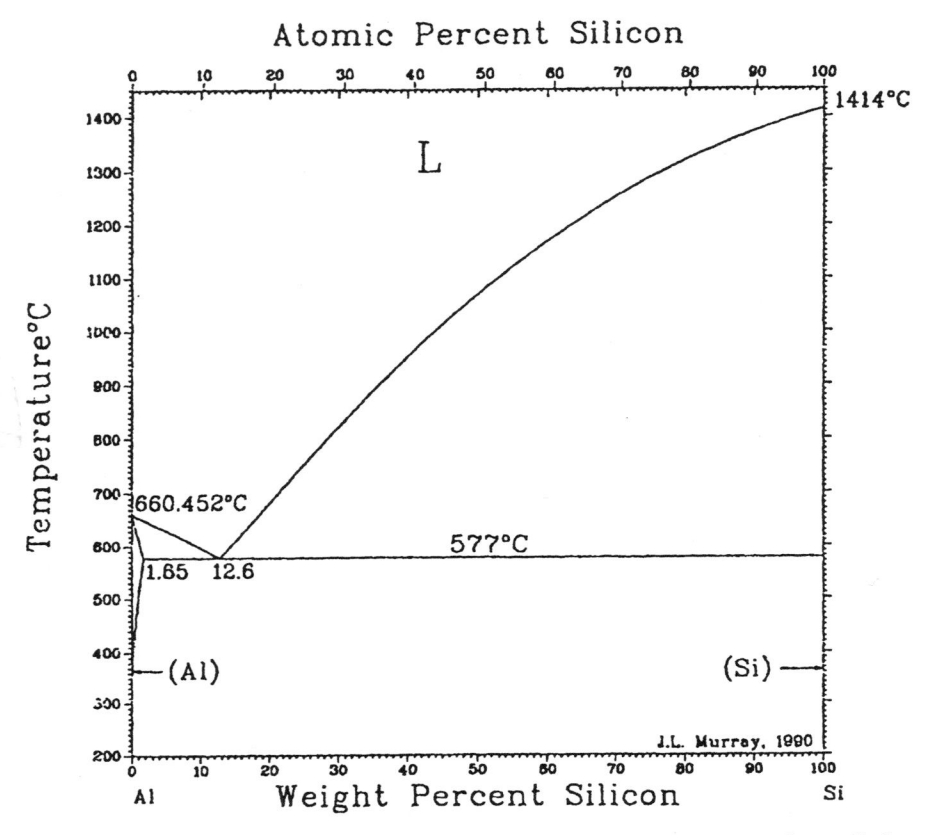

Fig. 6.3 The aluminum-silicon phase diagram. This system is used in 4xxx wrought aluminum alloys and in 3xx.x and 4xx.x cast aluminum alloys. Silicon content ranges from about 5 to 20 wt% in casting alloys.

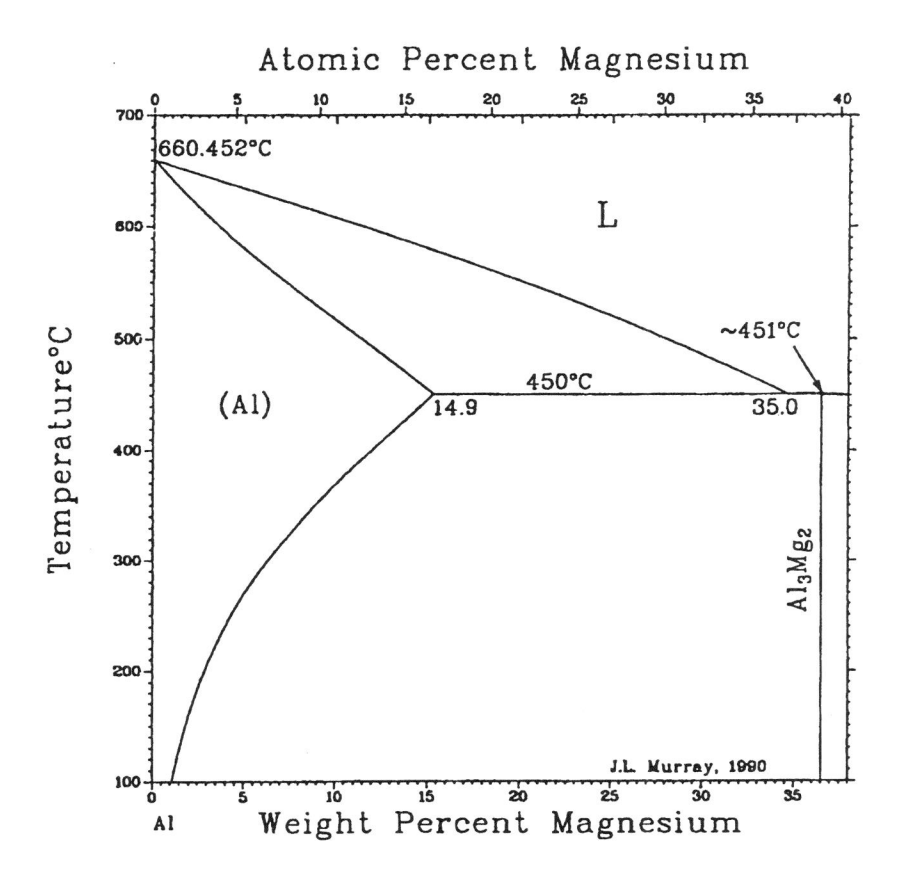

Fig. 6.4 The aluminum-rich portion of the aluminum-magnesium phase diagram. This system is the basis for the wrought 5xxx and cast 5xx.x nonheat-treatable aluminum alloys.

34 Concise Metals Engineering Data Book

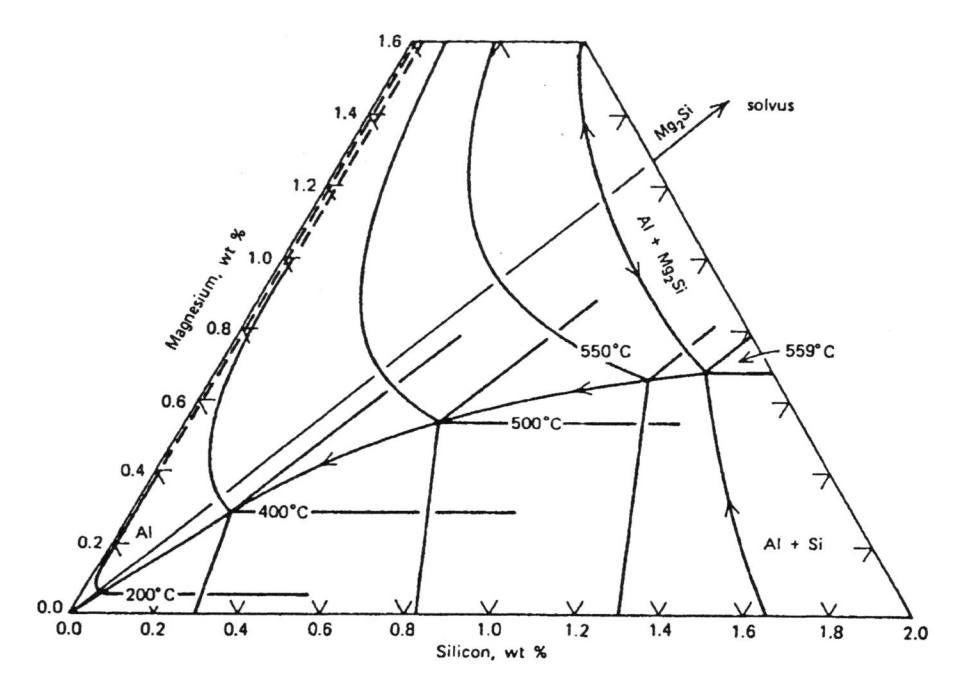

Fig. 6.5 Solvus for the aluminum-rich portion of the aluminum-magnesium-silicon phase diagram. This system is the basis for the wrought 6xxx aluminum alloys.

Phase Diagrams   35

(a)

(b)

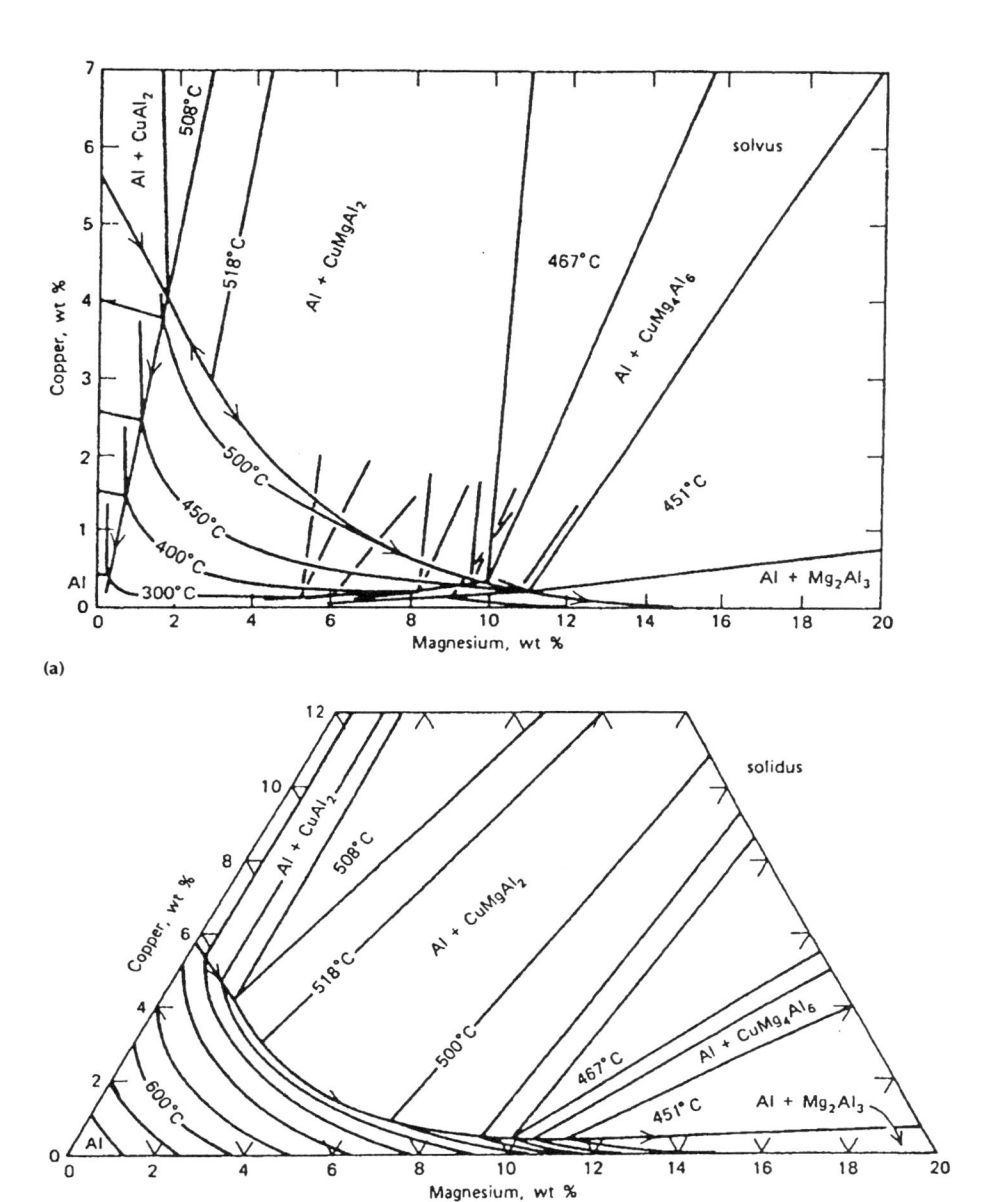

Fig. 6.6 Solvus (a) and solidus (b) for the aluminum-rich portion of the aluminum-copper-magnesium phase diagram. This system is the basis for magnesium-bearing wrought 2xxx aluminum-copper alloys such as 2014, 2024, and 2124.

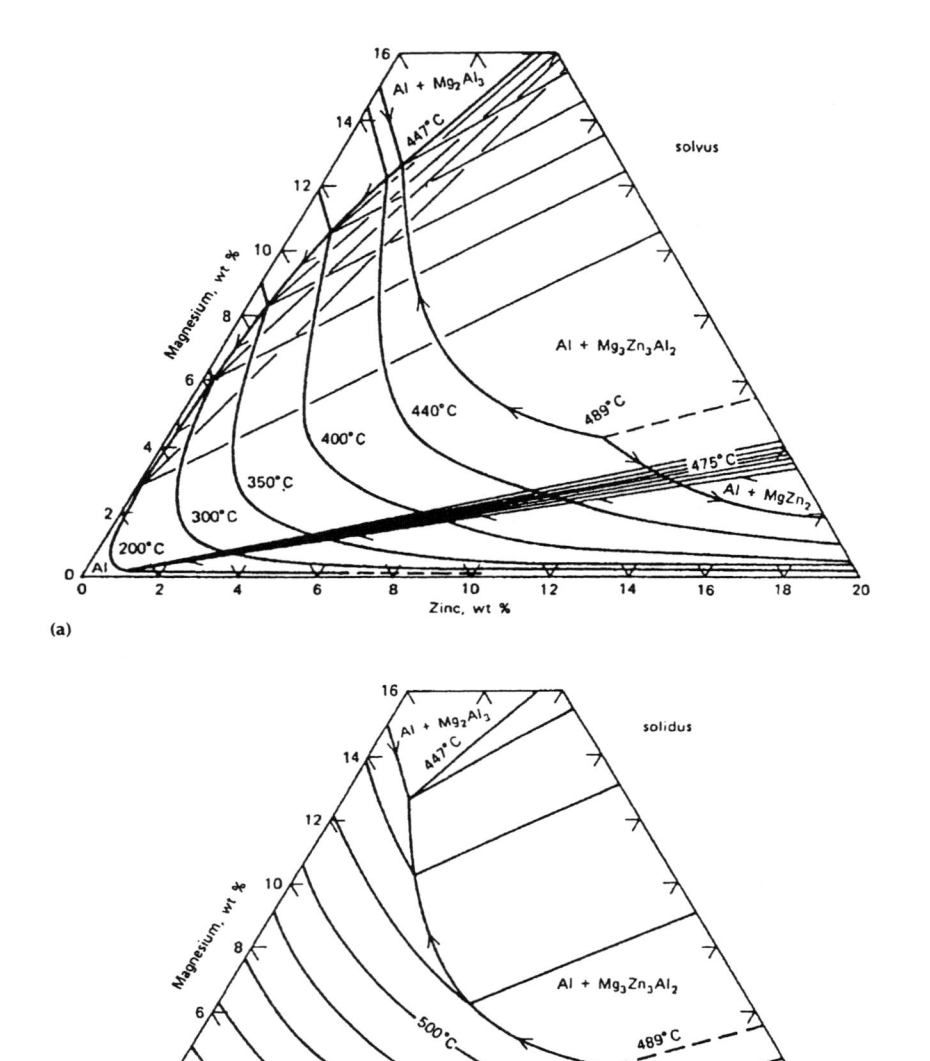

Fig. 6.7 Solvus (a) and solidus (b) for the aluminum-rich portion of the aluminum-magnesium-zinc phase diagram. This system is the basis for the wrought 7xxx aluminum alloys.

Phase Diagrams 37

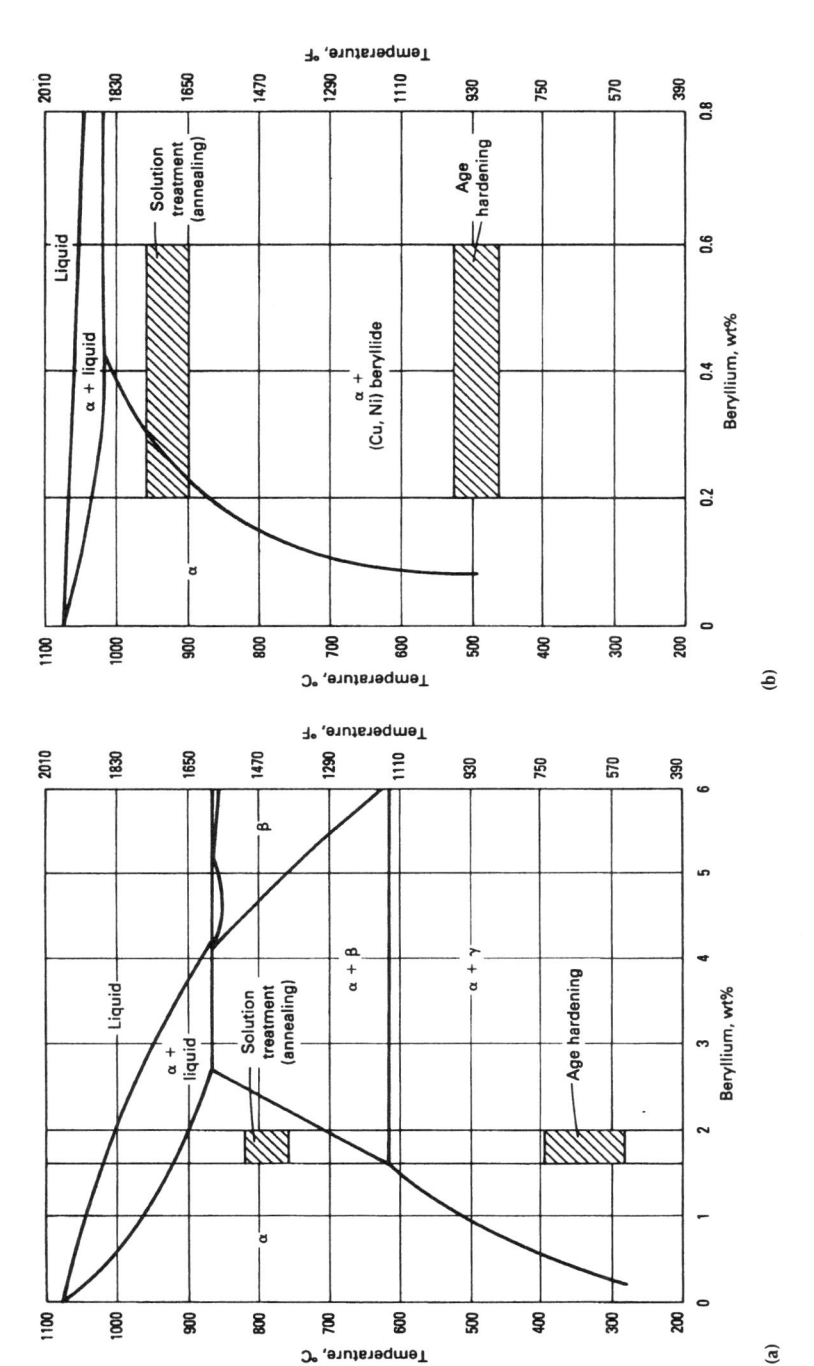

Fig. 6.8 The copper-rich portion of the beryllium-copper phase diagram. This system is the basis for the high-strength high-copper alloys containing from about 0.4 to 2.75 wt% Be. (a) Binary composition for beryllium-copper alloys such as C17200. (b) Pseudobinary composition for C17510, a high-conductivity alloy containing about 1.8 wt% Be.

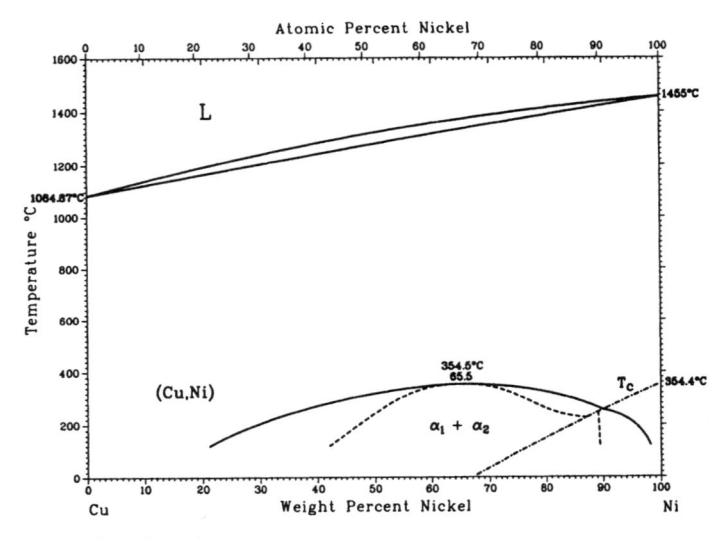

Fig. 6.9 The copper-nickel phase diagram

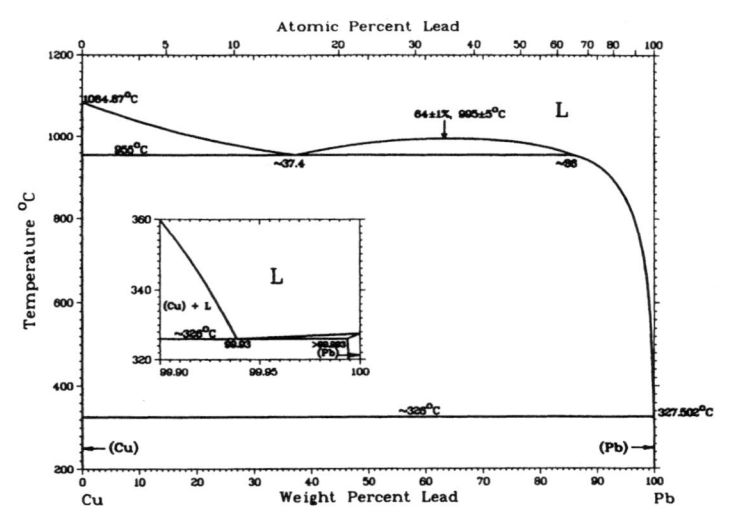

Fig. 6.10 The copper-lead phase diagram

Phase Diagrams 39

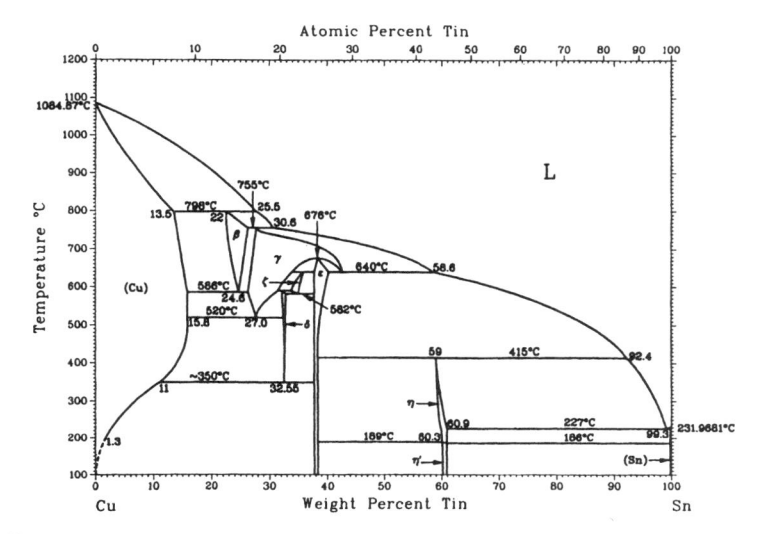

Fig. 6.11 The copper-tin phase diagram

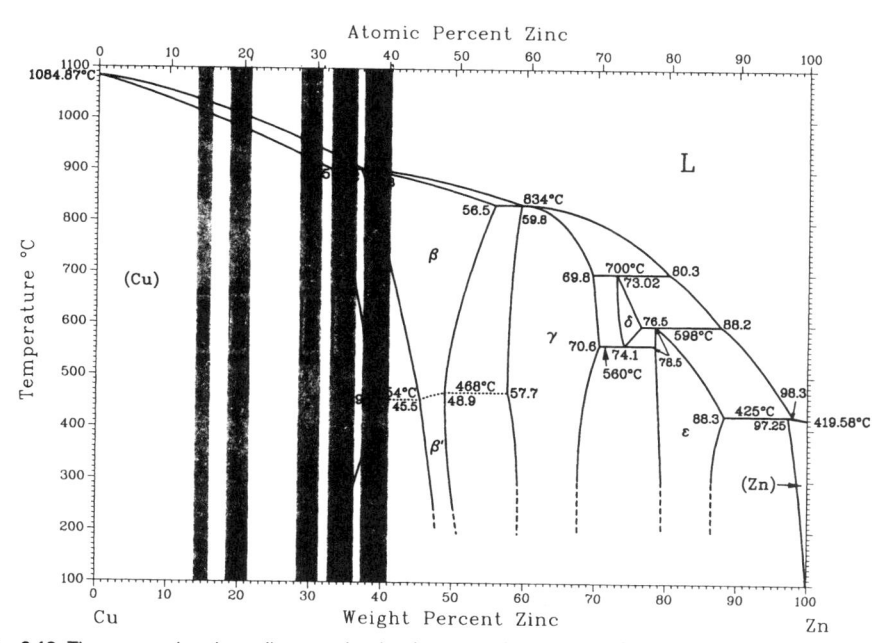

Fig. 6.12 The copper-zinc phase diagram, showing the composition range for five common brasses

40 Concise Metals Engineering Data Book

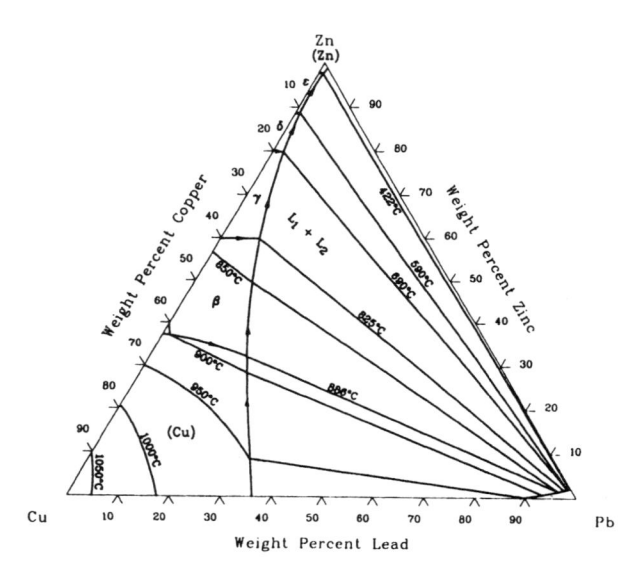

Fig. 6.13 The liquidus projection for the copper-lead-zinc system

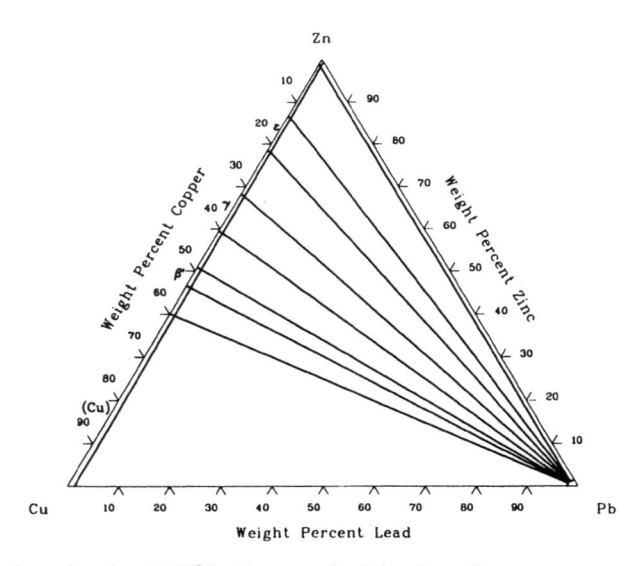

Fig. 6.14 The isothermal section at 25 °C for the copper-lead-zinc phase diagram

Phase Diagrams 41

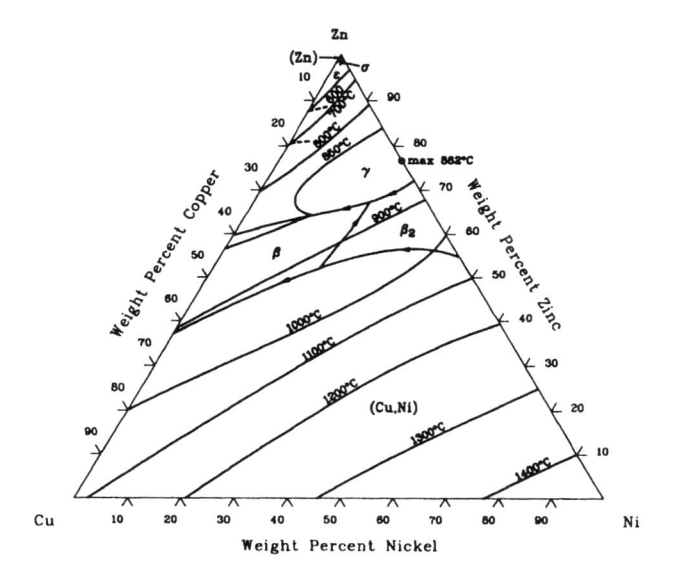

Fig. 6.15 The liquidus projection for the copper-nickel-zinc system

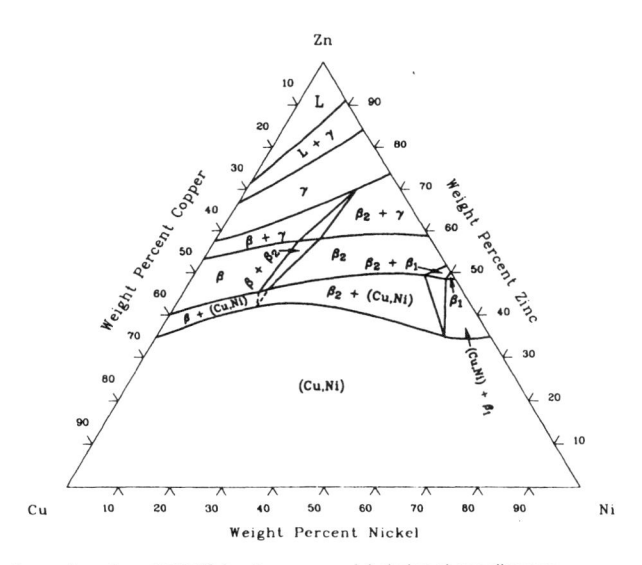

Fig. 6.16 The isothermal section at 775 °C for the copper-nickel-zinc phase diagram

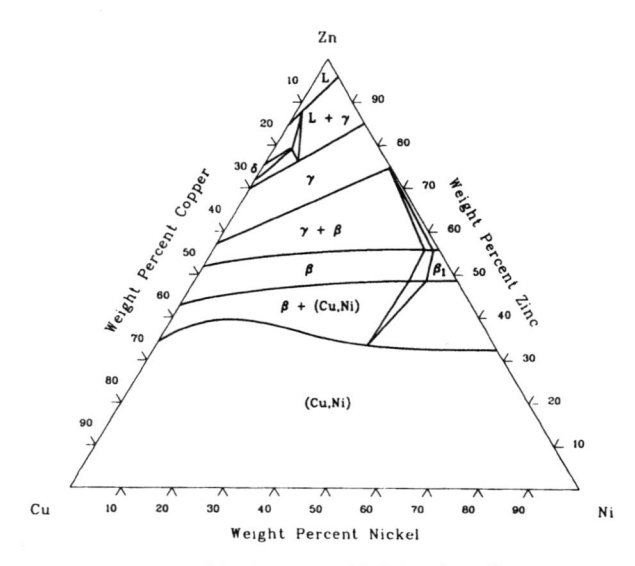

Fig. 6.17 The isothermal section at 650 °C for the copper-nickel-zinc phase diagram

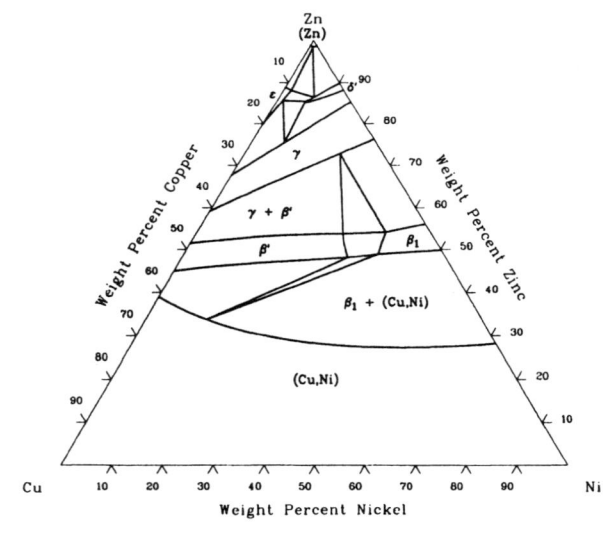

Fig. 6.18 The isothermal section at 20 °C for the copper-nickel-zinc phase diagram

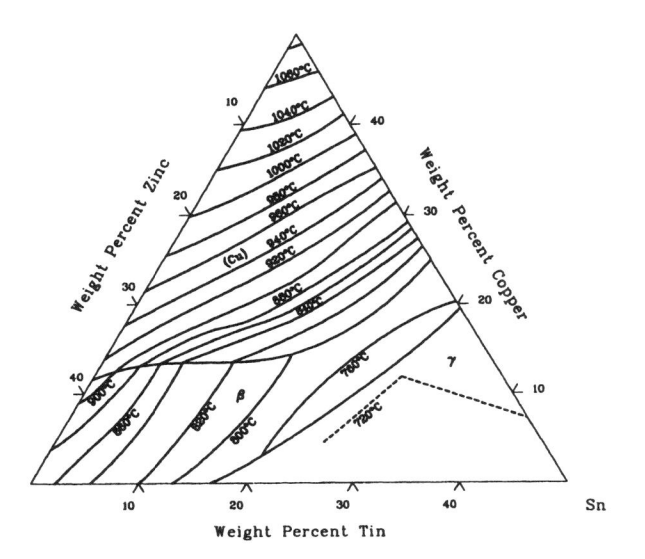

Fig. 6.19 The liquidus projection for the copper-tin-zinc system

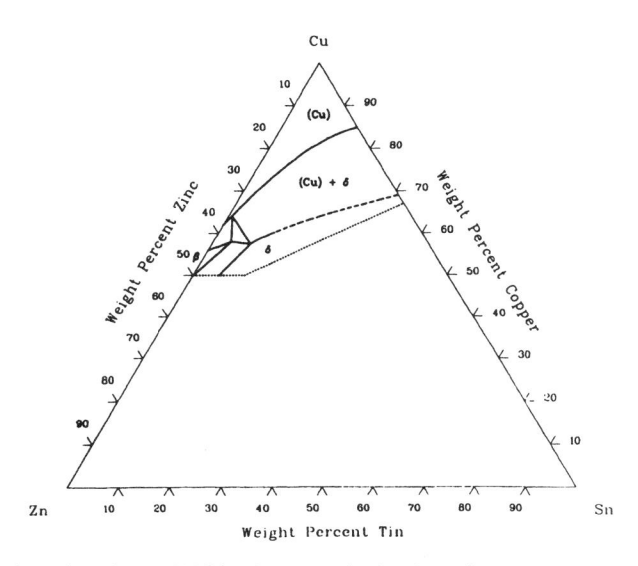

Fig. 6.20 The isothermal section at 500 °C for the copper-tin-zinc phase diagram

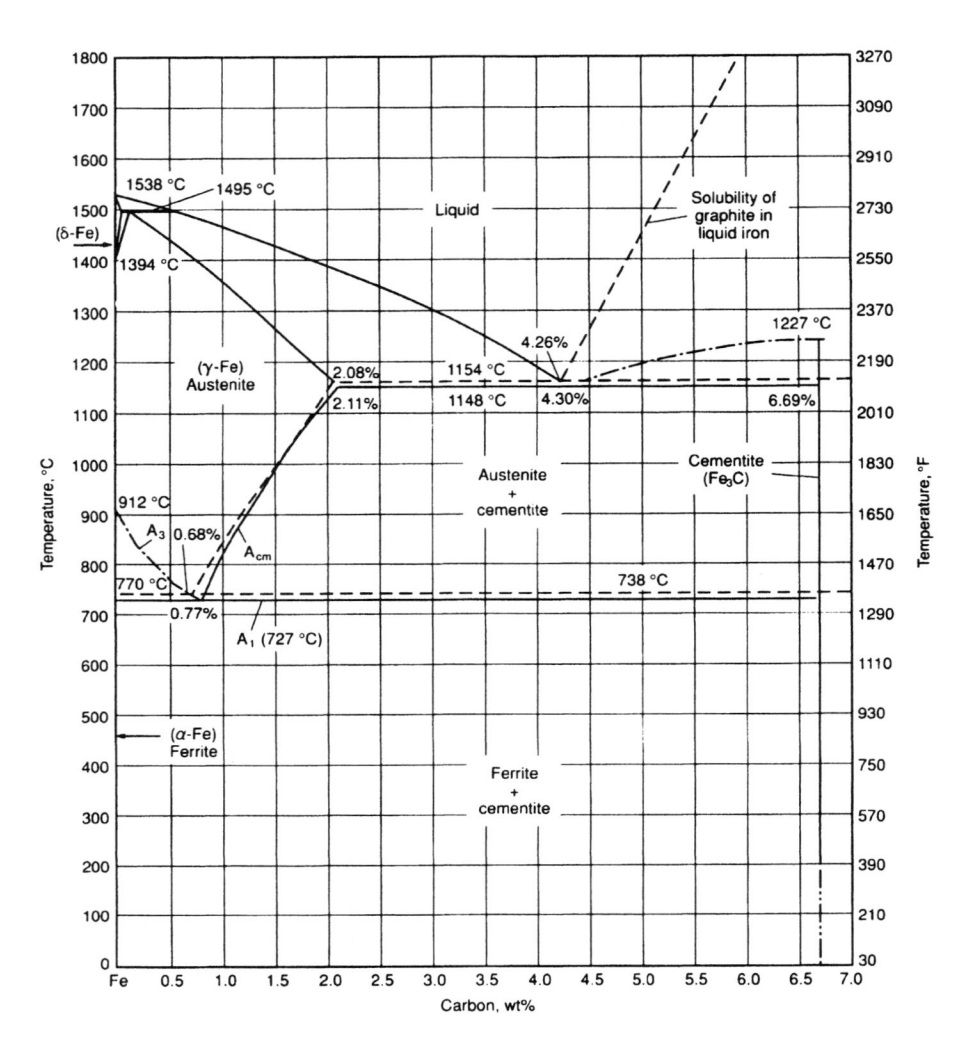

Fig. 6.21 The iron-rich portion of the iron-carbon phase diagram (up to 6.67 wt% C). Solid lines indicate Fe-Fe$_3$C (cementite) diagram; dashed lines indicate iron-graphite diagram.

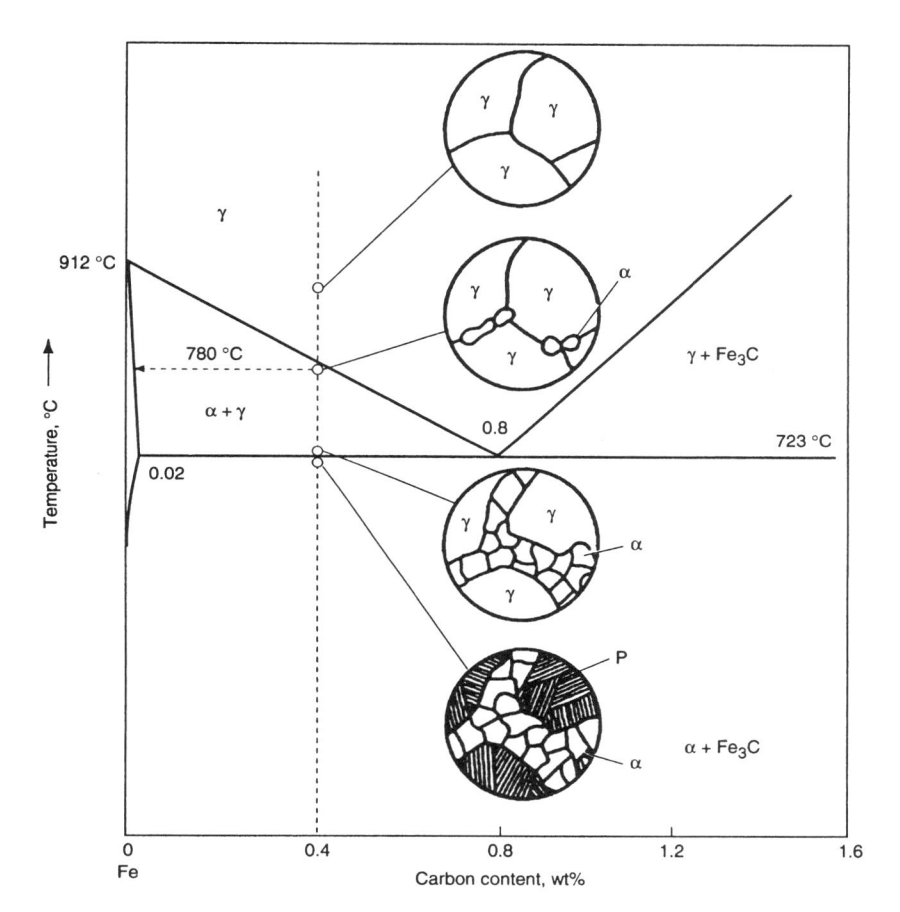

Fig. 6.22 The low-temperature iron-rich portion of the iron-carbon phase diagram illustrating the microstructural evolution of a 0.40 wt% C steel upon slow cooling to room temperature from the γ (austenite) phase field. When the steel is coooled below 780 °C, it enters a two-phase region of α (ferrite) and γ. As it cools, iron-rich α particles may precipitate from γ and, in many cases, become situated along γ grain boundaries. On further cooling, the amount of α-ferrite increases and, at a temperature slightly above 723 °C, the steel is now about 50% α and 50% γ. On cooling below 723 °C, the γ transforms to Fe$_3$C (cemenite). The steel structure now consists of a mixture of α-ferrite and Fe$_3$C referred to as pearlite (P). Such a microstructure is typical of the common ferritic-pearlitic steels.

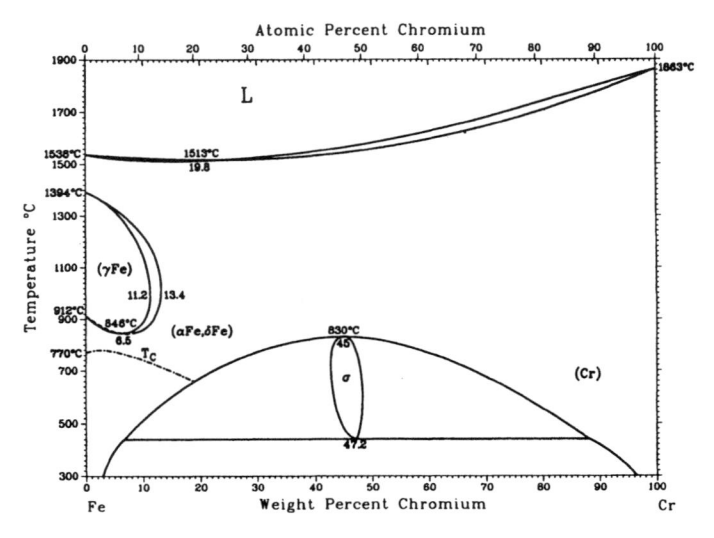

Fig. 6.23 The iron-chromium phase diagram. Chromium serves as a ferrite (α) stabilizer.

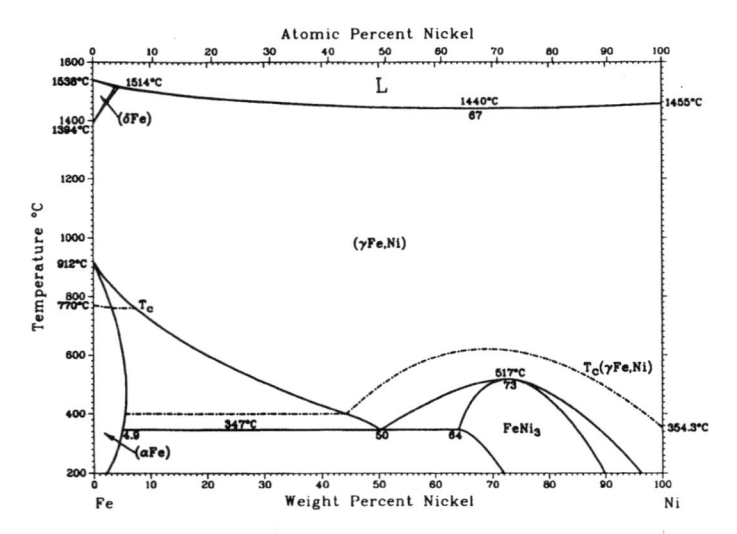

Fig. 6.24 The iron-nickel phase diagram. Nickel serves as an austenite (γ) stabilizer.

Phase Diagrams 47

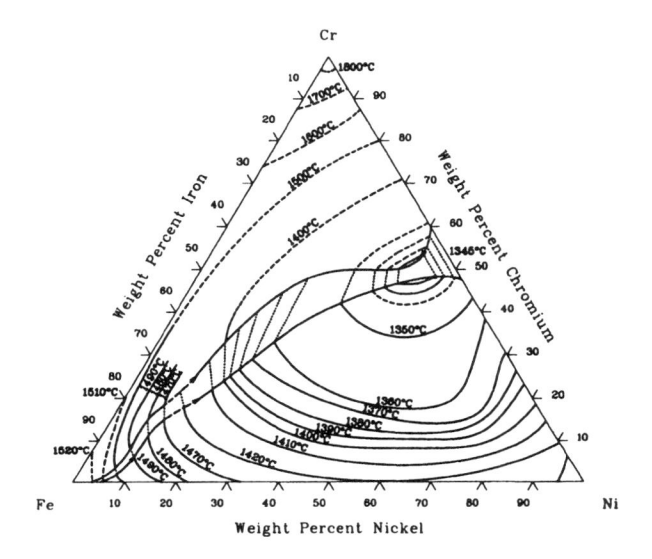

Fig. 6.25 The solidus projection for the iron-chromium-nickel system

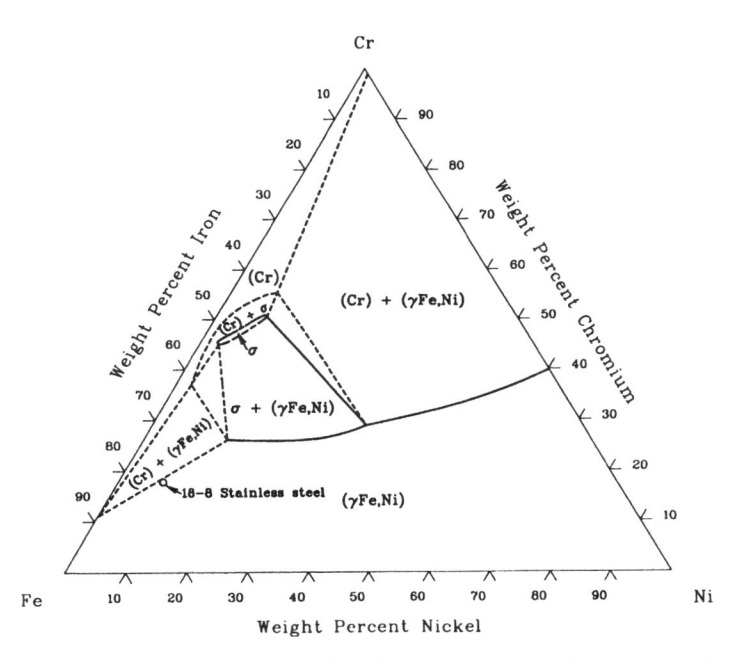

Fig. 6.26 The isothermal section at 900 °C (1652 °F) of the iron-chromium-nickel ternary phase diagram, showing the nominal composition of 18-8 stainless steel

7 Chemical Compositions of Metals and Alloys

Table 7.1 Guide to the Unified Numbering System (UNS) for metals and alloys
For additional details on the UNS, see the combined ASTM E 527/SAE J1086 standard, "Recommended Practice for Numbering Metals and Alloys."

UNS series	Metal/alloy	UNS series	Metal/alloy
Nonferrous metals and alloys		**Nonferrous metals and alloys (continued)**	
A00001–A99999	Aluminum and aluminum alloys	L01001–L01999	Cadmium
		L02001–L02999	Cesium
C00001–C99999	Copper and copper alloys	L03001–L03999	Gallium
E00001–E99999	Rare earth and rare earth-like metals and alloys	L04001–L04999	Indium
		L06001–L06999	Lithium
E00001–E00999	Actinium	L07001–L07999	Mercury
E01000–E20999	Cerium	L08001–L08999	Potassium
E21000–E45999	Mixed rare earths (e.g., mischmetal)	L09001–L09999	Rubidium
		L10001–L10999	Selenium
E46000–E47999	Dysprosium	L11001–L11999	Sodium
E48000–E49999	Erbium	L13001–L13999	Tin
E50000–E51999	Europium	L50001–L59999	Lead
E52000–E55999	Gadolinium	M00001–M99999	Miscellaneous nonferrous metals and alloys
E56000–E57999	Holmium		
E58000–E67999	Lanthanum	M00001–M00999	Antimony
E68000–E68999	Lutetium	M01001–M01999	Arsenic
E69000–E73999	Neodymium	M02001–M02999	Barium
E74000–E77999	Praseodymium	M03001–M03999	Calcium
E78000–E78999	Promethium	M04001–M04999	Germanium
E79000–E82999	Samarium	M05001–M05999	Plutonium
E83000–E84999	Scandium	M06001–M06999	Strontium
E85000–E86999	Terbium	M07001–M07999	Tellurium
E87000–E87999	Thulium	M08001–M08999	Uranium
E88000–E89999	Ytterbium	M10001–M19999	Magnesium
E90000–E99999	Yttrium	M20001–M29999	Manganese
L00001–L99999	Low-melting-point metals and alloys	M30001–M39999	Silicon
		P00001–P99999	Precious metals and alloys
L00001–L00999	Bismuth	P00001–P00999	Gold

(continued)

Chemical Compositions of Metals and Alloys 49

Table 7.1 (continued)

UNS series	Metal/alloy	UNS series	Metal/alloy
Nonferrous metals and alloys (continued)		**Ferrous metals and alloys (continued)**	
P01001–P01999	Iridium	G00001–G99999	AISI and SAE carbon and alloy steels (except tool steels)
P02001–P02999	Osmium		
P03001–P03999	Palladium		
P04001–P04999	Platinum	H00001–H99999	AISI and SAE H-steels (carbon, carbon-boron, and alloy H-steels
P05001–P05999	Rhodium		
P06001–P06999	Ruthenium		
P07001–P07999	Silver	J00001–J99999	Cast steels (except tool steels)
R00001–R99999	Reactive and refractory metals and alloys	K00001–K99999	Miscellaneous steels and ferrous alloys
R01001–R01999	Boron	S00001–S99999	Heat- and corrosion-resistant (stainless) steels
R02001–R02999	Hafnium		
R03001–R03999	Molybdenum	T00001–T99999	Tool steels
R04001–R04999	Niobium (Columbium)	**Welding filler metals**	
R05001–R05999	Tantalum	W00001–W99999	Welding filler metals, covered and tubular electrodes, classified by weld deposit composition
R06001–R06999	Thorium		
R07001–R07999	Tungsten		
R08001–R08999	Vanadium		
R10001–R19999	Beryllium	W00001–W09999	Carbon steel with no significant alloying elements
R20001–R29999	Chromium		
R30001–R39999	Cobalt		
R40001–R49999	Rhenium	W10000–W19999	Manganese-molybdenum low-alloys steels
R50001–R59999	Titanium		
R60001–R69999	Zirconium	W20000–W29999	Nickel low-alloy steels
Z00001–Z99999	Zinc and zinc alloys	W30000–W39999	Austenitic stainless steels
Ferrous metals and alloys		W40000–W49999	Ferritic stainless steels
D00001–D99999	Specified mechanical properties of steels	W50000–W59999	Chromium low-alloy steels
		W60000–W69999	Copper-base alloys
F00001–F99999	Cast irons (gray, malleable, and ductile irons)	W70000–W79999	Surfacing alloys
		W80000–W89999	Nickel-base alloys

Table 7.2 SAE-AISI system of designations for carbon and alloy steels

Numerals and digits	Type of steel and nominal alloy content, %	Numerals and digits	Type of steel and nominal alloy content, %
Carbon steels		**Nickel-chromium steels**	
10xx(a)	Plain carbon (Mn 1.00 max)	31xx	Ni 1.25; Cr 0.65 and 0.80
11xx	Resulfurized	32xx	Ni 1.75; Cr 1.07
12xx	Resulfurized and rephosphorized	33xx	Ni 3.50; Cr 1.50 and 1.57
15xx	Plain carbon (max Mn range: 1.00–1.65)	34xx	Ni 3.00; Cr 0.77
Manganese steels		**Molybdenum steels**	
13xx	Mn 1.75	40xx	Mo 0.20 and 0.25
Nickel steels		44xx	Mo 0.40 and 0.52
23xx	Ni 3.50	**Chromium-molybdenum steels**	
25xx	Ni 5.00	41xx	Cr 0.50, 0.80, and 0.95; Mo 0.12, 0.20, 0.25, and 0.30

(continued)

(a) The xx in the last two digits of these designations indicates that the carbon content (in hundredths of a percent) is to be inserted.

Table 7.2 (continued)

Numerals and digits	Type of steel and nominal alloy content, %
Nickel-chromium-molybdenum steels	
43xx	Ni 1.82; Cr 0.50 and 0.80; Mo 0.25
43BVxx	Ni 1.82; Cr 0.50; Mo 0.12 and 0.25; V 0.03 min
47xx	Ni 1.05; Cr 0.45; Mo 0.20 and 0.35
81xx	Ni 0.30; Cr 0.40; Mo 0.12
86xx	Ni 0.55; Cr 0.50; Mo 0.20
87xx	Ni 0.55; Cr 0.50; Mo 0.25
88xx	Ni 0.55; Cr 0.50; Mo 0.35
93xx	Ni 3.25; Cr 1.20; Mo 0.12
94xx	Ni 0.45; Cr 0.40; Mo 0.12
97xx	Ni 0.55; Cr 0.20; Mo 0.20
98xx	Ni 1.00; Cr 0.80; Mo 0.25
Nickel-molybdenum steels	
46xx	Ni 0.85 and 1.82; Mo 0.20 and 0.25
48xx	Ni 3.50; Mo 0.25
Chromium steels	
50xx	Cr 0.27, 0.40, 0.50, and 0.65
51xx	Cr 0.80, 0.87, 0.92, 0.95, 1.00, and 1.05

Numerals and digits	Type of steel and nominal alloy content, %
Chromium steels (continued)	
50xxx	Cr 0.50; C 1.00 min
51xxx	Cr 1.02; C 1.00 min
52xxx	Cr 1.45; C 1.00 min
Chromium-vanadium steels	
61xx	Cr 0.60, 0.80, and 0.95; V 0.10 and 0.15 min
Tungsten-chromium steel	
72xx	W 1.75; Cr 0.75
Silicon-manganese steels	
92xx	Si 1.40 and 2.00; Mn 0.65, 0.82, and 0.85; Cr 0 and 0.65
Boron steels	
xxBxx	B denotes boron steel
Leaded steels	
xxLxx	L denotes leaded steel
Vanadium steels	
xxVxx	V denotes vanadium steel

(a) The xx in the last two digits of these designations indicates that the carbon content (in hundredths of a percent) is to be inserted.

Table 7.3 Carbon steel compositions

Applicable to semifinished products for forging, hot-rolled and cold-finished bars, wire rods, and seamless tubing

UNS No.	SAE-AISI No.	C	Mn	P (max)	S (max)
G10050	1005	0.06 max	0.35 max	0.040	0.050
G10060	1006	0.08 max	0.25–0.40	0.040	0.050
G10080	1008	0.10 max	0.30–0.50	0.040	0.050
G10100	1010	0.08–0.13	0.30–0.60	0.040	0.050
G10120	1012	0.10–0.15	0.30–0.60	0.040	0.050
G10130	1013	0.11–0.16	0.50–0.80	0.040	0.050
G10150	1015	0.13–0.18	0.30–0.60	0.040	0.050
G10160	1016	0.13–0.18	0.60–0.90	0.040	0.050
G10170	1017	0.15–0.20	0.30–0.60	0.040	0.050
G10180	1018	0.15–0.20	0.60–0.90	0.040	0.050
G10190	1019	0.15–0.20	0.70–1.00	0.040	0.050
G10200	1020	0.18–0.23	0.30–0.60	0.040	0.050
G10210	1021	0.18–0.23	0.60–0.90	0.040	0.050
G10220	1022	0.18–0.23	0.70–1.00	0.040	0.050

(continued)

(a) When silicon ranges or limits are required for bar and semifinished products, the following ranges are commonly used: 0.10% max; 0.10–0.20%; 0.15–0.35%; 0.20–0.40%; or 0.30–0.60%. For rods, the following ranges are commonly used: 0.10 max; 0.07–0.15%; 0.10–0.20%; 0.15–0.35%; 0.20–0.40%; and 0.30–0.60%. Steels listed in this table can be produced with additions of lead or boron. Leaded steels typically contain 0.15–0.35% Pb and are identified by inserting the letter L in the designation (10L45); boron steels can be expected to contain 0.0005–0.003% B and are identified by inserting the letter B in the designation (10B46).

Table 7.3 (continued)

UNS No.	Designation SAE-AISI No.	C	Mn	P (max)	S (max)
G10230	1023	0.20–0.25	0.30–0.60	0.040	0.050
G10250	1025	0.22–0.28	0.30–0.60	0.040	0.050
G10260	1026	0.22–0.28	0.60–0.90	0.040	0.050
G10290	1029	0.25–0.31	0.60–0.90	0.040	0.050
G10300	1030	0.28–0.34	0.60–0.90	0.040	0.050
G10350	1035	0.32–0.38	0.60–0.90	0.040	0.050
G10370	1037	0.32–0.38	0.70–1.00	0.040	0.050
G10380	1038	0.35–0.42	0.60–0.90	0.040	0.050
G10390	1039	0.37–0.44	0.70–1.00	0.040	0.050
G10400	1040	0.37–0.44	0.60–0.90	0.040	0.050
G10420	1042	0.40–0.47	0.60–0.90	0.040	0.050
G10430	1043	0.40–0.47	0.70–1.00	0.040	0.050
G10440	1044	0.43–0.50	0.30–0.60	0.040	0 050
G10450	1045	0.43–0.50	0.60–0.90	0.040	0.050
G10460	1046	0.43–0.50	0.70–1.00	0.040	0.050
G10490	1049	0.46–0.53	0.60–0.90	0.040	0.050
G10500	1050	0.48–0.55	0.60–0.90	0.040	0.050
G10530	1053	0.48–0.55	0.70–1.00	0.040	0.050
G10550	1055	0.50–0.60	0.60–0.90	0.040	0.050
G10590	1059	0.55–0.65	0.50–0.80	0.040	0.050
G10600	1060	0.55–0.65	0.60–0.90	0.040	0.050
G10640	1064	0.60–0.70	0.50–0.80	0.040	0.050
G10650	1065	0.60–0.70	0.60–0.90	0.040	0.050
G10690	1069	0.65–0.75	0.40–0.70	0.040	0.050
G10700	1070	0.65–0.75	0.60–0.90	0.040	0.050
G10740	1074	0.70–0.80	0.50–0.80	0.040	0.050
G10750	1075	0.70–0.80	0.40–0.70	0.040	0.050
G10780	1078	0.72–0.85	0.30–0.60	0.040	0.050
G10800	1080	0.75–0.88	0.60–0.90	0.040	0.050
G10840	1084	0.80–0.93	0.60–0.90	0.040	0.050
G10850	1085	0.80–0.93	0.70–1.00	0.040	0.050
G10860	1086	0.80–0.93	0.30–0.50	0.040	0.050
G10900	1090	0.85–0.98	0.60–0.90	0.040	0.050
G10950	1095	0.90–1.03	0.30–0.50	0.040	0.050

(a) When silicon ranges or limits are required for bar and semifinished products, the following ranges are commonly used: 0.10% max; 0.10 – 0.20%; 0.15 – 0.35%; 0.20 – 0.40%; or 0.30 – 0.60%. For rods, the following ranges are commonly used: 0.10 max; 0.07–0.15%; 0.10–0.20%; 0.15–0.35%; 0.20–0.40%; and 0.30–0.60%. Steels listed in this table can be produced with additions of lead or boron. Leaded steels typically contain 0.15–0.35% Pb and are identified by inserting the letter L in the designation (10L45); boron steels can be expected to contain 0.0005–0.003% B and are identified by inserting the letter B in the designation (10B46).

Table 7.4 Carbon steel compositions

Applicable only to structural shapes, plates, strip, sheets, and welded tubing

UNS No.	Designation SAE-AISI No.	C	Mn	P max	S max
G10060	1006	0.08 max	0.45 max	0.040	0.050
G10080	1008	0.10 max	0.50 max	0.040	0.050
G10090	1009	0.15 max	0.60 max	0.040	0.050
G10100	1010	0.08–0.13	0.30–0.60	0.040	0.050
G10120	1012	0.10–0.15	0.30–0.60	0.040	0.050
G10150	1015	0.12–0.18	0.30–0.60	0.040	0.050
G10160	1016	0.12–0.18	0.60–0.90	0.040	0.050
G10170	1017	0.14–0.20	0.30–0.60	0.040	0.050
G10180	1018	0.14–0.20	0.60–0.90	0.040	0.050
G10190	1019	0.14–0.20	0.70–1.00	0.040	0.050
G10200	1020	0.17–0.23	0.30–0.60	0.040	0.050
G10210	1021	0.17–0.23	0.60–0.90	0.040	0.050
G10220	1022	0.17–0.23	0.70–1.00	0.040	0.050
G10230	1023	0.19–0.25	0.30–0.60	0.040	0.050
G10250	1025	0.22–0.28	0.30–0.60	0.040	0.050
G10260	1026	0.22–0.28	0.60–0.90	0.040	0.050
G10300	1030	0.27–0.34	0.60–0.90	0.040	0.050
G10330	1033	0.29–0.36	0.70–1.00	0.040	0.050
G10350	1035	0.31–0.38	0.60–0.90	0.040	0.050
G10370	1037	0.31–0.38	0.70–1.00	0.040	0.050
G10380	1038	0.34–0.42	0.60–0.90	0.040	0.050
G10390	1039	0.36–0.44	0.70–1.00	0.040	0.050
G10400	1040	0.36–0.44	0.60–0.90	0.040	0.050
G10420	1042	0.39–0.47	0.60–9.90	0.040	0.050
G10430	1043	0.39–0.47	0.70–1.00	0.040	0.050
G10450	1045	0.42–0.50	0.60–0.90	0.040	0.050
G10460	1046	0.42–0.50	0.70–1.00	0.040	0.050
G10490	1049	0.45–0.53	0.60–0.90	0.040	0.050
G10500	1050	0.47–0.55	0.60–0.90	0.040	0.050
G10550	1055	0.52–0.60	0.60–0.90	0.040	0.050
G10600	1060	0.55–0.66	0.60–0.90	0.040	0.050
G10640	1064	0.59–0.70	0.50–0.80	0.040	0.050
G10650	1065	0.59–0.70	0.60–0.90	0.040	0.050
G10700	1070	0.65–0.76	0.60–0.90	0.040	0.050
G10740	1074	0.69–0.80	0.50–0.80	0.040	0.050
G10750	1075	0.69–0.80	0.40–0.70	0.040	0.050
G10780	1078	0.72–0.86	0.30–0.60	0.040	0.050
G10800	1080	0.74–0.88	0.60–0.90	0.040	0.050
G10840	1084	0.80–0.94	0.60–0.90	0.040	0.050
G10850	1085	0.80–0.94	0.70–1.00	0.040	0.050
G10860	1086	0.80–0.94	0.30–0.50	0.040	0.050
G10900	1090	0.84–0.98	0.60–0.90	0.040	0.050
G10950	1095	0.90–1.04	0.30–0.50	0.040	0.050

(a) When silicon ranges or limits are required, the following ranges and limits are commonly used: up to SAE 1025 inclusive, 0.10% max, 0.10–0.25%, or 0.15–0.35%. Over SAE 1025, 0.10–0.25% or 0.15–0.35%.

Chemical Compositions of Metals and Alloys 53

Table 7.5 Free-cutting (resulfurized) carbon steel compositions

Applicable to semifinished products for forging, hot-rolled and cold-finished bars, wire rods, and seamless tubing

UNS No.	Designation SAE-AISI No.	C	Mn	P (max)	S
G11080	1108	0.08–0.13	0.50–0.80	0.040	0.08–0.13
G11100	1110	0.08–0.13	0.30–0.60	0.040	0.08–0.13
G11170	1117	0.14–0.20	1.00–1.30	0.040	0.08–0.13
G11180	1118	0.14–0.20	1.30–1.60	0.040	0.08–0.13
G11370	1137	0.32–0.39	1.35–1.65	0.040	0.08–0.13
G11390	1139	0.35–0.43	1.35–1.65	0.040	0.13–0.20
G11400	1140	0.37–0.44	0.70–1.00	0.040	0.08–0.13
G11410	1141	0.37–0.45	1.35–1.65	0.040	0.08–0.13
G11440	1144	0.40–0.48	1.35–1.65	0.040	0.24–0.33
G11460	1146	0.42–0.49	0.70–1.00	0.040	0.08–0.13
G11510	1151	0.48–0.55	0.70–1.00	0.040	0.08–0.13

(a) When lead ranges or limits are required, or when silicon ranges or limits are required for bars or semifinished products, the values in Table 7.3 apply. For rods, the following ranges and limits for silicon are commonly used: up to SAE 1110 inclusive, 0.10% max; SAE 1117 and over, 0.10% max, 0.10–0.20%, or 0.15–0.35%.

Table 7.6 Free-cutting (rephosphorized and resulfurized) carbon steel compositions

Applicable to semifinished products for forging, hot-rolled and cold-finished bars, wire rods, and seamless tubing

UNS No.	Designation SAE-AISI No.	C (max)	Mn	P	S	Pb
G12110	1211	0.13	0.60–0.90	0.07–0.12	0.10–0.15	...
G12120	1212	0.13	0.70–1.00	0.07–0.12	0.16–0.23	...
G12130	1213	0.13	0.70–1.00	0.07–0.12	0.24–0.33	...
G12150	1215	0.09	0.75–1.05	0.04–0.09	0.26–0.35	...
G12144	12L14	0.15	0.85–1.15	0.04–0.09	0.26–0.35	0.15–0.35

(a) When lead ranges or limits are required, the values in Table 7.3 apply. It is not common practice to produce the 12xx series of steels to specified limits for silicon because of its adverse effect on machinability.

Table 7.7 High-manganese carbon steel compositions

Applicable only to semifinished products for forging, hot-rolled and cold-finished bars, wire rods, and seamless tubing

UNS No.	Designation SAE-AISI No.	C	Mn	P (max)	S (max)
G15130	1513	0.10–0.16	1.10–1.40	0.040	0.050
G15220	1522	0.18–0.24	1.10–1.40	0.040	0.050
G15240	1524	0.19–0.25	1.35–1.65	0.040	0.050
G15260	1526	0.22–0.29	1.10–1.40	0.040	0.050
G15270	1527	0.22–0.29	1.20–1.50	0.040	0.050
G15360	1536	0.30–0.37	1.20–1.50	0.040	0.050
G15410	1541	0.36–0.44	1.35–1.65	0.040	0.050
G15480	1548	0.44–0.52	1.10–1.40	0.040	0.050
G15510	1551	0.45–0.56	0.85–1.15	0.040	0.050
G15520	1552	0.47–0.55	1.20–1.50	0.040	0.050
G15610	1561	0.55–0.65	0.75–1.05	0.040	0.050
G15660	1566	0.60–0.71	0.85–1.15	0.040	0.050

(a) When silicon, lead, and boron ranges or limits are required, the values in Table 7.3 apply.

Table 7.8 High-manganese carbon steel compositions

Applicable only to structural shapes, plates, strip, sheets, and welded tubing

UNS No.	Designation SAE-AISI No.	C (max)	Mn	P (max)	S (max)	Former SAE No.
G15240	1524	0.18–0.25	1.30–1.65	0.040	0.050	1024
G15270	1527	0.22–0.29	1.20–1.55	0.040	0.050	1027
G15360	1536	0.30–0.38	1.20–1.55	0.040	0.050	1036
G15410	1541	0.36–0.45	1.30–1.65	0.040	0.050	1041
G15480	1548	0.43–0.52	1.05–1.40	0.040	0.050	1048
G15520	1552	0.46–0.55	1.20–1.55	0.040	0.050	1052

(a) When silicon ranges or limits are required, the values shown in Table 7.4 apply.

Chemical Compositions of Metals and Alloys 55

Table 7.9 Low-alloy steel compositions applicable to billets, blooms, slabs, and hot-rolled and cold-finished bars

UNS No.	SAE No.	Corresponding AISI No.	C	Mn	P	S	Si	Ni	Cr	Mo	V
G13300	1330	1330	0.28–0.33	1.60–1.90	0.035	0.040	0.15–0.35	...	...	...	...
G13350	1335	1335	0.33–0.38	1.60–1.90	0.035	0.040	0.15–0.35	...	...	...	...
G13400	1340	1340	0.38–0.43	1.60–1.90	0.035	0.040	0.15–0.35	...	...	...	...
G13450	1345	1345	0.43–0.48	1.60–1.90	0.035	0.040	0.15–0.35	...	...	...	...
G40230	4023	4023	0.20–0.25	0.70–0.90	0.035	0.040	0.15–0.35	...	...	...	...
G40240	4024	4024	0.20–0.25	0.70–0.90	0.035	0.035–0.050	0.15–0.35	...	...	0.20–0.30	...
G40270	4027	4027	0.25–0.30	0.70–0.90	0.035	0.040	0.15–0.35	...	...	0.20–0.30	...
G40280	4028	4028	0.25–0.30	0.70–0.90	0.035	0.035–0.050	0.15–0.35	...	...	0.20–0.30	...
G40320	4032	...	0.30–0.35	0.70–0.90	0.035	0.040	0.15–0.35	...	...	0.20–0.30	...
G40370	4037	4037	0.35–0.40	0.70–0.90	0.035	0.040	0.15–0.35	...	...	0.20–0.30	...
G40420	4042	...	0.40–0.45	0.70–0.90	0.035	0.040	0.15–0.35	...	...	0.20–0.30	...
G40470	4047	4047	0.45–0.50	0.70–0.90	0.035	0.040	0.15–0.35	...	...	0.20–0.30	...
G41180	4118	4118	0.18–0.23	0.70–0.90	0.035	0.040	0.15–0.35	...	0.40–0.60	0.08–0.15	...
G41300	4130	4130	0.28–0.33	0.40–0.60	0.035	0.040	0.15–0.35	...	0.80–1.10	0.15–0.25	...
G41350	4135	...	0.33–0.38	0.70–0.90	0.035	0.040	0.15–0.35	...	0.80–1.10	0.15–0.25	...
G41370	4137	4137	0.35–0.40	0.70–0.90	0.035	0.040	0.15–0.35	...	0.80–1.10	0.15–0.25	...
G41400	4140	4140	0.38–0.43	0.75–1.00	0.035	0.040	0.15–0.35	...	0.80–1.10	0.15–0.25	...
G41420	4142	4142	0.40–0.45	0.75–1.00	0.035	0.040	0.15–0.35	...	0.80–1.10	0.15–0.25	...
G41450	4145	4145	0.41–0.48	0.75–1.00	0.035	0.040	0.15–0.35	...	0.80–1.10	0.15–0.25	...
G41470	4147	4147	0.45–0.50	0.75–1.00	0.035	0.040	0.15–0.35	...	0.80–1.10	0.15–0.25	...
G41500	4150	4150	0.48–0.53	0.75–1.00	0.035	0.040	0.15–0.35	...	0.80–1.10	0.15–0.25	...
G41610	4161	4161	0.56–0.64	0.75–1.00	0.035	0.040	0.15–0.35	...	0.70–0.90	0.25–0.35	...
G43200	4320	4320	0.17–0.22	0.45–0.65	0.035	0.040	0.15–0.35	1.65–2.00	0.40–0.60	0.20–0.30	...
G43400	4340	4340	0.38–0.43	0.60–0.80	0.035	0.040	0.15–0.35	1.65–2.00	0.70–0.90	0.20–0.30	...
G43406	E4340(b)	E4340	0.38–0.43	0.65–0.85	0.025	0.025	0.15–0.35	1.65–2.00	0.70–0.90	0.20–0.30	...
G44220	4422	...	0.20–0.25	0.70–0.90	0.035	0.040	0.15–0.35	...	...	0.35–0.45	...
G44270	4427	...	0.24–0.29	0.70–0.90	0.035	0.040	0.15–0.35	...	...	0.35–0.45	...
G46150	4615	4615	0.13–0.18	0.45–0.65	0.035	0.040	0.15–0.25	1.65–2.00	...	0.20–0.30	...
G46170	4617	...	0.15–0.20	0.45–0.65	0.035	0.040	0.15–0.35	1.65–2.00	...	0.20–0.30	...
G46200	4620	4620	0.17–0.22	0.45–0.65	0.035	0.040	0.15–0.35	1.65–2.00	...	0.20–0.30	...
G46260	4626	4626	0.24–0.29	0.45–0.65	0.035	0.04 max	0.15–0.35	0.70–1.00	...	0.15–0.25	...
G47180	4718	4718	0.16–0.21	0.70–0.90	...	...	...	0.90–1.20	0.35–0.55	0.30–0.40	...
G47200	4720	4720	0.17–0.22	0.50–0.70	0.035	0.040	0.15–0.35	0.90–1.20	0.35–0.55	0.15–0.25	...
G48150	4815	4815	0.13–0.18	0.40–0.60	0.035	0.040	0.15–0.35	3.25–3.75	...	0.20–0.30	...

(continued)

Table 7.9 (continued)

UNS No.	SAE No.	Corresponding AISI No.	C	Mn	P	S	Si	Ni	Cr	Mo	V
G48170	4817	4817	0.15–0.20	0.40–0.60	0.035	0.040	0.15–0.35	3.25–3.75	...	0.20–0.30	...
G48200	4820	4820	0.18–0.23	0.50–0.70	0.035	0.040	0.15–0.35	3.25–3.75	...	0.20–0.30	...
G50401	50B40(c)	...	0.38–0.43	0.75–1.00	0.035	0.040	0.15–0.35	...	0.40–0.60	...	...
G50441	50B44(c)	50B44	0.43–0.48	0.75–1.00	0.035	0.040	0.15–0.35	...	0.40–0.60	...	...
G50460	5046	...	0.43–0.48	0.75–1.00	0.035	0.040	0.15–0.35	...	0.20–0.35	...	...
G50461	50B46(c)	50B46	0.44–0.49	0.75–1.00	0.035	0.040	0.15–0.35	...	0.20–0.35	...	...
G50501	50B50(c)	50B50	0.48–0.53	0.75–1.00	0.035	0.040	0.15–0.35	...	0.40–0.60	...	...
G50600	5060	...	0.56–0.64	0.75–1.00	0.035	0.040	0.15–0.35	...	0.40–0.60	...	...
G50601	50B60(c)	50B60	0.56–0.64	0.75–1.00	0.035	0.040	0.15–0.35	...	0.40–0.60	...	...
G51150	5115	...	0.13–0.18	0.70–0.90	0.035	0.040	0.15–0.35	...	0.70–0.90	...	...
G51170	5117	5117	0.15–0.20	0.70–0.90	0.035	0.040	0.15–0.35	...	0.70–0.90	...	...
G51200	5120	5120	0.17–0.22	0.70–0.90	0.040	0.040	0.15–0.35	...	0.70–0.90	...	...
G51300	5130	5130	0.28–0.33	0.70–0.90	0.035	0.040	0.15–0.35	...	0.80–1.10	...	...
G51320	5132	5132	0.30–0.35	0.60–0.80	0.035	0.040	0.15–0.35	...	0.75–1.00	...	...
G51350	5135	5135	0.33–0.38	0.60–0.80	0.035	0.040	0.15–0.35	...	0.80–1.05	...	...
G51400	5140	5140	0.38–0.43	0.70–0.90	0.035	0.040	0.15–0.35	...	0.70–0.90	...	...
G51470	5147	5147	0.46–0.51	0.70–0.95	0.035	0.040	0.15–0.35	...	0.85–1.15	...	...
G51500	5150	5150	0.48–0.53	0.70–0.90	0.035	0.040	0.15–0.35	...	0.70–0.90	...	...
G51550	5155	5155	0.51–0.59	0.70–0.90	0.035	0.040	0.15–0.35	...	0.70–0.90	...	...
G51600	5160	5160	0.56–0.64	0.75–1.00	0.035	0.040	0.15–0.35	...	0.70–0.90	...	...
G51601	51B60(c)	51B60	0.56–0.64	0.75–1.00	0.035	0.040	0.15–0.35	...	0.40–0.60	...	...
G50986	50100(b)	...	0.98–1.10	0.25–0.45	0.025	0.025	0.15–0.35	...	0.40–0.60	...	...
G51986	51100(b)	E51100	0.98–1.10	0.25–0.45	0.025	0.025	0.15–0.35	...	0.90–1.15	...	0.10–0.15
G52986	52100(b)	E52100	0.98–1.10	0.25–0.45	0.025	0.025	0.15–0.35	...	1.30–1.60	...	0.15 min
G61180	6118	6118	0.16–0.21	0.50–0.70	0.035	0.040	0.15–0.35	...	0.50–0.70	...	...
G61500	6150	6150	0.48–0.53	0.70–0.90	0.035	0.040	0.15–0.35	...	0.80–1.10	...	...
G81150	8115	8115	0.13–0.18	0.70–0.90	0.035	0.040	0.15–0.35	0.20–0.40	0.30–0.50	0.08–0.15	...
G81451	81B45(c)	81B45	0.43–0.48	0.75–1.00	0.035	0.040	0.15–0.35	0.20–0.40	0.35–0.55	0.08–0.15	...
G86150	8615	8615	0.13–0.18	0.70–0.90	0.035	0.040	0.15–0.35	0.40–0.70	0.40–0.60	0.15–0.25	...
G86170	8617	8617	0.15–0.20	0.70–0.90	0.035	0.040	0.15–0.35	0.40–0.70	0.40–0.60	0.15–0.25	...
G86200	8620	8620	0.18–0.23	0.70–0.90	0.035	0.040	0.15–0.35	0.40–0.70	0.40–0.60	0.15–0.25	...
G86220	8622	8622	0.20–0.25	0.70–0.90	0.035	0.040	0.15–0.35	0.40–0.70	0.40–0.60	0.15–0.25	...
G86250	8625	8625	0.23–0.28	0.70–0.90	0.035	0.040	0.15–0.35	0.40–0.70	0.40–0.60	0.15–0.25	...
G86270	8627	8627	0.25–0.30	0.70–0.90	0.035	0.040	0.15–0.35	0.40–0.70	0.40–0.60	0.15–0.25	...

(continued)

Chemical Compositions of Metals and Alloys 57

Table 7.9 (continued)

UNS No.	SAE No.	Corresponding AISI No.	C	Mn	P	S	Si	Ni	Cr	Mo	V
G86300	8630	8630	0.28–0.33	0.70–0.90	0.035	0.040	0.15–0.35	0.40–0.70	0.40–0.60	0.15–0.25	...
G86370	8637	8637	0.35–0.40	0.75–1.00	0.035	0.040	0.15–0.35	0.40–0.70	0.40–0.60	0.15–0.25	...
G86400	8640	8640	0.38–0.43	0.75–1.00	0.035	0.040	0.15–0.35	0.40–0.70	0.40–0.60	0.15–0.25	...
G86420	8642	8642	0.40–0.45	0.75–1.00	0.035	0.040	0.15–0.35	0.40–0.70	0.40–0.60	0.15–0.25	...
G86450	8645	8645	0.43–0.48	0.75–1.00	0.035	0.040	0.15–0.35	0.40–0.70	0.40–0.60	0.15–0.25	...
G86451	86B45(c)	...	0.43–0.48	0.75–1.00	0.035	0.040	0.15–0.35	0.40–0.70	0.40–0.60	0.15–0.25	...
G86500	8650	...	0.48–0.53	0.75–1.00	0.035	0.040	0.15–0.35	0.40–0.70	0.40–0.60	0.15–0.25	...
G86550	8655	8655	0.51–0.59	0.75–1.00	0.035	0.040	0.15–0.35	0.40–0.70	0.40–0.60	0.15–0.25	...
G86600	8660	...	0.56–0.64	0.75–1.00	0.035	0.040	0.15–0.35	0.40–0.70	0.40–0.60	0.15–0.25	...
G87200	8720	8720	0.18–0.23	0.70–0.90	0.035	0.040	0.15–0.35	0.40–0.70	0.40–0.60	0.20–0.30	...
G87400	8740	8740	0.38–0.43	0.75–1.00	0.035	0.040	0.15–0.35	0.40–0.70	0.40–0.60	0.20–0.30	...
G88220	8822	8822	0.20–0.25	0.75–1.00	0.035	0.040	0.15–0.35	0.40–0.70	0.40–0.60	0.30–0.40	...
G92540	9254	...	0.51–0.59	0.60–0.80	0.035	0.040	1.20–1.60	...	0.60–0.80	...	...
G92600	9260	9260	0.56–0.64	0.75–1.00	0.035	0.040	1.80–2.20	...	...	...	...
G93106	9310(b)	...	0.08–0.13	0.45–0.65	0.025	0.025	0.15–0.35	3.00–3.50	1.00–1.40	0.08–0.15	...
G94151	94B15(c)	...	0.13–0.18	0.75–1.00	0.035	0.040	0.15–0.35	0.30–0.60	0.30–0.50	0.08–0.15	...
G94171	94B17(c)	94B17	0.15–0.20	0.75–1.00	0.035	0.040	0.15–0.35	0.30–0.60	0.30–0.50	0.08–0.15	...
G94301	94B30(c)	94B30	0.28–0.33	0.75–1.00	0.035	0.040	0.15–0.35	0.30–0.60	0.30–0.50	0.08–0.15	...

(a) Small quantities of certain elements that are not specified or required may be found in alloy steels. These elements are to be considered as incidental and are acceptable to the following maximum amount: copper to 0.35%, nickel to 0.25%, chromium to 0.20%, and molybdenum to 0.06%. (b) Electric furnace steel. (c) Boron content is 0.0005–0.003%.

Table 7.10 Composition ranges and limits for AISI-SAE standard low-alloy steel plate applicable for structural applications

Boron or lead can be added to these compositions. Small quantities of certain elements not required may be found. These elements are to be considered incidental and are acceptable to the following maximum amounts: copper to 0.35%, nickel to 0.25%, chromium to 0.20%, and molybdenum to 0.06%.

AISI-SAE No.	UNS No.	C	Mn	Si(b)	Cr	Ni	Mo
1330	G13300	0.27–0.34	1.50–1.90	0.15–0.30	...	...	...
1335	G13350	0.32–0.39	1.50–1.90	0.15–0.30	...	...	...
1340	G13400	0.36–0.44	1.50–1.90	0.15–0.30	...	...	...
1345	G13450	0.41–0.49	1.50–1.90	0.15–0.30	...	...	...
4118	G41180	0.17–0.23	0.60–0.90	0.15–0.30	0.40–0.65	...	0.08–0.15
4130	G41300	0.27–0.34	0.35–0.60	0.15–0.30	0.80–1.15	...	0.15–0.25
4135	G41350	0.32–0.39	0.65–0.95	0.15–0.30	0.80–1.15	...	0.15–0.25
4137	G41370	0.33–0.40	0.65–0.95	0.15–0.30	0.80–1.15	...	0.15–0.25
4140	G41400	0.36–0.44	0.70–1.00	0.15–0.30	0.80–1.15	...	0.15–0.25
4142	G41420	0.38–0.46	0.70–1.00	0.15–0.30	0.80–1.15	...	0.15–0.25
4145	G41450	0.41–0.49	0.70–1.00	0.15–0.30	0.80–1.15	...	0.15–0.25
4340	G43400	0.36–0.44	0.55–0.80	0.15–0.30	0.60–0.90	1.65–2.00	0.20–0.30
E4340(c)	G43406	0.37–0.44	0.60–0.85	0.15–0.30	0.65–0.90	1.65–2.00	0.20–0.30
4615	G46150	0.12–0.18	0.40–0.65	0.15–0.30	...	1.65–2.00	0.20–0.30
4617	G46170	0.15–0.21	0.40–0.65	0.15–0.30	...	1.65–2.00	0.20–0.30
4620	G46200	0.16–0.22	0.40–0.65	0.15–0.30	...	1.65–2.00	0.20–0.30
5160	G51600	0.54–0.65	0.70–1.00	0.15–0.30	0.60–0.90	...	...
6150(d)	G61500	0.46–0.54	0.60–0.90	0.15–0.30	0.80–1.15	...	...
8615	G86150	0.12–0.18	0.60–0.90	0.15–0.30	0.35–0.60	0.40–0.70	0.15–0.25
8617	G86170	0.15–0.21	0.60–0.90	0.15–0.30	0.35–0.60	0.40–0.70	0.15–0.25
8620	G86200	0.17–0.23	0.60–0.90	0.15–0.30	0.35–0.60	0.40–0.70	0.15–0.25
8622	G86220	0.19–0.25	0.60–0.90	0.15–0.30	0.35–0.60	0.40–0.70	0.15–0.25
8625	G86250	0.22–0.29	0.60–0.90	0.15–0.30	0.35–0.60	0.40–0.70	0.15–0.25
8627	G86270	0.24–0.31	0.60–0.90	0.15–0.30	0.35–0.60	0.40–0.70	0.15–0.25
8630	G86300	0.27–0.34	0.60–0.90	0.15–0.30	0.35–0.60	0.40–0.70	0.15–0.25
8637	G86370	0.33–0.40	0.70–1.00	0.15–0.30	0.35–0.60	0.40–0.70	0.15–0.25
8640	G86400	0.36–0.44	0.70–1.00	0.15–0.30	0.35–0.60	0.40–0.70	0.15–0.25
8655	G86550	0.49–0.60	0.70–1.00	0.15–0.30	0.35–0.60	0.40–0.70	0.15–0.25
8742	G87420	0.38–0.46	0.70–1.00	0.15–0.30	0.35–0.60	0.40–0.70	0.20–0.30

(a) Indicated ranges and limits apply to steels made by the open hearth or basic oxygen processes; maximum content for phosphorus is 0.035% and for sulfur 0.040%. For steels made by the electric furnace process, the ranges and limits are reduced as follows: C to 0.01%; Mn to 0.05%; Cr to 0.05% (<1.25%), 0.10% (>1.25%); maximum content for either phosphorus or sulfur is 0.025%. (b) Other silicon ranges may be negotiated. Silicon is available in ranges of 0.10–0.20%, 0.20–0.30%, and 0.35% maximum (when carbon deoxidized) when so specified by the purchaser. (c) Prefix "E" indicates that the steel is made by the electric furnace process. (d) Contains 0.15% V minimum

Chemical Compositions of Metals and Alloys 59

Table 7.11 Composition of carbon and carbon–boron H-steels

UNS No.	Designation SAE or AISI No.	C	Mn	Si	P(b), maximum	S(b), maximum
H10380	1038H	0.34/0.43	0.50/1.00	0.15/0.35	0.040	0.050
H10450	1045H	0.42/0.51	0.50/1.00	0.15/0.35	0.040	0.050
H15220	1522H	0.17/0.25	1.00/1.50	0.15/0.35	0.040	0.050
H15240	1524H	0.18/0.26	1.25/1.75	0.15/0.35	0.040	0.050
H15260	1526H	0.21/0.30	1.00/1.50	0.15/0.35	0.040	0.050
H15410	1541H	0.35/0.45	1.25/1.75	0.15/0.35	0.040	0.050
H15211	15B21H(a)	0.17/0.24	0.70/1.20	0.15/0.35	0.040	0.050
H15281	15B28H(a)	0.25/0.34	1.00/1.50	0.15/0.35	0.040	0.050
H15301	15B30H(a)	0.27/0.35	0.70/1.20	0.15/0.35	0.040	0.050
H15351	15B35H(a)	0.31/0.39	0.70/1.20	0.15/0.35	0.040	0.050
H15371	15B37H(a)	0.30/0.39	1.00/1.50	0.15/0.35	0.040	0.050
H15411	15B41H(a)	0.35/0.45	1.25/1.75	0.15/0.35	0.040	0.050
H15481	15B48H(a)	0.43/0.53	1.00/1.50	0.15/0.35	0.040	0.050
H15621	15B62H(a)	0.54/0.67	1.00/1.50	0.40/0.60	0.040	0.050

(a) These steels contain 0.005 to 0.003% B. (b) If electric furnace practice is specified or required, the limit for both phosphorus and sulfur is 0.025%, and the prefix E is added to the SAE or AISI number.

Table 7.12 Composition of standard alloy H-steels

UNS No.	H-steel SAE or AISI No.	C	Mn	Si	Ni	Cr	Mo	V
H13300	1330H	0.27/0.33	1.45/2.05	0.15/0.35	...	...	...	...
H13350	1335H	0.32/0.38	1.45/2.05	0.15/0.35	...	...	...	...
H13400	1340H	0.37/0.44	1.45/2.05	0.15/0.35	...	...	...	...
H13450	1345H	0.42/0.49	1.45/2.05	0.15/0.35	...	...	...	...
H40270	4027H	0.24/0.30	0.60/1.00	0.15/0.35	...	...	0.20/0.30	...
H40280(c)	4028H(c)	0.24/0.30	0.60/1.00	0.15/0.35	...	...	0.20/0.30	...
H40320	4032H	0.29/0.35	0.60/1.00	0.15/0.35	...	...	0.20/0.30	...
H40370	4037H	0.34/0.41	0.60/1.00	0.15/0.35	...	...	0.20/0.30	...
H40420	4042H	0.39/0.46	0.60/1.00	0.15/0.35	...	...	0.20/0.30	...
H40470	4047H	0.44/0.51	0.60/1.00	0.15/0.35	...	...	0.20/0.30	...
H41180	4118H	0.17/0.23	0.60/1.00	0.15/0.35	...	0.30/0.70	0.08/0.15	...
H41300	4130H	0.27/0.33	0.30/0.70	0.15/0.35	...	0.75/1.20	0.15/0.25	...
H41350	4135H	0.32/0.38	0.60/1.00	0.15/0.35	...	0.75/1.20	0.15/0.25	...
H41370	4137H	0.34/0.41	0.60/1.00	0.15/0.35	...	0.75/1.20	0.15/0.25	...
H41400	4140H	0.37/0.44	0.65/1.10	0.15/0.35	...	0.75/1.20	0.15/0.25	...
H41420	4142H	0.39/0.46	0.65/1.10	0.15/0.35	...	0.75/1.20	0.15/0.25	...
H41450	4145H	0.42/0.49	0.65/1.10	0.15/0.35	...	0.75/1.20	0.15/0.25	...
H41470	4147H	0.44/0.51	0.65/1.10	0.15/0.35	...	0.75/1.20	0.15/0.25	...
H41500	4150H	0.47/0.54	0.65/1.10	0.15/0.35	...	0.75/1.20	0.15/0.25	...
H41610	4161H	0.55/0.65	0.65/1.10	0.15/0.35	...	0.65/0.95	0.25/0.35	...
H43200	4320H	0.17/0.23	0.40/0.70	0.15/0.35	1.55/2.00	0.35/0.65	0.20/0.30	...
H43400	4340H	0.37/0.44	0.55/0.90	0.15/0.35	1.55/2.00	0.65/0.95	0.20/0.30	...
H43406(d)	E4340H(d)	0.37/0.44	0.60/0.95	0.15/0.35	1.55/2.00	0.65/0.95	0.20/0.30	...
H46200	4620H	0.17/0.23	0.35/0.75	0.15/0.35	1.55/2.00	...	0.20/0.30	...
H47180	4718H	0.15/0.21	0.60/0.95	0.15/0.35	0.85/1.25	0.30/0.60	0.30/0.40	...
H47200	4720H	0.17/0.23	0.45/0.75	0.15/0.35	0.85/1.25	0.30/0.60	0.15/0.25	...
H48150	4815H	0.12/0.18	0.30/0.70	0.15/0.35	3.20/3.80	...	0.20/0.30	...
H48170	4817H	0.14/0.20	0.30/0.70	0.15/0.35	3.20/3.80	...	0.20/0.30	...

(continued)

Table 7.12 (continued)

UNS No.	H-steel SAE or AISI No.	C	Mn	Si	Ni	Cr	Mo	V
H48200	4820H	0.17/0.23	0.40/0.80	0.15/0.35	3.20/3.80	...	0.20/0.30	...
H50401(e)	50B40H(e)	0.37/0.44	0.65/1.10	0.15/0.35	...	0.30/0.70	...	...
H50441(e)	50B44H(e)	0.42/0.49	0.65/1.10	0.15/0.35	...	0.30/0.70	...	...
H50460	5046H	0.43/0.50	0.65/1.10	0.15/0.35	...	0.13/0.43	...	...
H50461(e)	50B46H(e)	0.43/0.50	0.65/1.10	0.15/0.35	...	0.13/0.43	...	...
H50501(e)	50B50H(e)	0.47/0.54	0.65/1.10	0.15/0.35	...	0.30/0.70	...	...
H50601(e)	50B60H(e)	0.55/0.65	0.65/1.10	0.15/0.35	...	0.30/0.70	...	...
H51200	5120H	0.17/0.23	0.60/1.00	0.15/0.35	...	0.60/1.00	...	...
H51300	5130H	0.27/0.33	0.60/1.10	0.15/0.35	...	0.75/1.20	...	...
H51320	5132H	0.29/0.35	0.50/0.90	0.15/0.35	...	0.65/1.10	...	...
H51350	5135H	0.32/0.38	0.50/0.90	0.15/0.35	...	0.70/1.15	...	...
H51400	5140H	0.37/0.44	0.60/1.00	0.15/0.35	...	0.60/1.00	...	...
H51470	5147H	0.45/0.52	0.60/1.05	0.15/0.35	...	0.80/1.25	...	...
H51500	5150H	0.47/0.54	0.60/1.00	0.15/0.35	...	0.60/1.00	...	...
H51550	5155H	0.50/0.60	0.60/1.00	0.15/0.35	...	0.60/1.00	...	...
H51600	5160H	0.55/0.65	0.65/1.10	0.15/0.35	...	0.60/1.00	...	...
H51601(e)	51B60H(e)	0.55/0.65	0.65/1.10	0.15/0.35	...	0.60/1.00	...	...
H61180	6118H	0.15/0.21	0.40/0.80	0.15/0.35	...	0.40/0.80	...	0.10/0.15
H61500	6150H	0.47/0.54	0.60/1.00	0.15/0.35	...	0.75/1.20	...	0.15
H81451(e)	81B4S5(e)	0.42/0.49	0.70/1.05	0.15/0.35	0.15/0.45	0.30/0.60	0.08/0.15	...
H86170	8617H	0.14/0.20	0.60/0.95	0.15/0.35	0.35/0.75	0.35/0.65	0.15/0.25	...
H86200	8620H	0.17/0.23	0.60/0.95	0.15/0.35	0.35/0.75	0.35/0.65	0.15/0.25	...
H86220	8622H	0.19/0.25	0.60/9.95	0.15/0.35	0.35/0.75	0.35/0.65	0.15/0.25	...
H86250	8625H	0.22/0.28	0.60/0.95	0.15/0.35	0.35/0.75	0.35/0.65	0.15/0.25	...
H86270	8627H	0.24/0.30	0.60/0.95	0.15/0.35	0.35/0.75	0.35/0.65	0.15/0.25	...
H86300	8630H	0.27/0.33	0.60/0.95	0.15/0.35	0.35/0.75	0.35/0.65	0.15/0.25	...
H86301(e)	86B30H(e)	0.27/0.33	0.60/0.95	0.15/0.35	0.35/0.75	0.35/0.65	0.15/0.25	...
H86370	8637H	0.34/0.41	0.70/1.05	0.15/0.35	0.35/0.75	0.35/0.65	0.15/0.25	...
H86400	8640H	0.37/0.44	0.70/1.05	0.15/0.35	0.35/0.75	0.35/0.65	0.15/0.25	...
H86420	8642H	0.39/0.46	0.70/1.05	0.15/0.35	0.35/0.75	0.35/0.65	0.15/0.25	...
H86450	8645H	0.42/0.49	0.70/1.05	0.15/0.35	0.35/0.75	0.35/0.65	0.15/0.25	...
H86451(e)	86B45H9(e)	0.42/0.49	0.70/1.05	0.15/0.35	0.35/0.75	0.35/0.65	0.15/0.25	...
H86500	8650H	0.47/0.54	0.70/1.05	0.15/0.35	0.35/0.75	0.35/0.65	0.15/0.25	...
H86550	8655H	0.50/0.60	0.70/1.05	0.15/0.35	0.35/0.75	0.35/0.65	0.15/0.25	...
H86600	8660H	0.55/0.65	0.70/1.05	0.15/0.35	0.35/0.75	0.35/0.65	0.15/0.25	...
H87200	8720H	0.17/0.23	0.60/0.95	0.15/0.35	0.35/0.75	0.35/0.65	0.20/0.30	...
H87400	8740H	0.37/0.44	0.70/1.05	0.15/0.35	0.35/0.75	0.35/0.65	0.20/0.30	...
H88220	8822H	0.19/0.25	0.70/1.05	0.15/0.35	0.35/0.75	0.35/0.65	0.30/0.40	...
H92600	9260H	0.55/0.65	0.65/1.10	1.70/2.20	...	...	...	...
H93100(d)	9310H(d)	0.07/0.13	0.40/0.70	0.15/0.35	2.95/3.55	1.00/1.45	0.08/0.15	...
H94151(e)	94B15H(e)	0.12/0.18	0.70/1.05	0.15/0.35	0.25/0.65	0.25/0.55	0.08/0.15	...
H94171(e)	94B17H(e)	0.14/0.20	0.70/1.05	0.15/0.35	0.25/0.65	0.25/0.55	0.08/0.15	...
H94301(e)	94B30H(e)	0.27/0.33	0.70/1.05	0.15/0.35	0.25/0.65	0.25/0.55	0.08/0.15	...

(a) Small quantities of certain elements may be found in alloy steel that are not specified or required. These elements are to be considered incidental and acceptable to the following maximum amounts: copper to 0.35%, nickel to 0.25%, chromium to 0.20%, and molybdenum to 0.06%. (b) For open hearth and basic oxygen steels, maximum sulfur content is to be 0.040%, and maximum phosphorus content is to be 0.035%. Maximum phosphorus and sulfur in basic electric furnace steels are to be 0.025% each. (c) Sulfur content range is 0.035/0.050%. (d) Electric furnace steel. (e) These steels contain 0.0005 to 0.003% B.

Table 7.13 Composition limits of principal types of tool steels

AISI type	Designation UNS No.	C	Mn	Si	Cr	Ni	Mo	W	V	Co
Molybdenum high-speed steels										
M1	T11301	0.78–0.88	0.15–0.40	0.20–0.50	3.50–4.00	0.30 max	8.20–9.20	1.40–2.10	1.00–1.35	...
M2	T11302	0.78–0.88; 0.95–1.05	0.15–0.40	0.20–0.45	3.75–4.50	0.30 max	4.50–5.50	5.50–6.75	1.75–2.20	...
M3, class 1	T11313	1.00–1.10	0.15–0.40	0.20–0.45	3.75–4.50	0.30 max	4.75–6.50	5.00–6.75	2.25–2.75	...
M3, class 2	T11323	1.15–1.25	0.15–0.40	0.20–0.45	3.75–4.50	0.30 max	4.75–6.50	5.00–6.75	2.75–3.75	...
M4	T11304	1.25–1.40	0.15–0.40	0.20–0.45	3.75–4.75	0.30 max	4.25–5.50	5.25–6.50	3.75–4.50	...
M7	T11307	0.97–1.05	0.15–0.40	0.20–0.55	3.50–4.00	0.30 max	8.20–9.20	1.40–2.10	1.75–2.25	...
M10	T11310	0.84–0.94; 0.95–1.05	0.10–0.40	0.20–0.45	3.75–4.50	0.30 max	7.75–8.50	...	1.80–2.20	...
M30	T11330	0.75–0.85	0.15–0.40	0.20–0.45	3.50–4.25	0.30 max	7.75–9.00	1.30–2.30	1.00–1.40	4.50–5.50
M33	T11333	0.85–0.92	0.15–0.40	0.15–0.50	3.50–4.00	0.30 max	9.00–10.00	1.30–2.10	1.00–1.35	7.75–8.75
M34	T11334	0.85–0.92	0.15–0.40	0.20–0.45	3.50–4.00	0.30 max	7.75–9.20	1.40–2.10	1.90–2.30	7.75–8.75
M35	T11335	0.82–0.88	0.15–0.40	0.20–0.45	3.75–4.50	0.30 max	4.50–5.50	5.50–6.75	1.75–2.20	4.50–5.50
M36	T11336	0.80–0.90	0.15–0.40	0.20–0.45	3.75–4.50	0.30 max	4.50–5.50	5.50–6.50	1.75–2.25	7.75–8.75
M41	T11341	1.05–1.15	0.20–0.60	0.15–0.50	3.75–4.50	0.30 max	3.25–4.25	6.25–7.00	1.75–2.25	4.75–5.75
M42	T11342	1.05–1.15	0.15–0.40	0.15–0.65	3.50–4.25	0.30 max	9.00–10.00	1.15–1.85	0.95–1.35	7.75–8.75
M43	T11343	1.15–1.25	0.20–0.40	0.15–0.65	3.50–4.25	0.30 max	7.50–8.50	2.25–3.00	1.50–1.75	7.75–8.75
M44	T11344	1.10–1.20	0.20–0.40	0.30–0.55	4.00–4.75	0.30 max	6.00–7.00	5.00–5.75	1.85–2.20	11.00–12.25
M46	T11346	1.22–1.30	0.20–0.40	0.40–0.65	3.70–4.20	0.30 max	8.00–8.50	1.90–2.20	3.00–3.30	7.80–8.80
M47	T11347	1.05–1.15	0.15–0.40	0.20–0.45	3.50–4.00	0.30 max	9.25–10.00	1.30–1.80	1.15–1.35	4.75–5.25
M48	T11348	1.42–1.52	0.15–0.40	0.15–0.40	3.50–4.00	0.30 max	4.75–5.50	9.50–10.50	2.75–3.25	8.00–10.00
M62	T11362	1.25–1.35	0.15–0.40	0.15–0.40	3.50–4.00	0.30 max	10.00–11.00	5.75–6.50	1.80–2.10	...
Tungsten high-speed steels										
T1	T12001	0.65–0.80	0.10–0.40	0.20–0.40	3.75–4.50	0.30 max	...	17.25–18.75	0.90–1.30	...
T2	T12002	0.80–0.90	0.20–0.40	0.20–0.40	3.75–4.50	0.30 max	1.0 max	17.50–19.00	1.80–2.40	...
T4	T12004	0.70–0.80	0.10–0.40	0.20–0.40	3.75–4.50	0.30 max	0.40–1.00	17.50–19.00	0.80–1.20	4.25–5.75
T5	T12005	0.75–0.85	0.20–0.40	0.20–0.40	3.75–5.00	0.30 max	0.50–1.25	17.50–19.00	1.80–2.40	7.00–9.50
T6	T12006	0.75–0.85	0.20–0.40	0.20–0.40	4.00–4.75	0.30 max	0.40–1.00	18.50–21.00	1.50–2.10	11.00–13.00
T8	T12008	0.75–0.85	0.20–0.40	0.20–0.40	3.75–4.50	0.30 max	0.40–1.00	13.25–14.75	1.80–2.40	4.25–5.75
T15	T12015	1.50–1.60	0.15–0.40	0.15–0.40	3.75–5.00	0.30 max	1.00 max	11.75–13.00	4.50–5.25	4.75–5.25

(continued)

Table 7.13 (continued)

AISI type	Designation UNS No.	C	Mn	Si	Cr	Ni	Mo	W	V	Co
Intermediate high-speed steels										
M50	T11350	0.78–0.88	0.15–0.45	0.20–0.60	3.75–4.50	0.30 max	3.90–4.75	...	0.80–1.25	...
M52	T11352	0.85–0.95	0.15–0.45	0.20–0.60	3.50–4.30	0.30 max	4.00–4.90	0.75–1.50	1.65–2.25	...
Chromium hot-work steels										
H10	T20810	0.35–0.45	0.25–0.70	0.80–1.20	3.00–3.75	0.30 max	2.00–3.00	...	0.25–0.75	...
H11	T20811	0.33–0.43	0.20–0.50	0.80–1.20	4.75–5.50	0.30 max	1.10–1.60	...	0.30–0.60	...
H12	T20812	0.30–0.40	0.20–0.50	0.80–1.20	4.75–5.50	0.30 max	1.25–1.75	1.00–1.70	0.50 max	...
H13	T20813	0.32–0.45	0.20–0.50	0.80–1.20	4.75–5.50	0.30 max	1.10–1.75	...	0.80–1.20	...
H14	T20814	0.35–0.45	0.20–0.50	0.80–1.20	4.75–5.50	0.30 max	...	4.00–5.25	...	...
H19	T20819	0.32–0.45	0.20–0.50	0.20–0.50	4.00–4.75	0.30 max	0.30–0.55	3.75–4.50	1.75–2.20	4.00–4.50
Tungsten hot-work steels										
H21	T20821	0.26–0.36	0.15–0.40	0.15–0.50	3.00–3.75	0.30 max	...	8.50–10.00	0.30–0.60	...
H22	T20822	0.30–0.40	0.15–0.40	0.15–0.40	1.75–3.75	0.30 max	...	10.00–11.75	0.25–0.50	...
H23	T20823	0.25–0.35	0.15–0.40	0.15–0.60	11.00–12.75	0.30 max	...	11.00–12.75	0.75–1.25	...
H24	T20824	0.42–0.53	0.15–0.40	0.15–0.40	2.50–3.50	0.30 max	...	14.00–16.00	0.40–0.60	...
H25	T20825	0.22–0.32	0.15–0.40	0.15–0.40	3.75–4.50	0.30 max	...	14.00–16.00	0.40–0.60	...
H26	T20826	0.45–0.55(b)	0.15–0.40	0.15–0.40	3.75–4.50	0.30 max	...	17.25–19.00	0.75–1.25	...
Molybdenum hot-work steels										
H42	T20842	0.55–0.70(b)	0.15–0.40	...	3.75–4.50	0.30 max	4.50–5.50	5.50–6.75	1.75–2.20	...
Air-hardening, medium-alloy, cold-work steels										
A2	T30102	0.95–1.05	1.00 max	0.50 max	4.75–5.50	0.30 max	0.90–1.40	...	0.15–0.50	...
A3	T30103	1.20–1.30	0.40–0.60	0.50 max	4.75–5.50	0.30 max	0.90–1.40	...	0.80–1.40	...
A4	T30104	0.95–1.05	1.80–2.20	0.50 max	0.90–2.20	0.30 max	0.90–1.40	...	...	...
A6	T30106	0.65–0.75	1.80–2.50	0.50 max	0.90–1.20	0.30 max	0.90–1.40	...	...	...
A7	T30107	2.00–2.85	0.80 max	0.50 max	5.00–5.75	0.30 max	0.90–1.40	0.50–1.50	3.90–5.15	...
A8	T30108	0.50–0.60	0.50 max	0.75–1.10	4.75–5.50	0.30 max	1.15–1.65	1.00–1.50	...	...
A9	T30109	0.45–0.55	0.50 max	0.95–1.15	4.75–5.50	1.25–1.75	1.30–1.80	...	0.80–1.40	...
A10	T30110	1.25–1.50(c)	1.60–2.10	1.00–1.50	...	1.55–2.05	1.25–1.75	...	...	...
High-carbon, high-chromium, cold-work steels										
D2	T30402	1.40–1.60	0.60 max	0.60 max	11.00–13.00	0.30 max	0.70–1.20	...	1.10 max	...

(continued)

Table 7.13 (continued)

AISI type	UNS No.	C	Mn	Si	Cr	Ni	Mo	W	V	Co
High-carbon, high-chromium, cold-work steels (continued)										
D3	T30403	2.00–2.35	0.60 max	0.60 max	11.00–13.50	0.30 max	...	1.00 max	1.00 max	...
D4	T30404	2.05–2.40	0.60 max	0.60 max	11.00–13.00	0.30 max	0.70–1.20	...	1.00 max	...
D5	T30405	1.40–1.60	0.60 max	0.60 max	11.00–13.00	0.30 max	0.70–1.20	...	1.00 max	2.50–3.50
D7	T30407	2.15–2.50	0.60 max	0.60 max	11.50–13.50	0.30 max	0.70–1.20	...	3.80–4.40	...
Oil-hardening cold-work steels										
O1	T31501	0.85–1.00	1.00–1.40	0.50 max	0.40–0.60	0.30 max	...	0.40–0.60	0.30 max	...
O2	T31502	0.85–0.95	1.40–1.80	0.50 max	0.50 max	0.30 max	0.30 max	...	0.30 max	...
O6	T31506	1.25–1.55(c)	0.30–1.10	0.55–1.50	0.30 max	0.30 max	0.20–0.30	...	...	...
O7	T31507	1.10–1.30	1.00 max	0.60 max	0.35–0.85	0.30 max	0.30 max	1.00–2.00	0.40 max	...
Shock-resisting steels										
S1	T41901	0.40–0.55	0.10–0.40	0.15–1.20	1.00–1.80	0.30 max	0.50 max	1.50–3.00	0.15–0.30	...
S2	T41902	0.40–0.55	0.30–0.50	0.90–1.20	...	0.30 max	0.30–0.60	...	0.50 max	...
S5	T41905	0.50–0.65	0.60–1.00	1.75–2.25	0.50 max	...	0.20–1.35	...	0.35 max	...
S6	T41906	0.40–0.50	1.20–1.50	2.00–2.50	1.20–1.50	...	0.30–0.50	...	0.20–0.40	...
S7	T41907	0.45–0.55	0.20–0.90	0.20–1.00	3.00–3.50	...	1.30–1.80	...	0.20–0.30(d)	...
Low-alloy special-purpose tool steels										
L2	T61202	0.45–1.00(b)	0.10–0.90	0.50 max	0.70–1.20	...	0.25 max	...	0.10–0.30	...
L6	T61206	0.65–0.75	0.25–0.80	0.50 max	0.60–1.20	1.25–2.00	0.50 max	...	0.20–0.30(d)	...
Low-carbon mold steels										
P2	T51602	0.10 max	0.10–0.40	0.10–0.40	0.75–1.25	0.10–0.50	0.15–0.40	...	...	...
P3	T51603	0.10 max	0.20–0.60	0.40 max	0.40–0.75	1.00–1.50	...	...	...	...
P4	T51604	0.12 max	0.20–0.60	0.10–0.40	4.00–5.25	...	0.40–1.00	...	...	...
P5	T51605	0.10 max	0.20–0.60	0.40 max	2.00–2.50	...	...	...	...	...
P6	T51606	0.05–0.15	0.35–0.70	0.10–0.40	1.25–1.75	0.35 max	...	...	...	...
P20	T51620	0.28–0.40	0.60–1.00	0.20–0.80	1.40–2.00	3.25–3.75	0.30–0.55	...	...	...
P21	T51621	0.18–0.22	0.20–0.40	0.20–0.40	0.50 max	3.90–4.25	...	...	0.15–0.25	1.05–1.25Al
Water-hardening tool steels										
W1	T72301	0.70–1.50(e)	0.10–0.40	0.10–0.40	0.15 max	0.20 max	0.10 max	0.15 max	0.10 max	...
W2	T72302	0.85–1.50(e)	0.10–0.40	0.10–0.40	0.15 max	0.20 max	0.10 max	0.15 max	0.15–0.35	...
W5	T72305	1.05–1.15	0.10–0.40	0.10–0.40	0.40–0.60	0.20 max	0.10 max	0.15 max	0.10 max	...

(a) All steels except group W contain 0.25% max Cu, 0.03 max P, and max sincerely; group W steels contain 0.20 max Cu, 0.025 max P, and 0.025 max sincerely. Where specified, sulfur may be increased to 0.06–0.15% to improve machinability of group A, D, H, M, and T steels. (b) Available in several carbon ranges. (c) Contains free graphite in the microstructures. (d) Optional. (e) Specified carbon ranges are designated by duffix numbers.

Table 7.14 Composition of selected standard and special stainless steels

UNS No.	AISI type	C	Mn	Si	P	S	Cr	Ni	Mo	N	Others
Ferritic alloys											
S40500	405	0.08	1.00	1.00	0.040	0.030	11.50–14.50	...	...	...	0.10–0.30 Al
S40900	409	0.08	1.00	1.00	0.045	0.045	10.50–11.75	0.50	...	...	6×C–0.75 Ti
S43000	430	0.12	1.00	1.00	0.040	0.030	16.00–18.00	...	...	...	...
S43020	430F	0.12	1.25	1.00	0.060	0.15(a)	16.00–18.00	...	0.60	...	...
S43023	430FSe	0.12	1.25	1.00	0.060	0.060	16.00–18.00	...	...	...	0.15 min Se
S43400	434	0.12	1.00	1.00	0.040	0.030	16.00–18.00	...	0.75–1.25	...	...
S44200	442	0.20	1.00	1.00	0.040	0.030	18.00–23.00	...	...	...	...
S44300	443(b)	0.20	1.00	1.00	0.040	0.030	18.00–23.00	0.50	...	...	0.90–1.25 Cu
S44400	444(b)	0.025	1.00	1.00	0.040	0.030	17.50–19.50	1.00	1.75–2.50	0.025	[0.20 + 4 (C + N)]–0.80 Ti + Nb
S44600	446(b)	0.20	1.50	1.00	0.040	0.030	23.00–27.00	...	...	0.25	...
S18200	18-2FM(c)	0.08	1.25–2.50	1.00	0.040	0.15(a)	17.50–19.50	...	1.50–2.50	...	...
Martensitic alloys											
S40300	403	0.15	1.00	0.50	0.040	0.030	11.50–13.00	...	...	...	...
S41000	410	0.15	1.00	1.00	0.040	0.030	11.50–13.00	...	...	...	...
S41400	414	0.15	1.00	1.00	0.040	0.030	11.50–13.50	1.25–2.50	...	...	...
S41600	416	0.15	1.25	1.00	0.060	0.15(a)	12.00–14.00	...	0.60	...	...
S41610	416 Plus X(d)	0.15	1.50–2.50	1.00	0.060	0.15(a)	12.00–14.00	...	0.60	...	...
S41623	416Se	0.15	1.25	1.00	0.060	0.060	12.00–14.00	...	...	...	0.15 min Se
S42000	420	0.15(a)	1.00	1.00	0.040	0.030	12.00–14.00	...	...	...	...
S42010	Trim Rite(e)	0.15–0.30	1.00	1.00	0.040	0.030	13.50–15.00	0.25–1.00	0.40–1.00	...	...
S42020	420F	0.15(a)	1.25	1.00	0.060	0.15(a)	12.00–14.00	...	0.60	...	...
S42023	420FSe(b)	0.30–0.40	1.25	1.00	0.060	0.060	12.00–14.00	...	0.60	...	0.15 min Se; 0.60 Zr or Cu
S43100	431	0.20	1.00	1.00	0.040	0.030	15.00–17.00	1.25–2.50	...	...	...
S44002	440A	0.60–0.75	1.00	1.00	0.040	0.030	16.00–18.00	...	0.75	...	...
S44003	440B	0.75–0.95	1.00	1.00	0.040	0.030	16.00–18.00	...	0.75	...	...
S44004	440C	0.95–1.20	1.00	1.00	0.040	0.030	16.00–18.00	...	0.75	...	...
S44020	440F(b)	0.95–1.20	1.25	1.00	0.040	0.10–0.35	16.00–18.00	0.75	0.40–0.60	0.08	...
S44023	440FSe(b)	0.95–1.20	1.25	1.00	0.040	0.030	16.00–18.00	0.75	0.60	0.08	0.15 min Se
Austenitic alloys											
S20100	201	0.15	5.50–7.50	1.00	0.060	0.030	16.00–18.00	3.50–5.50	...	0.25	...
S20161	Gall-Tough(e)	0.15	4.00–6.00	3.00–4.00	0.040	0.040	15.00–18.00	4.00–6.00	...	0.08–0.20	...
S20300	203EZ(f)	0.08	5.00–6.50	1.00	0.040	0.18–0.35	16.00–18.00	5.00–6.50	0.50	...	1.75–2.25 Cu

(continued)

Chemical Compositions of Metals and Alloys

Table 7.14 (continued)

UNS No.	AISI type	C	Mn	Si	P	S	Cr	Ni	Mo	N	Others
Austenitic alloys (continued)											
S20910	22-13-5(c)	0.06	4.00–6.00	1.00	0.040	0.030	20.50–23.50	11.50–13.50	1.50–3.00	0.20–0.40	0.10–0.30Nb; 0.10–0.30 V
S21000	SCF19(e)	0.10	4.00–7.00	0.60	0.030	0.030	18.00–23.00	16.00–20.00	4.00–6.00	0.15	2.00 Cu
S21300	15-15LC(e)	0.25	15.00–18.00	1.00	0.050	0.050	16.00–21.00	3.00	0.50–3.00	0.20–0.80	0.50–2.00 Cu
S21800	Nitronic 60(g)	0.10	7.00–9.00	3.50–4.50	0.040	0.030	16.00–18.00	7.00–9.00	...	0.08–0.20	...
S21904	21-6-9LC(c)	0.04	8.00–10.00	1.00	0.060	0.030	19.00–21.50	5.50–7.50	...	0.15–0.40	...
S24100	18-2Mn(c)	0.15	11.00–14.00	1.00	0.060	0.030	16.50–19.50	0.50–2.50	...	0.20–0.45	...
S28200	18-18 Plus(e)	0.15	17.00–19.00	1.00	0.045	0.030	17.00–19.00	...	0.50–1.50	0.40–0.60	0.50–1.50 Cu
...	Nitronic 30(g)	0.10	7.00–9.00	1.00	...	...	15.00–17.00	1.50–3.00	...	0.15–0.30	1.00 Cu
S30100	301	0.15	2.00	1.00	0.045	0.030	16.00–18.00	6.00–8.00	...	...	...
S30200	302	0.15	2.00	1.00	0.045	0.030	17.00–19.00	8.00–10.00	...	...	...
S30300	303	0.15	2.00	1.00	0.20	0.15(a)	17.00–19.00	8.00–10.00	0.60	...	...
S30310	303 Plus X(d)	0.15	2.50–4.50	1.00	0.20	0.25(a)	17.00–19.00	7.00–10.00	0.75	...	...
S30323	303Se	0.15	2.00	1.00	0.20	0.060	17.00–19.00	8.00–10.00	...	...	0.15 min Se
S30330	303 Cu(b)	0.15	2.00	1.00	0.15	0.10(a)	17.00–19.00	6.00–10.00	...	...	2.50–4.00 Cu; 0.10 Se
S30400	304	0.08	2.00	1.00	0.045	0.030	18.00–20.00	8.00–10.50	...	...	...
S30403	304L	0.03	2.00	1.00	0.045	0.030	18.00–20.00	8.00–12.00	...	...	...
S30430	302 HQ(b)	0.10	2.00	1.00	0.045	0.030	17.00–19.00	8.00–10.00	...	...	3.00–4.00 Cu
S30431	302 HQ-FM(e)	0.06	2.00	1.00	0.040	0.14	16.00–19.00	9.00–11.00	...	...	1.30–2.40 Cu
S30452	304 HN(b)	0.08	2.00	1.00	0.045	0.030	18.00–20.00	8.00–10.50	...	0.16–0.30	...
S30500	305	0.12	2.00	1.00	0.045	0.030	17.00–19.00	10.00–13.00	...	...	...
S30900	309	0.20	2.00	1.00	0.045	0.030	22.00–24.00	12.00–15.00	...	...	...
S30908	309S	0.08	2.00	1.00	0.045	0.030	22.00–24.00	12.00–15.00	...	...	...
S31000	310	0.25	2.00	1.50	0.045	0.030	24.00–26.00	19.00–22.00	...	...	...
S31008	310S	0.08	2.00	1.50	0.045	0.030	24.00–26.00	19.00–22.00	...	...	...
S31600	316	0.08	2.00	1.00	0.045	0.030	16.00–18.00	10.00–14.00	2.00–3.00	...	...
S31603	316L	0.030	2.00	1.00	0.045	0.030	16.00–18.00	10.00–14.00	2.00–3.00	...	...
S31620	316F	0.08	2.00	1.00	0.20	0.10(a)	17.00–19.00	12.00–14.00	1.75–2.50	...	...
S31700	317	0.08	2.00	1.00	0.045	0.30	18.00–20.00	11.00–15.00	3.00–4.00	...	...
S31703	317L	0.030	2.00	1.00	0.045	0.030	18.00–20.00	11.00–15.00	3.00–4.00	...	...
S32100	321	0.08	2.00	1.00	0.045	0.030	17.00–19.00	9.00–12.00	...	...	5 × C min Ti
S34700	347	0.08	2.00	1.00	0.045	0.030	17.00–19.00	9.00–13.00	...	...	10 × C min Nb
S34720	347F(b)	0.08	2.00	1.00	0.045	0.18–0.35	17.00–19.00	9.00–12.00	...	...	10 × C–1.10Nb
S34723	347FSe(b)	0.08	2.00	1.00	0.11–0.17	0.030	17.00–19.00	9.00–12.00	...	...	10 × C–1.10 Nb; 0.15–0.35 Se

(continued)

Table 7.14 (continued)

UNS No.	AISI type	C	Mn	Si	P	S	Cr	Ni	Mo	N	Others
Austenitic alloys (continued)											
S38400	384	0.08	2.00	1.00	0.045	0.030	15.00–17.00	17.00–19.00	...	...	...
N08020	20Cb-3(e)	0.07	2.00	1.00	0.045	0.035	19.00–21.00	32.00–38.00	2.00–3.00	...	8 × C–1.00 Nb; 3.00–4.00 Cu
Duplex alloys											
S31803	2205(c)	0.030	2.00	1.00	0.030	0.020	21.0–23.0	4.50–6.50	2.50–3.50	0.08–0.20	...
S32550	Alloy 255(c)	0.04	1.50	1.00	0.04	0.03	24.0–27.0	4.50–6.50	2.00–4.00	0.10–0.25	1.50–2.50 Cu
S32900	329	0.20	1.00	0.75	0.040	0.030	23.00–28.00	2.50–5.00	1.00–2.00	...	...
S32950	7-Mo Plus(e)	0.03	2.00	0.60	0.035	0.010	26.0–29.0	3.50–5.20	1.00–2.50	0.15–0.35	...
Precipitation-hardenable alloys											
S13800	PH13-8 Mo(g)	0.05	0.20	0.10	0.010	0.008	12.25–13.25	7.50–8.50	2.00–2.50	0.01	0.90–1.35 Al
S15500	15-5PH(g)	0.07	1.00	1.00	0.040	0.030	14.00–15.50	3.50–5.50	...	...	0.15–0.45 Nb; 2.50–4.50 Cu
S15700	15-7PH(g)	0.09	1.00	1.00	0.040	0.030	14.00–16.00	6.50–7.25	2.00–3.00	...	0.75–1.50 Al
S17400	17-4PH(g)	0.07	1.00	1.00	0.040	0.030	15.50–17.50	3.00–5.00	...	...	0.15–0.45 Nb; 3.00–5.00 Cu
S17700	PH 17-7(g)	0.09	1.00	1.00	0.040	0.040	16.00–18.00	6.50–7.75	...	...	0.75–1.50 Al
S35000	633(b)	0.07–0.11	0.50–1.25	0.50	0.040	0.030	16.00–17.00	4.00–5.00	2.50–3.25	0.07–0.13	...
S35500	634(b)	0.10–0.15	0.50–1.25	0.50	0.040	0.030	15.00–16.00	4.00–5.00	2.50–3.25	0.07–0.13	...
S44000	Custom 450(e)	0.05	1.00	1.00	0.030	0.030	14.00–16.00	5.00–7.00	0.50–1.00	...	8 × C min; 1.25–1.75 Cu
S45500	Custom 455(e)	0.05	0.50	0.50	0.040	0.030	11.00–12.50	7.50–9.50	0.50	...	0.10–0.50 Nb; 1.50–2.50 Cu; 0.80–1.40 Ti
S66286	A286(c)	0.08	2.00	1.00	0.040	0.030	13.50–16.00	24.0–27.0	1.00–1.50	...	0.35 Al; 0.0010–0.010 B; 1.90–2.35 Ti; 0.10–0.50 V

Note: All compositions include Fe as balance. (a) Minimum, rather than maximum wt%. (b) Designation resembles AISI type, but is not used in that system. (c) Common trade name, rather than AISI type. (d) Trade name of Crucible Inc. (e) Trade name of Carpenter Technology Corporation. (f) Trade name of Al-Tech Corporation (g) Trade name of Armco Inc.

Table 7.15 Compositions of maraging and high fracture toughness steels

Grade	C	Cr	Ni	Co	Mo	Other
Standard (cobalt-bearing) maraging steel grades						
18Ni (200)	(a)	...	18	8.5	3.3	0.2Ti; 0.1Al
18Ni (250)	(a)	...	18	8.5	5.0	0.4Ti; 0.1Al
18Ni (300)	(a)	...	18	9.0	5.0	0.7Ti; 0.1Al
18Ni (350)	(a)	...	18	12.5	4.2(b)	1.6Ti; 0.1Al
18Ni (Cast)	(a)	...	17	10.0	4.6	0.3Ti; 0.1Al
12-5-3 (180)	(a)	5.0	12	...	3.0	0.2Ti; 0.3Al
Cobalt-free and low-cobalt bearing maraging steel grades						
Co-free 18Ni (200)	(a)	...	18.5	...	3.0	0.7Ti; 0.1Al
Co-free 18Ni (250)	(a)	...	18.5	...	3.0	1.4Ti; 0.1Al
Low-Co 18Ni (250)	(a)	...	18.5	2.0	2.6	1.2Ti; 0.1Al; 0.1Nb
Co-free 18Ni (300)	(a)	...	18.5	...	4.0	1.85Ti; 0.1Al
High fracture toughness steels						
AF1410	0.13–0.17	1.80–2.20	9.50–10.50	13.50–14.50	0.90–1.10	...
HP 9-4-30	0.29–0.34	0.90–1.10	7.0–8.0	4.25–4.75	0.90–1.10	0.06–0.12V
AerMet 100	0.23	3.10	11.10	13.40	1.20	...

(a) All maraging steel grades contain no more than 0.03% C. (b) Some producers use a combination of 4.8% Mo and 1.4% Ti, nominal.

Table 7.16 Four-digit numerical system used to identify wrought aluminum and aluminum alloys

Aluminum, ≥99.00%	1xxx
Aluminum alloys grouped by major alloying element(s):	
Copper	2xxx
Manganese	3xxx
Silicon	4xxx
Magnesium	5xxx
Magnesium and silicon	6xxx
Zinc	7xxx
Other elements	8xxx
Unused series	9xxx

Table 7.17 Four-digit numerical system used to identify cast aluminum and aluminum alloys

The last digit, which is separated from the others by a decimal point, indicates the product form, whether casting or ingot.

Aluminum, ≥99.00%.	1xx.x
Aluminum alloys grouped by major alloying element(s):	
Copper	2xx.x
Silicon, with added copper and/or magnesium	3xx.x
Silicon	4xx.x
Magnesium	5xx.x
Zinc	7xx.x
Tin	8xx.x
Other elements	9xx.x
Unused series	6xx.x

Table 7.18 Designations and nominal compositions of common wrought aluminum and aluminum alloys

AA No.	Product(a)	Al	Si	Cu	Mn	Mg	Cr	Zn	Others
1050	DT	99.50 min	...	...	...	...	...	...	...
1060	S, P, ET, DT	99.60 min	...	...	...	...	...	...	...
1100	S, P, F, E, ES, ET, C, DT, FG	99.00 min	...	0.12	...	...	...	...	...
1145	S, P, F	99.45 min	...	...	...	...	...	...	...
1199	F	99.99 min	...	...	...	...	...	...	...
1350	S, P, E, ES, ET, C	99.50 min	...	...	...	...	...	...	...
2011	E, ES, ET, C, DT	93.7	...	5.5	...	...	...	...	0.4 Bi; 0.4 Pb
2014	S, P, E, ES, ET, C, DT, FG	93.5	0.8	4.4	0.8	0.5	...	...	...
2024	S, P, E, ES, ET, C, DT	93.5	...	4.4	0.6	1.5	...	...	...
2036	S	96.7	...	2.6	0.25	0.45	...	...	...
2048	S, P	94.8	...	3.3	0.4	1.5	...	...	...
2124	P	93.5	...	4.4	0.6	1.5	...	...	...
2218	FG	92.5	...	4.0	...	1.5	...	...	2.0 Ni
2219	S, P, E, ES, ET, C, FG	93.0	...	6.3	0.3	...	...	...	0.6 Ti; 0.10 V; 0.18 Zr
2319	C	93.0	...	6.3	0.3	...	...	...	0.18 Zn; 0.15 Ti; 0.10 V
2618	FG	93.7	0.18	2.3	...	1.6	...	...	1.1 Fe; 1.0 Ni; 0.07 Ti
3003	S, P, F, E, ES, ET, C, DT, FG	98.6	...	0.12	1.2	...	...	...	...
3004	S, P, ET, DT	97.8	...	...	1.2	1.0	...	...	...
3105	S	99.0	...	...	0.55	0.50	...	...	...
4032	FG	85.0	12.2	0.9	...	1.0	...	...	0.9 Ni
4043	C	94.8	5.2	...	...	...	...	...	...
5005	S, P, C	99.2	...	...	...	0.8	...	...	...
5050	S, P, C, DT	98.6	...	...	...	1.4	...	...	...
5052	S, P, F, C, DT	97.2	...	...	...	2.5	0.25	...	...
5056	F, C	95.0	...	...	0.12	5.0	0.12	...	...
5083	S, P, E, ES, ET, FG	94.7	...	...	0.7	4.4	0.15	...	...

(continued)

Table 7.18 (continued)

AA No.	Product(a)	Al	Si	Cu	Mn	Mg	Cr	Zn	Others
5086	S, P, E, ES, ET, DT	95.4	...	...	0.4	4.0	0.15	...	...
5154	S, P, E, ES, ET, C, DT	96.2	...	...	...	3.5	0.25	...	...
5182	S	95.2	...	...	0.35	4.5	...	...	...
5252	S	97.5	...	...	...	2.5	...	...	...
5254	S, P	96.2	...	...	...	3.5	0.25	...	...
5356	C	94.6	...	...	0.12	5.0	0.12	...	0.13 Ti
5454	S, P, E, ES, ET	96.3	...	...	0.8	2.7	0.12	...	...
5456	S, P, E, ES, ET, DT, FG	93.9	...	...	0.8	5.1	0.12	...	...
5457	S	98.7	...	...	0.3	1.0	...	...	...
5652	S, P	97.2	...	...	...	2.5	0.25	...	...
5657	S	99.2	...	...	...	0.8	...	...	...
6005	E, ES, ET	98.7	0.8	...	...	0.5	...	...	...
6009	S	97.7	0.8	0.35	0.5	0.6	...	...	...
6010	S	97.3	1.0	0.35	0.5	0.8	...	...	...
6061	S, P, E, ES, ET, C, DT, FG	97.9	0.6	0.28	...	1.0	0.2	...	...
6063	E, ES, ET, DT	98.9	0.4	...	...	0.7	...	...	...
6066	E, ES, ET, DT, FG	95.7	1.4	1.0	0.8	1.1	...	...	...
6070	E, ES, ET	96.8	1.4	0.28	0.7	0.8	...	...	...
6101	E, ES, ET	98.9	0.5	...	...	0.6	...	...	...
6151	FG	98.2	0.9	...	...	0.6	0.25	...	...
6201	C	98.5	0.7	...	...	0.8	...	...	...
6205	E, ES, ET	98.4	0.8	...	0.1	0.5	0.1	...	0.1 Zr
6262	E, ES, ET, C, DT	96.8	0.6	0.28	...	1.0	0.09	...	0.6 Bi; 0.6 Pb
6351	E, ES	97.8	1.0	...	0.6	0.6	...	...	...
6463	E, ES	98.9	0.4	...	...	0.7	...	...	...
7005	E, ES	93.3	...	...	0.45	1.4	0.13	4.5	0.04 Ti; 0.14 Zr
7049	P, E, ES, FG	88.2	...	1.5	...	2.5	0.15	7.6	...
7050	P, E, ES, FG	89.0	...	2.3	...	2.3	...	6.2	0.12 Zr
7072	S, F	99.0	...	...	...	...	...	1.0	...
7075	S, P, E, ES, ET, C, DT, FG	90.0	...	1.6	...	2.5	0.23	5.6	...
7175	S, P, FG	90.0	...	1.6	...	2.5	0.23	5.6	...
7178	S, P, E, ES, C	88.1	...	2.0	...	2.7	0.26	6.8	...
7475	S, P, FG	90.3	1.5	...	...	2.3	0.22	5.7	...

(a) S, sheet; P, plate; F, foil; E, extruded rod, bar and wire; ES, extruded shapes; ET, extruded tubes; C, cold-finished rod, bar, and wire; DT, drawn tube; FG, forgings

Table 7.19 Designations and nominal compositions of common aluminum alloys used for casting

AA No.	Former AA designation	Former ASTM No.	Product(a)	Cu	Mg	Mn	Si	Others
201.0	...	...	S	4.6	0.35	0.35	...	0.7 Ag, 0.25 Ti
206.0	...	...	S or P	4.6	0.25	0.35	0.10(b)	0.22 Ti, 0.15 Fe(b)
A206.0	...	...	S or P	4.6	0.25	0.35	0.05(b)	0.22 Ti, 0.10 Fe(b)
208.0	108	CS43A	S	4.0	...	...	3.0	...
242.0	142	CN42A	S or P	4.0	1.5	...	...	2.0 Ni
295.0	195	C4A	S	4.5	...	...	0.8	...
296.0	B295.0, B195	...	P	4.5	...	...	2.5	...
308.0	A108	SC64A	S or P	4.5	...	...	5.5	...
319.0	319, Allcast	SC64D	S or P	3.5	...	...	6.0	...
336.0	A332.0, A132	SN122A	P	1.0	1.0	...	12.0	2.5 Ni
354.0	354	SC92A	P	1.8	0.50	...	9.0	...
355.0	355	SC51A	S or P	1.2	0.50	0.50(b)	5.0	0.6 Fe(b), 0.35 Zn(b)
C355.0	C355	SC51B	S or P	1.2	0.50	0.10(b)	5.0	0.20 Fe(b), 0.10 Zn(b)
356.0	356	SG70A	S or P	0.25(b)	0.32	0.35(b)	7.0	0.6 Fe(b), 0.35 Zn(b)
A356.0	A356	SG70B	S or P	0.20(b)	0.35	0.10(b)	7.0	0.20 Fe(b), 0.10 Zn(b)
357.0	357	...	S or P	...	0.50	...	7.0	...
A357.0	A357	...	S or P	...	0.6	...	7.0	0.15 Ti, 0.005 Be
359.0	359	SG91A	S or P	...	0.6	...	9.0	...
360.0	360	SG100B	D	...	0.50	...	9.5	2.0 Fe(b)
A360.0	A360	SG100A	D	...	0.50	...	9.5	1.3 Fe(b)
380.0	380	SC84B	D	3.5	...	...	8.5	2.0 Fe(b)
A380.0	A380	SC84A	D	3.5	...	...	8.5	1.3 Fe(b)
383.0	...	SC102A	D	2.5	...	...	10.5	...
384.0	384	SC114A	D	3.8	...	...	11.2	3.0 Zn(b)
A384.0	384	SC114A	D	3.8	...	...	11.2	1.0 Zn(b)
390.0	A390	...	S or P	4.5	0.6	...	17.0	0.5 Zn(b)
A390.0	A390	...	S or P	4.5	0.6	...	17.0	0.5 Zn(b)
413.0	13	S12B	D	...	...	...	12.0	2.0 Fe(b)
A413.0	A13	S12A	D	...	...	...	12.0	1.3 Fe(b)
4430	43	S5B	S	0.6(b)	...	...	5.2	...
A443.0	43	...	S	0.30(b)	...	...	5.2	...
B443.0	43	S5A	S or P	0.15(b)	...	...	5.2	...
C443.0	A43	S5C	D	0.6(b)	...	...	5.2	2.0 Fe(b)
514.0	214	G4A	S	...	4.0	...	...	...
518.0	218	G8A	D	...	8.0	...	...	...
520.0	220	G10A	S	...	10.0	...	...	...
535.0	Almag 35	GM70B	S	...	6.8	0.18	...	0.18 Ti
A535.0	A218	...	S	...	7.0	0.18	...	...
B535.0	B218	...	S	...	7.0	...	...	0.18 Ti
712.0	D712.0, D612, 40E	ZG61A	S or P	0.7	0.35	...	...	7.5 Zn, 0.7 Cu
771.0	Precedent 71A	ZG71B	S	...	0.9	...	...	7.0 Zn, 0.13 Cr, 0.15 Ti
850.0	750	...	S or P	1.0	...	...	...	6.2 Sn, 1.0 Ni

(a) S, sand casting; P, permanent mold casting; D, die casting. (b) Maximum

Chemical Compositions of Metals and Alloys

Table 7.20 Generic classification of copper alloys

Generic name	UNS No.	Composition
Wrought alloys		
Coppers	C10100–C15760	>99% Cu
High-copper alloys	C16200–C19600	>96% Cu
Brasses	C20500–C28580	Cu-Zn
Leaded brasses	C31200–C38590	Cu-Zn-Pb
Tin brasses	C40400–C49080	Cu-Zn-Sn-Pb
Phosphor bronzes	C50100–C52400	Cu-Sn-P
Leaded phosphor bronzes	C53200–C54800	Cu-Sn-Pb-P
Copper-phosphorus and copper-silver-phosphorus alloys	C55180–C55284	Cu-P-Ag
Aluminum bronzes	C60600–C64400	Cu-Al-Ni-Fe-Si-Sn
Silicon bronzes	C64700–C66100	Cu-Si-Sn
Other copper-zinc alloys	C66400–C69900	...
Copper-nickels	C70000–C79900	Cu-Ni-Fe
Nickel silvers	C73200–C79900	Cu-Ni-Zn
Cast alloys		
Coppers	C80100–C81100	>99% Cu
High-copper alloys	C81300–C82800	>94% Cu
Red and leaded red brasses	C83300–C85800	Cu-Zn-Sn-Pb (75–89% Cu)
Yellow and leaded yellow brasses	C85200–C85800	Cu-Zn-Sn-Pb (57–74% Cu)
Manganese bronzes and leaded manganese bronzes	C86100–C86800	Cu-Zn-Mn-Fe-Pb
Silicon bronzes, silicon brasses	C87300–C87900	Cu-Zn-Si
Tin bronzes and leaded tin bronzes	C90200–C94500	Cu-Sn-Zn-Pb
Nickel-tin bronzes	C94700–C94900	Cu-Ni-Sn-Zn-Pb
Aluminum bronzes	C95200–C95810	Cu-Al-Fe-Ni
Copper-nickels	C96200–C96800	Cu-Ni-Fe
Nickel silvers	C97300–C97800	Cu-Ni-Zn-Pb-Sn
Leaded coppers	C98200–C98800	Cu-Pb
Miscellaneous alloys	C99300–C99750	...

Table 7.21 Nominal compositions of wrought copper and copper alloys

UNS No. (and name)	Nominal composition, %	Commercial form(a)
Coppers		
C10100 (oxygen-free electronic copper)	99.99 Cu	F, R, W, T, P, S
C10200 (oxygen-free copper)	99.95 Cu	F, R, W, T, P, S
C10300 (oxygen-free extra-low-phosphorus copper)	99.95 Cu, 0.003 P	F, R, T, P, S
C10400, C10500, C10700 (oxygen-free silver-bearing copper)	99.95 Cu(b)	F, R, W, S
C10800 (oxygen-free low-phosphorus copper)	99.95 Cu, 0.009 P	F, R, T, P
C11000 (electrolytic tough pitch copper)	99.90 Cu, 0.04 O	F, R, W, T, P, S
C11100 (electrolytic tough pitch anneal-resistant copper)	99.90 Cu, 0.04 O, 0.01 Cd	W
C11300, C11400, C11500, C11600 (silver-bearing tough pitch copper)	99.90 Cu, 0.04 O, Ag(c)	F, R, W, T, S
C12000, C12100	99.9 Cu(d)	F, T, P
C12200 (phosphorus-deoxidized copper, high residual phosphorus)	99.90 Cu, 0.02 P	F, R, T, P
C12500, C12700, C12800, C12900, C13000 (fire-refined tough pitch with silver)	99.88 Cu(e)	F, R, W, S
C14200 (phosphorus-deoxidized arsenical copper)	99.68 Cu, 0.3 As, 0.02 P	F, R, T
(continued)		

Table 7.21 (continued)

UNS No. (and name)	Nominal composition, %	Commercial form(a)
Coppers (continued)		
C14300	99.9 Cu, 0.1 Cd	F
C14310	99.8 Cu, 0.2 Cd	F
C14500 (phosphorus-deoxidized tellurium-bearing copper)	99.5 Cu, 0.50 The, 0.008 P	F, R, W, T
C14700 (sulfur-bearing copper)	99.6 Cu, 0.40 S	R, W
C15000 (zirconium-copper)	99.8 Cu, 0.15 Zr	R, W
C15100	99.82 Cu, 0.1 Zr	F
C15500	99.75 Cu, 0.06 P, 0.11 Mg, Ag(f)	F
C15710	99.8 Cu, 0.2 Al_2O_3	R, W
C15720	99.6 Cu, 0.4 Al_2O_3	F, R
C15735	99.3 Cu, 0.7 Al_2O_3	R
C15760	98.9 Cu, 1.1 Al_2O_3	F, R
High-copper alloys		
C16200 (cadmium-copper)	99.0 Cu, 1.0 Cd	F, R, W
C16500	98.6 Cu, 0.8 Cd, 0.6 Sn	F, R, W
C17000 (beryllium-copper)	99.5 Cu, 1.7 Be, 0.20 Co	F, R
C17200 (beryllium-copper)	99.5 Cu, 1.9 Be, 0.20 Co	F, R, W, T, P, S
C17300 (beryllium-copper)	99.5 Cu, 1.9 Be, 0.40 Pb	R
C17400	99.5 Cu, 0.3 Be, 0.25 Co	F
C17500 (copper-cobalt-beryllium alloy)	99.5 Cu, 2.5 Co, 0.6 Be	F, R
C18200, C18400, C18500 (chromium-copper)	99.5 Cu(g)	F, W, R, S, T
C18700 (leaded copper)	99.0 Cu, 1.0 Pb	R
C18900	98.75 Cu, 0.75 Sn, 0.3 Si, 0.20 Mn	R, W
C19000 (copper-nickel-phosphorus alloy)	98.7 Cu, 1.1 Ni, 0.25 P	F, R, W
C19100 (copper-nickel-phosphorus-tellurium alloy)	98.15 Cu, 1.1 Ni, 0.50 The, 0.25 P	R, F
C19200	98.97 Cu, 1.0 Fe, 0.03 P	F, T
C19400	97.5 Cu, 2.4 Fe, 0.13 Zn, 0.03 P	F
C19500	97.0 Cu, 1.5 Fe, 0.6 Sn, 0.10 P, 0.80 Co	F
C19700	99 Cu, 0.6 Fe, 0.2 P, 0.05 Mg	F
Brasses		
C21000 (gilding, 95%)	95.0 Cu, 5.0 Zn	F, W
C22000 (commercial bronze, 90%)	90.0 Cu, 10.0 Zn	F, R, W, T
C22600 (jewelry bronze, 87.5%)	87.5 Cu, 12.5 Zn	F, W
C23000 (red brass, 85%)	85.0 Cu, 15.0 Zn	F, W, T, P
C24000 (low brass, 80%)	80.0 Cu, 20.0 Zn	F, W
C26000 (cartridge brass, 70%)	70.0 Cu, 30.0 Zn	F, R, W, T
C26800, C27000 (yellow brass)	65.0 Cu, 35.0 Zn	F, R, W
C28000 (Muntz metal)	60.0 Cu, 40.0 Zn	F, R, T
Leaded brasses		
C31400 (leaded commercial bronze)	89.0 Cu, 1.75 Pb, 9.25 Zn	F, R
C31600 (leaded commercial bronze, nickel-bearing)	89.0 Cu, 1.9 Pb, 1.0 Ni, 8.1 Zn	F, R
C33000 (low-leaded brass tube)	66.0 Cu, 0.5 Pb, 33.5 Zn	T
C33200 (high-leaded brass tube)	66.0 Cu, 1.6 Pb, 32.4 Zn	T
C33500 (low-leaded brass)	65.0 Cu, 0.5 Pb, 34.5 Zn	F
C34000 (medium-leaded brass)	65.0 Cu, 1.0 Pb, 34.0 Zn	F, R, W, S
C34200 (high-leaded brass)	64.5 Cu, 2.0 Pb, 33.5 Zn	F, R
C34900	62.2 Cu, 0.35 Pb, 37.45 Zn	R, W
C35000 (medium-leaded brass)	62.5 Cu, 1.1 Pb, 36.4 Zn	F, R
C35300 (high-leaded brass)	62.0 Cu, 1.8 Pb, 36.2 Zn	F, R
C35600 (extra-high-leaded brass)	63.0 Cu, 2.5 Pb, 34.5 Zn	F
C36000 (free-cutting brass)	61.5 Cu, 3.0 Pb, 35.5 Zn	F, R, S
C36500 to C36800 (leaded Muntz metal)	60.0 Cu(h), 0.6 Pb, 39.4 Zn	F

(continued)

Table 7.21 (continued)

UNS No. (and name)	Nominal composition, %	Commercial form(a)
Leaded brasses (continued)		
C37000 (free-cutting Muntz metal)	60.0 Cu, 1.0 Pb, 39.0 Zn	T
C37700 (forging brass)	59.0 Cu, 2.0 Pb, 39.0 Zn	R, S
C38500 (architectural bronze)	57.0 Cu, 3.0 Pb, 40.0 Zn	R, S
Tin brasses		
C40500	95 Cu, 1 Sn, 4 Zn	F
C40800	95 Cu, 2 Sn, 3 Zn	F
C41100	91 Cu, 0.5 Sn, 8.5 Zn	F, W
C41300	90.0 Cu, 1.0 Sn, 9.0 Zn	F, R, W
C41500	91 Cu, 1.8 Sn, 7.2 Zn	F
C42200	87.5 Cu, 1.1 Sn, 11.4 Zn	F
C42500	88.5 Cu, 2.0 Sn, 9.5 Zn	F
C43000	87.0 Cu, 2.2 Sn, 10.8 Zn	F
C43400	85.0 Cu, 0.7 Sn, 14.3 Zn	F
C43500	81.0 Cu, 0.9 Sn, 18.1 Zn	F, T
C44300, C44400, C44500 (inhibited admiralty)	71.0 Cu, 28.0 Zn, 1.0 Sn	F, W, T
C46400 to C46700 (naval brass)	60.0 Cu, 39.25 Zn, 0.75 Sn	F, R, T, S
C48200 (naval brass, medium-leaded)	60.5 Cu, 0.7 Pb, 0.8 Sn, 38.0 Zn	F, R, S
C48500 (leaded naval brass)	60.0 Cu, 1.75 Pb, 37.5 Zn, 0.75 Sn	F, R, S
Phosphor bronzes		
C50500 (phosphor bronze, 1.25% E)	98.75 Cu, 1.25 Sn, trace P	F, W
C51000 (phosphor bronze, 5% A)	95.0 Cu, 5.0 Sn, trace P	F, R, W, T
C51100	95.6 Cu, 4.2 Sn, 0.2 P	F
C52100 (phosphor bronze, 8% C)	92.0 Cu, 8.0 Sn, trace P	F, R, W
C52400 (phosphor bronze, 10% D)	90.0 Cu, 10.0 Sn, trace P	F, R, W
Leaded phosphor bronzes		
C54400 (free-cutting phosphor bronze)	88.0 Cu, 4.0 Pb, 4.0 Zn, 4.0 Sn	F, R
Aluminum bronzes		
C60800 (aluminum bronze, 5%)	95.0 Cu, 5.0 Al	T
C61000	92.0 Cu, 8.0 Al	R, W
C61300	92.65 Cu, 0.35 Sn, 7.0 Al	F, R, T, P, S
C61400 (aluminum bronze, D)	91.0 Cu, 7.0 Al, 2.0 Fe	F, R, W, T, P, S
C61500	90.0 Cu, 8.0 Al, 2.0 Ni	F
C61800	89.0 Cu, 1.0 Fe, 10.0 Al	R
C61900	86.5 Cu, 4.0 Fe, 9.5 Al	F
C62300	87.0 Cu, 3.0 Fe, 10.0 Al	F, R
C62400	86.0 Cu, 3.0 Fe, 11.0 Al	F, R
C62500	82.7 Cu, 4.3 Fe, 13.0 Al	F, R
C63000	82.0 Cu, 3.0 Fe, 10.0 Al, 5.0 Ni	F, R
C63200	82.0 Cu, 4.0 Fe, 9.0 Al, 5.0 Ni	F, R
C63600	95.5 Cu, 3.5 Al, 1.0 Si	R, W
C63800	95.0 Cu, 2.8 Al, 1.8 Si, 0.40 Co	F
C64200	91.2 Cu, 7.0 Al	F, R
Silicon bronzes		
C65100 (low-silicon bronze, B)	98.5 Cu, 1.5 Si	R, W, T
C65400	95.44 Cu, 3 Si, 1.5 Sn, 0.06 Cr	F
C65500 (high-silicon bronze, A)	97.0 Cu, 3.0 Si	F, R, W, T
Other copper-zinc alloys		
C66700 (manganese brass)	70.0 Cu, 28.8 Zn, 1.2 Mn	F, W
C67400	58.5 Cu, 36.5 Zn, 1.2 Al, 2.8 Mn, 1.0 Sn	F, R

(continued)

Table 7.21 (continued)

UNS No. (and name)	Nominal composition, %	Commercial form(a)
Other copper-zinc alloys (continued)		
C67500 (manganese bronze, A)	58.5 Cu, 1.4 Fe, 39.0 Zn, 1.0 Sn, 0.1 Mn	R, S
C68700 (aluminum brass, arsenical)	77.5 Cu, 20.5 Zn, 2.0 Al, 0.1 As	T
C68800	73.5 Cu, 22.7 Zn, 3.4 Al, 0.40 Co	F
C69000	73.3 Cu, 3.4 Al, 0.6 Ni, 22.7 Zn	F
C69400 (silicon red brass)	81.5 Cu, 14.5 Zn, 4.0 Si	R
Copper-nickels		
C70250	96.2 Cu, 3 Ni, 0.65 Si, 0.15 Mg	F
C70400	92.4 Cu, 1.5 Fe, 5.5 Ni, 0.6 Mn	F, T
C70600 (copper-nickel, 10%)	88.7 Cu, 1.3 Fe, 10.0 Ni	F, T
C71000 (copper-nickel, 20%)	79.0 Cu, 21.0 Ni	F, W, T
C71300	75 Cu, 25 Ni	F
C71500 (copper-nickel, 30%)	70.0 Cu, 30.0 Ni	F, R, T
C71700	67.8 Cu, 0.7 Fe, 31.0 Ni, 0.5 Be	F, R, W
C72500	88.2 Cu, 9.5 Ni, 2.3 Sn	F, R, W, T
Nickel-silvers		
C73500	72.0 Cu, 10.0 Zn, 18.0 Ni	F, R, W, T
C74500 (nickel silver, 65-10)	65.0 Cu, 25.0 Zn, 10.0 Ni	F, W
C75200 (nickel silver, 65-18)	65.0 Cu, 17.0 Zn, 18.0 Ni	F, R, W
C75400 (nickel silver, 65-15)	65.0 Cu, 20.0 Zn, 15.0 Ni	F
C75700 (nickel silver, 65-12)	65.0 Cu, 23.0 Zn, 12.0 Ni	F, W
C76200	59.0 Cu, 29.0 Zn, 12.0 Ni	F, T
C77000 (nickel silver, 55-18)	55.0 Cu, 27.0 Zn, 18.0 Ni	F, R, W
C78200 (leaded nickel silver, 65-8-2)	65.0 Cu, 2.0 Pb, 25.0 Zn, 8.0 Ni	F

(a) F, flat products; R, rod; W, wire; T, tube; P, pipe; S, shapes. (b) C10400, 250 g/Mg (8 oz/ton) Ag; C10500, 310 g/Mg (10 oz/ton); C10700, 780 g/Mg (25 oz/ton). (c) C11300, 250 g/Mg (8 oz/ton) Ag; C11400, 310 g/Mg (10 oz/ton); C11500, 500 g/Mg (16 oz/ton); C11600, 780 g/Mg (25 oz/ton). (d) C12000, 0.008 P; C12100, 0.008 P and 125 g/Mg (4 oz/ton) Ag. (e) C12700, 250 g/Mg (8 oz/ton) Ag; C12800, 500 g/Mg (10 oz/ton); C12900, 500 g/Mg (16 oz/ton); C13000, 780 g/Mg (25 oz/ton). (f) 260 g/Mg (8.30 oz/ton) Ag. (g) C18200, 0.9 Cr; C18400, 0.8 Cr; C18500, 0.7 Cr. (h) Rod, 61.0 Cu min. Source: Copper Development Association Inc.

Chemical Compositions of Metals and Alloys

Table 7.22 Nominal compositions of cast copper and copper alloys

UNS designation(a)	Nominal composition(a), %	Casting type(b)
Coppers		
C80100	99.95 Cu + Ag min, 0.05 other max	C, T, I, M, P, S
C80300	99.95 Cu + Ag min, 0.034 Ag min, 0.05 other max	C, T, I, M, P, S
C80500	99.75 Cu + Ag min, 0.034 Ag min, 0.02 B max, 0.23 other max	C, T, I, M, P, S
C80700	99.75 Cu + Ag min, 0.02 B max, 0.23 other max	C, T, I, M, P, S
C80900	99.70 Cu + Ag min, 0.034 Ag min, 0.30 other max	C, T, I, M, P, S
C81100	99.70 Cu + Ag min, 0.30 other max	C, T, I, M, P, S
High-copper alloys		
C81300	98.5 Cu min, 0.06 Be, 0.80 Co, 0.40 other max	C, T, I, M, P, S
C81400	98.5 Cu min, 0.06 Be, 0.80 Cr, 0.40 other max	C, T, I, M, P, S
C81500	98.0 Cu min, 1.0 Cr, 0.50 other max	C, T, I, M, P, S
C81700	94.25 Cu min, 1.0 Ag, 0.4 Be, 0.9 Co, 0.9 Ni	C, T, I, M, P, S
C81800	95.6 Cu min, 1.0 Ag, 0.4 Be, 1.6 Co	C, T, I, M, P, S
C82000	96.8 Cu, 0.6 Be, 2.6 Co	C, T, I, M, P, S(c)
C82100	97.7 Cu, 0.5 Be, 0.9 Co, 0.9 Ni	C, T, I, M, P, S
C82200	96.5 Cu min, 0.6 Be, 1.5 Ni	C, T, I, M, P, S
C82400	96.4 Cu min, 1.70 Be, 0.25 Co	C, I, M, P, S(c)
C82500	97.2 Cu, 2.0 Be, 0.5 Co, 0.25 Si	C, I, M, P, S(c)
C82600	95.2 Cu min, 2.3 Be, 0.5 Co, 0.25 Si	C, I, M, P, S(c)
C82700	96.3 Cu, 2.45 Be, 1.25 Ni	C, I, M, P, S
C82800	96.6 Cu, 2.6 Be, 0.5 Co, 0.25 Si	C, I, M, P, S(c)
Red brasses and leaded red brasses		
C83300	93 Cu, 1.5 Sn, 1.5 Pb, 4 Zn	S
C83400	90 Cu, 10 Zn	C, S
C83600	85 Cu, 5 Sn, 5 Pb, 5 Zn	C, T, I, S
C83800	83 Cu, 4 Sn, 6 Pb, 7 Zn	C, T, S
Semired brasses and leaded semired brasses		
C84200	80 Cu, 5 Sn, 2.5 Pb, 12.5 Zn	C, T, S
C84400	81 Cu, 3 Sn, 7 Pb, 9 Zn	C, T, S
C84500	78 Cu, 3 Sn, 7 Pb, 12 Zn	C, T, S
C84800	76 Cu, 3 Sn, 6 Pb, 15 Zn	C, S
Yellow brasses and leaded yellow brasses		
C85200	72 Cu, 1 Sn, 3 Pb, 24 Zn	C, T
C85400	67 Cu, 1 Sn, 3 Pb, 29 Zn	C, T, M, P, S
C85500	61 Cu, 0.8 Al, bal Zn	C, S
C85700	63 Cu, 1 Sn, 1 Pb, 34.7 Zn, 0.3 Al	C, M, P, S
C85800	58 Cu, 1 Sn, 1 Pb, 40 Zn	D

UNS designation(a)	Nominal composition(a), %	Casting type(b)
Manganese and leaded manganese bronze alloys		
C86100	67 Cu, 21 Zn, 3 Fe, 5 Al, 4 Mn	C, I, P, S
C86200	64 Cu, 26 Zn, 3 Fe, 4 Al, 3 Mn	C, T, D, I, P, S
C86300	63 Cu, 25 Zn, 3 Fe, 6 Al, 3 Mn	C, I, P, S
C86400	59 Cu, 1 Pb, 40 Zn	C, D, M, P, S
C86500	58 Cu, 0.5 Sn, 39.5 Zn, 1 Fe, 1 Al	C, I, P, S
C86700	58 Cu, 1 Pb, 41 Zn	C, S
C86800	55 Cu, 37 Zn, 3 Ni, 2 Fe, 3 Mn	S
Silicon bronzes and silicon brasses		
C87200	89 Cu min, 4 Si	C, I, M, P, S
C87400	83 Cu, 14 Zn, 3 Si	C, D, I, M, P, S
C87500	82 Cu, 14 Zn, 4 Si	C, D, I, M, P, S
C87600	90 Cu, 5.5 Zn, 4.5 Si	S
C87800	82 Cu, 14 Zn, 4 Si	D
C87900	65 Cu, 34 Zn, 1 Si	D
Tin bronzes		
C90200	93 Cu, 7 Sn	C, S
C90300	88 Cu, 8 Sn, 4 Zn	C, T, I, P, S
C90500	88 Cu, 10 Sn, 2 Zn	C, T, I, S
C90700	89 Cu, 11 Sn	C, T, I, M, S
C90800	87 Cu, 12 Sn	
C90900	87 Cu, 13 Sn	C, S
C91000	85 Cu, 14 Sn, 1 Zn	C, T, I, S
C91100	84 Cu, 16 Sn	S
C91300	81 Cu, 19 Sn	C, T, M, S
C91600	88 Cu, 10.5 Sn, 1.5 Ni	C, T, M, S
C91700	86.5 Cu, 12 Sn, 1.5 Ni	C, T, I, M, S
Leaded tin bronzes		
C92200	88 Cu, 6 Sn, 1.5 Pb, 4.5 Zn	C, T, I, M, P, S
C92300	87 Cu, 8 Sn, 4 Zn	C, T, S
C92400	88 Cu, 10 Sn, 2 Pb, 2 Zn	...
C92500	87 Cu, 11 Sn, 1 Pb, 1 Ni	C, T, M, S
C92600	87 Cu, 10 Sn, 1 Pb, 2 Zn	C, T, S
C92700	88 Cu, 10 Sn, 2 Pb	C, T, S
C92800	79 Cu, 16 Sn, 5 Pb	C, S
C92900	84 Cu, 10 Sn, 2.5 Pb, 3.5 Ni	C, T, M, S
High-leaded tin bronzes		
C93200	83 Cu, 7 Sn, 7 Pb, 3 Zn	C, T, M, S
C93400	84 Cu, 8 Sn, 8 Pb	C, T, S
C93500	85 Cu, 5 Sn, 9 Pb	C, T, S
C93700	80 Cu, 10 Sn, 10 Pb	C, T, M, S
C93800	78 Cu, 7 Sn, 15 Pb	C, T, M, S
C93900	79 Cu, 6 Sn, 15 Pb	T
C94000	70.5 Cu, 13.0 Sn, 15.0 Pb, 0.50 Zn, 0.75 Ni, 0.25 Fe, 0.05 P, 0.35 Sb	...
C94100	70.0 Cu, 5.5 Sn, 18.5 Pb, 3.0 Zn, 1.0 other max	...
C94300	70 Cu, 5 Sn, 25 Pb	C, S
C94400	81 Cu, 8 Sn, 11 Pb	C, T, S
C94500	73 Cu, 7 Sn, 20 Pb	C, S

(continued)

Table 7.22 (continued)

UNS designation(a)	Nominal composition(a), %	Casting type(b)
Nickel-tin bronzes		
C94700	88 Cu, 5 Sn, 2 Zn, 5 Ni	C, T, I, M, S
C94800	87 Cu, 5 Sn, 5 Ni	M, S
C94900	80 Cu, 5 Sn, 5 Pb, 5 Zn, 5 Ni	...
Aluminum bronzes		
C95200	88 Cu, 3 Fe, 9 Al	C, T, M, P, S
C95300	89 Cu, 1 Fe, 10 Al	C, T, M, P, S
C95400	85 Cu, 4 Fe, 11 Al	C, T, M, P, S
C95410	85 Cu, 4 Fe, 11 Al, 2 Ni	
C95500	81 Cu, 4 Ni, 4 Fe, 11 Al	C, T, M, P, S
C95600	91 Cu, 7 Al, 2 Si	C, T, M, P, S
C95700	75 Cu, 2 Ni, 3 Fe, 8 Al, 12 Mn	C, T, M, P, S
C95800	81 Cu, 5 Ni, 4 Fe, 9 Al, 1 Mn	C, T, M, P, S
Copper-nickels		
C96200	88.6 Cu, 10 Ni, 1.4 Fe	C, S
C96300	79.3 Cu, 20 Ni, 0.7 Fe	C, S
C96400	69.1 Cu, 30 Ni, 0.9 Fe	C, T, S
C96600	68.5 Cu, 30 Ni, 1 Fe, 0.5 Be	C, T, I, M, S
C96700	67.6 Cu, 30 Ni, 0.9 Fe, 1.15 Be, 0.15 Zr, 0.15 Ti	I, M, S
Nickel silvers		
C97300	56 Cu, 2 Sn, 10 Pb, 12 Ni, 20 Zn	I, M, S
C97400	59 Cu, 3 Sn, 5 Pb, 17 Ni, 16 Zn	C, I, S
C97600	64 Cu, 4 Sn, 4 Pb, 20 Ni, 8 Zn	C, I, S
C97800	66 Cu, 5 Sn, 2 Pb, 25 Ni, 2 Zn	I, M, S
Leaded coppers		
C98200	76.0 Cu, 24.0 Pb	...
C98400	70.5 Cu, 28.5 Pb, 1.5 Ag	...
C98600	65.0 Cu, 35.0 Pb, 1.5 Ag	...
C98800	59.5 Cu, 40.0 Pb, 5.5 Ag	...
Special alloys		
C99300	71.8 Cu, 15 Ni, 0.7 Fe, 11 Al, 1.5 Co	T, S
C99400	90.4 Cu, 2.2 Ni, 2.0 Fe, 1.2 Al, 1.2 Si, 3.0 Zn	C, T, I, S
C99500	87.9 Cu, 4.5 Ni, 4.0 Fe, 1.2 Al, 1.2 Si, 1.2 Zn	C, T, S
C99600	58 Cu, 2 Al, 40 Mn	C, T, M, S
C99700	56.5 Cu, 1 Al, 1.5 Pb, 12 Mn, 5 Ni, 24 Zn	C, D, I, M, P, S
C99750	58 Cu, 1 Al, 1 Pb, 20 Mn, 20 Zn	D, I, M, P, S

(a) Nominal composition, unless otherwise noted. For seldom-used alloys, only compositions are available. (b) C, centrifugal; T, continuous; D, die; I, investment; M, permanent mold; P, plaster; S, sand. (c) Also pressure cast. Source: Copper Development Association Inc.

Chemical Compositions of Metals and Alloys 77

Table 7.23 Nominal compositions of common zinc alloy die castings and zinc alloy ingot for die casting

UNS No.	Alloy ASTM designation	Common designation	Cu	Al	Mg	Fe max	Pb max	Cd max	Sn max	Ni	Zn
Castings (ASTM B 86)											
Z33520(a)	AG40A	No. 3	0.25 max(c)	3.5–4.3	0.020–0.05(d)	0.100	0.005	0.004	0.003	...	bal
Z33523(a)	AG40B	No. 7	0.25 max	3.5–4.3	0.005–0.020	0.075	0.0030	0.0020	0.0010	0.005–0.020	bal
Z35531(a)	AC41A	No. 5	0.75–1.25	3.5–4.3	0.03–0.08(d)	0.100	0.005	0.004	0.003	...	bal
Z35541	AC43A	No. 2	2.5–3.0	3.5–4.3	0.020–0.050	0.100	0.005	0.004	0.003	...	bal
Ingot form (ASTM B 240)											
Z33521(b)	AG40A	No. 3	0.10 max	3.9–4.3	0.025–0.05	0.075	0.004	0.003	0.002	...	bal
Z33522(b)	AG40B	No. 7	0.10 max	3.9–4.3	0.010–0.02	0.075	0.002	0.002	0.001	0.005–0.020	bal
Z35530(b)	AC41A	No. 5	0.75–1.25	3.9–4.3	0.03–0.06	0.075	0.004	0.003	0.002	...	bal
Z35540	AC43A	No. 2	2.6–2.9	3.9–4.3	0.025–0.05	0.075	0.004	0.003	0.002	...	bal

(a) Zinc alloy die castings may contain nickel, chromium, silicon, and manganese in amounts of 0.02, 0.02, 0.035, and 0.06%, respectively. (b) Zinc alloy ingot for die casting may contain nickel, chromium, silicon, and manganese in amounts of up to 0.02, 0.02, 0.035, and 0.05%, respectively. No harmful effects have ever been noted from the presence of these elements up to these concentrations; therefore, analyses are not required for these elements, except that nickel analysis is required for Z33522. (c) For the majority of commercial applications, a copper content in the range of 0.25–0.75% will not adversely affect the serviceability of die castings and should not serve as a basis for rejection. (d) Magnesium may be as low as 0.015% provided that the lead, cadmium, and tin do not exceed 0.003, 0.003, and 0.001%, respectively.

Table 7.24 Nominal compositions of zinc-aluminum foundry and die casting alloys directly poured to produce castings and in ingot form for remelting to produce castings

Alloy				Additions		Composition, %		Impurities(a)		
Common designation	UNS No.	Al	Cu	Mg	Zn	Fe max	Pb max	Cd max	Sn max

Castings (ASTM B 791)

ZA-8	Z35636	8.0–8.8	0.8–1.3	0.015–0.030	bal	0.075	0.006	0.006	0.003
ZA-12	Z35631	10.5–11.5	0.5–1.2	0.015–0.030	bal	0.075	0.006	0.006	0.003
ZA-27	Z35841	25.0–28.0	2.0–2.5	0.010–0.020	bal	0.075	0.006	0.006	0.003

Ingot form (ASTM B 669)

ZA-8	Z35635	8.2–8.8	0.8–1.3	0.020–0.030	bal	0.065	0.005	0.005	0.002
ZA-12	Z35630	10.8–11.5	0.5–1.2	0.020–0.030	bal	0.065	0.005	0.005	0.002
ZA-27	Z35840	25.5–28.0	2.0–2.5	0.012–0.020	bal	0.072	0.005	0.005	0.002

(a) Zinc-aluminum ingot for foundry and pressure die casting may contain chromium, manganese, or nickel in amounts of up to 0.01% each or 0.03% total. No harmful effects have ever been noted from the presence of these elements in these concentrations; therefore, analyses are not required for these elements.

Table 7.25 Nominal compositions of zinc casting alloys used for sheet metal forming dies and for slush casting alloys in ingot form

Alloy					Composition, %				
Common designation	UNS No.	Al	Cd max	Cu	Fe max	Pb max	Mg	Sn max	Zn

Forming die alloys (ASTM B 793)

Alloy A	Z35543	3.5–4.5	0.005	2.5–3.5	0.100	0.007	0.02–0.10	0.005	bal
Alloy B	Z35542	3.9–4.3	0.003	2.5–2.9	0.075	0.003	0.02–0.05	0.001	bal

Slush casting alloys (ASTM B 792)

Alloy A	Z34510	4.50–5.00	0.005	0.2–0.3	0.100	0.007	...	0.005	bal
Alloy B	Z30500	5.25–5.75	0.005	0.1 max	0.100	0.007	...	0.005	bal

Chemical Compositions of Metals and Alloys 79

Table 7.26 Nominal compositions of rolled zinc alloys per ASTM B 69

Common designation	Alloy UNS No.	Cu	Pb	Cd	Fe max	Al max	Other max	Zn
Zn-0.08Pb	Z21210	0.001 max	0.10 max	0.005 max	0.012	0.001	0.001 Sn	bal
Zn-0.06Pb-0.06Cd	Z21220	0.005 max	0.05–0.10	0.05–0.08	0.012	0.001	0.001 Sn	bal
Zn-0.3Pb-0.3Cd	Z21540	0.005 max	0.25–0.50	0.25–0.45	0.002	0.001	0.001 Sn	bal
Zn-1Cu	Z44330	0.85–1.25	0.10 max	0.005 max	0.012	0.001	0.001 Sn	bal
Zn-1Cu-0.010Mg	Z45330	0.85–1.25	0.15 max	0.04 max	0.015	0.001	0.006–0.016 Mg 0.001 Sn	bal
Zn-0.8Cu-0.15Ti	Z41320	0.50–1.50	0.10 max	0.05 max	0.012	0.001	0.12–0.50 Ti 0.001 Sn	bal
Zn-0.8Cu	Z40330	0.70–0.90	0.02 max	0.02 max	0.01	0.005	0.02 Ti	bal

Table 7.27 Nominal compositions of magnesium casting alloys

ASTM No.	UNS No.	Al	Mn(a)	Zn	Th	Zr	Rare earths	Other
Sand and permanent mold castings								
AM100A	M10100	10.0	0.1	0.3	...	...	...	...
AZ63A	M11630	6.0	0.15	3.0	...	...	...	...
AZ81A	M11810	7.6	0.13	0.7	...	...	...	...
AZ91C	M11914	8.7	0.13	0.7	...	...	...	...
AZ91E	M11918	8.7	0.13	0.7	...	...	...	0.005 Fe(b)
AZ92A	M11920	9.0	0.10	2.0	...	...	...	...
EQ21A	M12210	...	...	...	...	0.7	2.25(c)	1.5 Ag
EZ33A	M12330	...	...	2.55	...	0.75	3.25	...
HK31A	M13310	...	...	0.3	3.25	0.7	...	...
HZ32A	M13320	...	...	2.1	3.25	0.75	0.1	...
K1A	M18010	...	...	...	...	0.7	...	...
QE22A	M18220	...	...	...	...	0.7	2.15(c)	2.5 Ag
QH21A	M18210	...	...	0.2	1.1(d)	0.7	1.05 (c)(d)	2.5 Ag
WE43A	M18430	...	0.15	0.2	...	0.7	3.4(e)	4.0 Y
WE54A	M18410	...	0.15	0.2	...	0.7	2.75(e)	5.0 Y
ZC63A	M16331	...	0.25	6.0	...	...	...	2.7 Cu
ZE41A	M16410	...	0.15	4.25	...	0.7	1.25	...
ZE63A	M16630	...	...	5.75	...	0.7	2.55	...
ZH62A	M16620	...	...	5.7	1.8	0.75	...	...
ZK51A	M16510	...	...	4.55	...	0.75	...	...
ZK61A	M16610	...	...	6.0	...	0.8	...	...
Die castings								
AM60A	M10600	6.0	0.13	0.22	...	...	...	0.5 Si; 0.35 Cu
AS41A	M10410	4.25	0.20	0.12	...	...	...	1.0 Si
AS41B	M10412	4.25	0.35	0.12	...	...	...	1.0 Si
AZ91A	M11910	9.0	0.13	0.7	...	...	...	0.5 Si
AZ91B	M11912	9.0	0.13	0.7	...	...	...	0.5 Si; 0.35 Cu
AZ91D	M11916	9.0	0.15	0.7	...	...	...	...
AM60B	M10602	6.0	0.24	0.22	...	...	...	...
AM50A	M10500	4.9	0.26	0.22	...	...	...	...

(a) Minimum. (b) If iron exceeds 0.005%, the iron to manganese ratio shall not exceed 0.032. (c) Rare earth elements are in the form of didymium (a mixture of rare earth elements made up chiefly of neodymium and praseodymium). (d) Thorium and didymium total is 1.5 to 2.4%. (e) Rare earths are 2.0–2.5% and 1.5–2.0% Nd for WE43A and WE54A, respectively, the remainder being heavy rare earths.

Table 7.28 Nominal compositions of wrought magnesium alloys

Alloy ASTM No.	Alloy UNS No.	Product form(a)	Composition, wt % Al	Mn (min)	Zn	Th	Zr	Other
AZ31B	M11310	F, S, E	3.0	0.20	1.0	...	...	...
AZ31C	M11312	S, E	3.0	0.15	1.0	...	...	...
AZ61A	M11610	F, E	6.5	0.15	0.95	...	...	...
AZ80A	M11800	F, E	8.5	0.12	0.5	...	...	...
HK31A	M13310	S	3.0	...	0.3	3.25	0.7	...
HM21A	M13210	F, S	...	0.45	...	2.0	...	...
LA141A	M14141	S	1.25	0.15	...	...	...	14 Li
M1A	M15100	E	...	1.6	...	...	...	0.3 Ca
ZE10A	M16100	S	...	0.15	1.25	...	...	0.17 RE(b)
ZK40A	M16400	E	...	...	4.0	...	0.45	...
ZK60A	M16600	F, E	...	...	5.5	...	0.45	...

(a) S, sheet and plate; F, forging; E, extruded bar, shape, tube, and wire. (b) RE, rare earths

Table 7.29 UNS categories and nominal compositions of various lead grades and lead-base alloys

Lead alloy type(a)	UNS No.
Pure leads (UNS L50000–L50099)	
Zone-refined lead (99.9999% Pb min)	L50001
Refined soft lead (99.999% Pb min)	L50005
Refined soft lead (99.99% Pb min)	L50011, L50012, L50013, L50014
Corroding lead (99.94% Pb min)	L50042
Common lead (99.94% Pb min)	L50045
Lead-silver alloys (UNS L50100–L50199)	
Cable-sheathing alloy (0.2% Ag, 99.8% Pb)	L50101
Electrowinning alloys (0.5–1.0% Ag, 99.5–99% Pb)	L50110, L50115, L50120
Electrowinning alloy (1.0% Ag, 1.0% As, 98% Pb)	L50122
Cathodic protection anode alloy (2.0% Ag, 98% Pb)	L50140
Solder alloys (1.0–1.5% Ag, 1.0 Sn, bal Pb)	L50121, L50131
Solder alloys (1.5–2.5% Ag, with no tin)	L50132, L50150, L50151
Solder alloy (1.5% Ag, 5.0% Sn, 93.5% Pb)	L50134
Solder alloy (2.5% Ag, 2.0% Sn, 95.5% Pb)	L50152
Solder alloy (5.0% Ag, 95% Pb)	L50170
Solder alloys (5.0% Ag, with 5% Sn or 5% In)	L50171, L50172
Solder alloy (5.5% Ag)	L50180
Lead-arsenic alloys (UNS L50300–L50399)	
Arsenical lead cable-sheathing alloy (0.15% As, 0.10% Bi, 0.10% Sn, 99.6% Pb)	L50310
Lead-barium alloys (UNS L50500–L50599)	
Lead-barium alloy (0.05% Ba, 99.9% Pb)	L50510
Lead-tin-barium alloys (0.05–0.10% Ba, 1.0–2.0% Sn, 97.9–99% Pb)	L50520–L50522, L50530, L50535
Frary metal (0.4–1.2% Ba, 0.5–0.8% Ca, 97.2–98.8% Pb)	L50540–L50543
Lead-calcium alloys (UNS L50700–L50899)	
Lead-calcium alloys (99.9% Pb, 0.008–0.03% Ca)	L50710, L50720

(continued)

(a) Unless otherwise specified as a minimum (min) or balance (bal), the listed compositions represent nominal values (or the range of nominal values when several alloy designations are grouped together).

Chemical Compositions of Metals and Alloys

Table 7.29 (continued)

Lead alloy type(a)	UNS No.
Lead-calcium alloys (UNS L50700–L50899) (continued)	
Cable-sheathing alloys (0.025% Ca, 99.7–99.9% Pb, 0.0–0.025% Sn)	L50712, L50713
Lead-copper-calcium alloy (99.9% Pb, 0.06% Cu, 0.03% Ca)	L50722
Electrowinning anode alloy (0.5% Ag, 99.4% Pb, 0.05% Ca)	L50730
Battery grid alloy (99.9% Pb, 0.06% Ca)	L50735
Battery grid alloys (0.065% Ca, 0.2–1.5% Sn, 99.7–98.4% Pb)	L50736, L50737, L50740, L50745, L50750, L50755
Battery grid alloys (0.07% Ca, 0.0–0.7% Sn, 99.2–99.9% Pb)	L50760, L50765
Battery grid alloys (0.10% Ca, 0.0–1.0% Sn, 98.9–99.9% Pb)	L50770, L50775, L50780, L50790
Battery grid alloys (0.12% Ca, 0.3% Sn, 99.6% Pb)	L50795, L50800
Bearing metal (0.02% Al, 0.04% Li, 0.7% Ca, 0.6% Na, 98.7% Pb)	L50810
Bearing metal (0.02% Al, 0.04% Li, 0.7% Ca, 0.2% Na, 0.4% Ba, 98.7% Pb)	L50820
Lead-calcium alloys (1.0–6.0% Ca, 94.0–99.0% Pb)	L50840, L50850, L50880
Lead-cadmium alloys (UNS L50900–L50999)	
Lead-cadmium eutectic alloy (17.0% Cd, 83.0% Pb)	L50940
Lead-copper alloys (UNS L51100–L51199)	
Copperized lead (0.05% Cu, 99.9% Pb)	L51110
Chemical lead (see Table 2)	L51120
Copper-bearing lead (0.06% Cu, 99.90% Pb min)	L51121
Lead-tellurium-copper alloys (0.06% Cu, 0.045–0.055% The, 99.82–99.85% Pb min)	L51123, L51124
Copperized soft lead (0.06% Cu, 99.9% Pb min)	L51125
Copper-bearing alloy (51% Pb, 3.0% Sn, other 0.8% max, bal Cu) (alloy 485 in SAE J460)	L51180
Lead-indium alloys (UNS L51500–L51599)	
Lead-indium-silver solder alloys (2.38–2.5% Ag, 4.76–5.0% In, 92.5–92.8% Pb)	L51510, L51512
Lead-indium solder alloys (5.0% In, 95.0% Pb)	L51511
Lead-indium alloys (19.0–70% In, 30–81% Pb)	L51530, L51532, L51535, L51540, L51550, L51560, L51570
Indium-tin-lead alloy (40% In, 40% Sn, 20% Pb)	L51545
Indium-silver-lead alloy (80% In, 5% Ag, 15% Pb)	L51585
Lead-lithium alloys (UNS L51700–L51799)	
Lead-lithium alloys (0.01–0.07% Li, 99.9% Pb)	L51705, L51708, L51710, L51720, L51730
Lead-tin-lithium alloys (0.02–0.04% Li, 0.35–0.7% Sn, 99.2–99.9% Pb)	L51740, L51748
Lead-tin-lithium-calcium alloys (0.08–0.065% Li, 1–2% Sn, 0.02–0.15% Ca, 97.8–99.6% Pb)	L51770, L51775, L51778, L51780, L51790
Lead-antimony alloys (UNS L52500–L53799)	
Lead-antimony alloys (<1.0% Sb)	L52500–L52599
Lead-antimony alloys (1.0–1.99% Sb)	L52600–L52699
Lead-antimony alloys (2.0–2.99% Sb)	L52700–L52799
Lead-antimony alloys (3.0–3.99% Sb)	L52800–L52899
Lead-antimony alloys (4.0–4.99% Sb)	L52900–L52999
Lead-antimony alloys (5.0–5.99% Sb)	L53000–L53099
Lead-antimony alloys (6.0–6.99% Sb)	L53100–L53199
Lead-antimony alloys (7.0–8.99% Sb)	L53200–L53299
Lead-antimony alloys (9.0–10.99% Sb)	L53300–L53399
Lead-antimony alloys (11.0–12.99% Sb)	L53400–L53499
Lead-antimony alloys (13.0–15.99% Sb)	L33500–L33599
Lead-antimony alloys (16.0–19.99% Sb)	L53600–L53699
Lead-antimony alloys (>20% Sb)	L53700–L53799

(continued)

(a) Unless otherwise specified as a minimum (min) or balance (bal), the listed compositions represent nominal values (or the range of nominal values when several alloy designations are grouped together).

Table 7.29 (continued)

Lead alloy type(a)	UNS No.
Lead-tin alloys (UNS L54000–L55099)	
Lead-tin alloys (<1.0% Sn)	L54000–L54099
Lead-tin alloys (1.0–1.99% Sn)	L54100–L54199
Lead-tin alloys (2.0–3.99% Sn)	L54200–L54299
Lead-tin alloys (4.0–7.99% Sn)	L54300–L54399
Lead-tin alloys (8.0–11.99% Sn)	L54400–L54499
Lead-tin alloys (12.0–15.99% Sn)	L54500–L54599
Lead-tin alloys (16.0–19.99% Sn)	L54600–L54699
Lead-tin alloys (20.0–27.99% Sn)	L54700–L54799
Lead-tin alloys (28.0–37.99% Sn)	L54800–L54899
Lead-tin alloys (38.0–47.99% Sn)	L54900–L54999
Lead-tin alloys (48.0–57.99% Sn)	L55000–L55099
Lead-strontium alloys (UNS L55200–L55299)	
Battery alloys (0.06–0.2% Sr, 0.0–0.03% Al, 0.0–0.08% Sn, 0.0–0.6% Ca, 99–99.8% Pb)	L55210, L55230, L55260
Lead-strontium alloy (2% Sr, 98% Pb)	L55290

(a) Unless otherwise specified as a minimum (min) or balance (bal), the listed compositions represent nominal values (or the range of nominal values when several alloy designations are grouped together).

Table 7.30 Chemical compositions of common titanium and titanium alloys

Designation	N	C	H	Fe	O	Al	Sn	Zr	Mo	Others
Unalloyed grades										
ASTM Grade 1	0.03	0.08	0.015	0.20	0.18	...	...	...	...	...
ASTM Grade 2	0.03	0.08	0.015	0.30	0.25	...	...	...	...	...
ASTM Grade 3	0.05	0.08	0.015	0.30	0.35	...	...	...	...	...
ASTM Grade 4	0.05	0.08	0.015	0.50	0.40	...	...	...	...	...
ASTM Grade 7	0.03	0.08	0.015	0.30	0.25	...	...	...	...	0.2 Pd
ASTM Grade 11	0.03	0.08	0.015	0.20	0.18	...	...	...	...	0.2 Pd
Alpha and near-alpha alloys										
Ti-0.3Mo-0.8Ni	0.03	0.10	0.015	0.30	0.25	...	...	...	0.3	0.8 Ni
Ti-5Al-2.5Sn	0.05	0.08	0.02	0.50	0.20	5	2.5	...	...	...
Ti-5Al-2.5Sn-ELI	0.07	0.08	0.0125	0.25	0.12	5	2.5	...	...	...
Ti-8Al-1Mo-1V	0.05	0.08	0.015	0.30	0.12	8	...	...	1	1 V
Ti-6Al-2Sn-4Zr-2Mo	0.05	0.05	0.0125	0.25	0.15	6	2	4	2	0.08 Si
Ti-6Al-2Nb-1Ta-0.8Mo	0.02	0.03	0.0125	0.12	0.10	6	...	...	1	2 Nb; 1 Ta
Ti-2.25Al-11Sn-5Zr-1Mo	0.04	0.04	0.008	0.12	0.17	2.25	11	5	1	0.2 Si
Ti-5.8Al-4Sn-3.5Zr-0.7Nb-0.5Mo-0.35Si	0.03	0.08	0.006	0.05	0.15	5.8	4	3.5	0.5	0.7 Nb; 0.35 Si
Alpha-beta alloys										
Ti-6Al-4V	0.05	0.10	0.0125	0.30	0.20	6	...	...	...	4 V
Ti-6Al-4V-ELI	0.05	0.08	0.0125	0.25	0.13	6	...	...	...	4 V
Ti-6Al-6V-2Sn	0.04	0.05	0.015	1.0	0.20	6	2	...	...	0.75 Ca; 6 V
Ti-6Al-2Sn-4Zr-6Mo	0.04	0.04	0.0125	0.15	0.15	6	2	4	6	...
Ti-5Al-2Sn-2Zr-4Mo-4Cr	0.04	0.05	0.0125	0.30	0.13	5	2	2	4	4 Cr
Ti-6Al-2Sn-2Zr-2Mo-2Cr	0.03	0.05	0.0125	0.25	0.14	5.7	2	2	2	2 Cr; 0.25 Si
Ti-3Al-2.5V	0.015	0.05	0.015	0.30	0.12	3	...	...	...	2.5 V
Ti-4Al-4Mo-2Sn-0.5Si	(a)	0.02	0.0125	0.20	(a)	4	2	...	4	0.5 Si
Beta alloys										
Ti-10V-2Fe-3Al	0.05	0.05	0.015	2.5	0.16	3	...	...	...	10 V
Ti-3Al-8V-6Cr-4Mo-4Zr	0.03	0.05	0.020	0.25	0.12	3	...	4	4	6 Cr; 8 V
Ti-15V-3Cr-3Al-3Sn	0.05	0.05	0.015	0.25	0.13	3	3	...	...	15 V; 3 Cr
Ti-15Mo-3Al-2.7Nb-0.2Si	0.05	0.05	0.015	0.25	0.13	3	...	...	15	2.7 Nb; 0.2 Si

(a) Combined $O_2 + 2N_2 = 0.27\%$

84 Concise Metals Engineering Data Book

Table 7.31 Compositions of selected nickel and nickel-base alloys

Alloy	Ni	Cr	Fe	Cu	Mo	W	Fe	Mn	C	Si	S	Other
Commercially pure and low-alloy nickels												
Nickel 200	99.0 min			0.25			0.40	0.35	0.15	0.35	0.01	...
Nickel 201	99.0 min			0.25			0.40	0.35	0.02	0.35	0.01	...
Nickel 205	99.0 min(b)			0.15			0.20	0.35	0.15	0.15	0.008	0.01–0.08 Mg, 0.01–0.05 Ti
Nickel 211	93.7 min(b)			0.25			0.75	4.25–5.25	0.20	0.15	0.015	...
Nickel 212	97.0 min			0.20			0.25	1.5–2.5	0.10	0.20	...	0.20 Mg
Nickel 222	99.0 min(b)			0.10			0.10	0.30	...	0.10	0.008	0.01–0.10 Mg, 0.005 Ti
Nickel 270	99.9 min			0.01			0.05	0.003	0.02	0.005	0.003	0.005 Mg, 0.005 Ti
Duranickel 301	93.00 min			0.25			0.60	0.50	0.30	1.00	0.01	4.00–4.75 Al, 0.25–1.00 Ti
Nickel-copper alloys												
Alloy 400	63.0 min(b)			28.0–34.0			2.5	0.20	0.3	0.5	0.024	...
Alloy 401	40.0–45.0(b)			bal			0.75	2.25	0.10	0.25	0.015	...
Alloy R-405	63.0 min(b)			28.0–34.0			2.5	2.0	0.3	0.5	0.025–0.060	...
Alloy 450	29.0–33.0			bal			0.4–1.0	1.0	...	...	0.02	1.0 Zn, 0.05 Pb, 0.02 P
Alloy K-500	63.0 min(b)			27.0–33.0			2.0	1.5	0.25	0.5	0.01	2.30–3.15 Al, 0.35–0.85 Ti

Alloy	Ni	Cr	Fe	Co	Mo	W	Fe	Nb	Ti	Al	C	Mn	Si	B	Other
Nickel-chromium and nickel-chromium-iron alloys															
Alloy 230	bal	22.0	3.0	5.0	2.0	14.0		...	...	0.3	0.10	0.5	0.4	0.005	0.02 La
Alloy 600	72.0 min(b)	14.0–17.0	6.0–10.0	...	...			...	...		0.15	1.0	0.5	...	0.5 Cu
Alloy 601	58.0–63.0	21.0–25.0	bal	10.0–15.0	...			...	...	1.0–1.7	0.10	1.0	0.50	...	1.0 Cu
Alloy 617	44.5 min	20.0–24.0	3.0	1.0	8.0–10.0			...	0.6	0.8–1.5	0.05–0.15	1.0	1.0	0.006	0.5 Cu
Alloy 625	58.0 min	20.0,23.0	5.0		8.0–10.0			3.15–4.15(c)	0.40	0.40	0.10	0.50	0.50	...	...
Alloy 690	58.0 min	27.0–31.0	7.0–11.0		...			...	...	...	0.05	0.50	0.50	...	0.50 Cu
Alloy 718	50.0–55.0(b)	17.0–21.0	bal	1.0	2.80–3.30			4.75–5.50(c)	0.65–1.15	0.20–0.80	0.08	0.35	0.35	0.006	0.30 Cu
Alloy X750	70.0 min(b)	14.0–17.0	5.0–9.0	1.0	...			0.70–1.20(c)	2.25–2.75	0.40–1.00	0.08	1.00	0.50	...	0.50 Cu
Alloy 751	70.0 min(b)	14.0–17.0	5.0–9.0		...			0.7–1.2(c)	2.0–2.6	...	0.10	1.0	0.5	...	0.5 Cu
Alloy MA 754(d)	78.0	20	1.0		...			...	0.5	0.3	0.05	...	...	...	0.6 Y$_2$O$_3$
Alloy C-22	51.6	21.5	5.5	2.5	13.5	4.0		...	...	...	0.01	1.0	0.1	...	0.3 V
Alloy C-276	bal	14.5–16.5	4.0–7.0	2.5	15.0–17.0	3.0–4.5		...	...	...	0.01	1.0	0.08	...	0.35 V
Alloy G3	bal	21.0–23.5	18.0–21.0	5.0	6.0–8.0	1.5		0.50(c)	...	...	0.015	1.0	1.0	...	1.5–2.5 Cu

(continued)

Table 7.31 (continued)

Alloy	Ni	Cr	Fe	Co	Mo	W	Nb	Ti	Al	C	Mn	Si	B	Other
Nickel-chromium and nickel-chromium-iron alloys (continued)
Alloy HX	bal	20.5–23.0	17.0–20.0	0.5–2.5	8.0–10.0	0.2–1.0	...	...	...	0.05–0.15	1.0	1.0	...	0.01–0.10
Alloy S	bal	14.5–17.0	3.0	2.0	14.0–16.5	1.0	...	...	0.10–0.50	0.02	0.30–1.0	0.20–0.75	0.015	La, 0.35 Cu
Alloy W	63.0	5.0	6.0	2.5	24.0	...	...	...	...	0.12	1.0	1.0	...	...
Alloy X	bal	20.50–23.00	17.0–20.0	0.5–2.5	8.0–10.0	0.2–1.0	...	0.15	0.50	0.05–0.15	1.0	1.0	0.008	0.5 Cu
Iron-nickel-chromium alloys
Alloy 556	20.0	22.0	bal	18.0	3.0	2.5	...	...	0.2	0.10	1.0	0.4	...	0.6 Ta, 0.02 La, 0.02 Zr
Alloy 800	30.0–35.0	19.0–23.0	39.5 min	...	...	...	...	0.15–0.60	0.15–0.60	0.10	1.5	1.0	...	...
Alloy 800HT	30.0–35.0	19.0–23.0	39.5 min	...	...	...	...	0.15–0.60	0.15–0.60	0.06–0.10	1.5	1.0	...	0.85–1.20 Al + Ti
Alloy 825	38.0–44.0	19.5–23.5	22.0 min	...	2.5–3.5	...	...	0.6–1.2	0.2	0.05	1.0	0.5	...	...
Alloy 925	44.0	21.0	28.0	...	3.0	...	...	2.1	0.3	0.01	...	...	...	...
20Cb3	32.0–38.0	19.0–21.0	bal	...	2.0–3.0	...	1.0	...	...	0.07	2.0	1.0	...	3.0–4.0 Cu
20Mo-4	35.0–40.0	22.5–25.0	bal	...	3.5–5.0	...	0.15–0.35	...	...	0.03	1.0	0.5	...	0.5–1.5 Cu
20Mo-6	33.0–37.20	22.0–26.0	bal	...	5.0–6.7	...	...	...	...	0.03	1.0	0.5	...	2.0–4.0 Cu
Controlled-expansion alloys (Fe-Ni-Cr, Fe-Ni-Co)
Alloy 902	41.0–43.5(b)	4.9–5.75	bal	...	...	...	...	2.2–2.75	0.3–0.8	0.06	0.8	1.0	...	...
Alloy 903	38.0	...	42.0	15.0	...	...	3.0	1.4	0.9	...	...	...	...	...
Alloy 907	38.0	...	42.0	13.0	...	...	4.7	1.5	0.03	...	...	0.15	...	...
Alloy 909	38.0	...	42.0	13.0	...	...	4.7	1.5	0.03	0.01	...	0.4	...	...
Nickel-iron alloys
Alloy 36	35.0–38.0	0.50	bal	1.0	0.5	...	...	...	...	0.10	0.60	0.35	...	...
Alloy 42	42.0(e)	0.50	bal	1.0	0.5	...	...	...	0.15	0.05	0.80	0.30	...	...
Alloy 48	48.0(e)	0.25	bal	1.0	...	...	...	...	0.10	0.05	0.80	0.30	...	...

(a) Single values are maximum values unless otherwise indicated. (b) Nickel plus cobalt content. (c) Niobium plus tantalum content. (d) Mechanically alloyed, dispersion-strengthened, powder metallurgy alloy. (e) Nominal value; adjusted to meet expansion requirements

Table 7.32 Nominal compositions of various cobalt-base alloys

Alloy trade name	Co	Cr	W	Mo	C	Fe	Ni	Si	Mn	Others
Cobalt-base wear-resistant alloys										
Stellite 1	bal	31	12.5	1 (max)	2.4	3 (max)	3 (max)	2 (max)	1 (max)	...
Stellite 6	bal	28	4.5	1 (max)	1.2	3 (max)	3 (max)	2 (max)	1 (max)	...
Stellite 12	bal	30	8.3	1 (max)	1.4	3 (max)	3 (max)	2 (max)	1 (max)	...
Stellite 21	bal	28	...	5.5	0.25	2 (max)	2.5	2 (max)	1 (max)	...
Haynes alloy 6B	bal	30	4	1	1.1	3 (max)	2.5	0.7	1.5	...
Tribaloy T-800	bal	17.5	...	29	0.08 (max)	...	...	3.5	...	...
Stellite F	bal	25	12.3	1 (max)	1.75	3 (max)	22	2 (max)	1 (max)	...
Stellite 4	bal	30	14.0	1 (max)	0.57	3 (max)	3 (max)	2 (max)	1 (max)	...
Stellite 190	bal	26	14.5	1 (max)	3.3	3 (max)	3 (max)	2 (max)	1 (max)	...
Stellite 306	bal	25	2.0	...	0.4	...	5	...	...	6 Nb
Stellite 6K	bal	31	4.5	1.5 (max)	1.6	3 (max)	3 (max)	2 (max)	2 (max)	...
Cobalt-base high-temperature alloys										
Haynes alloy 25 (L605)	bal	20	15	...	0.10	3 (max)	10	1 (max)	1.5	...
Haynes alloy 188	bal	22	14	...	0.10	3 (max)	22	0.35	1.25	0.05 La
MAR-M alloy 509	bal	22.5	7	...	0.60	1.5 (max)	10	0.4 (max)	0.1 (max)	3.5 Ta, 0.2 Ti, 0.5 Zr
Cobalt-base corrosion-resistant alloys										
MP35N, Multiphase alloy	bal	20	...	10	...	...	35	...	...	...
Ultimet (1233)	bal	25.5	2	5	0.08 (max)	3	9	...	...	0.1 N (max)

bal, balance

8 Physical Properties of Metals and Alloys

Table 8.1 Density of metals and alloys

Metal or alloy	Density g/cm³	Density lb/in.³	Metal or alloy	Density g/cm³	Density lb/in.³
Aluminum and aluminum alloys			**Copper and copper alloys**		
Aluminum (99.996%)	2.6989	0.0975	**Wrought coppers**		
Wrought alloys			Pure copper	8.96	0.324
			Electrolytic tough pitch copper (ETP)	8.89	0.321
EC, 1060 alloys	2.70	0.098	Deoxidized copper, high residual phosphorus (DHP)	8.94	0.323
1100	2.71	0.098			
2011	2.82	0.102	Free-machining copper		
2014	2.80	0.101	0.5% Te	8.94	0.323
2024	2.77	0.100	1.0% Pb	8.94	0.323
2218	2.81	0.101			
3003	2.73	0.099	**Wrought alloys**		
4032	2.69	0.097	Gilding, 95%	8.86	0.320
5005	2.70	0.098	Commercial bronze, 90%	8.80	0.318
5050	2.69	0.097	Jewelry bronze, 87.5%	8.78	0.317
5052	2.68	0.097	Red brass, 85%	8.75	0.316
5056	2.64	0.095	Low brass, 80%	8.67	0.313
5083	2.66	0.096	Cartridge brass, 70%	8.53	0.308
5086	2.65	0.096	Yellow brass	8.47	0.306
5154	2.66	0.096	Muntz metal	8.39	0.303
5357	2.70	0.098	Leaded commercial bronze	8.83	0.319
5456	2.66	0.096	Low-leaded brass (tube)	8.50	0.307
6061, 6063	2.70	0.098	Medium-leaded brass	8.47	0.306
6101, 6151	2.70	0.098	High-leaded brass (tube)	8.53	0.308
7075	2.80	0.101	High-leaded brass	8.50	0.307
7079	2.74	0.099	Extra-high-leaded brass	8.50	0.307
7178	2.82	0.102	Free-cutting brass	8.50	0.307
			Leaded Muntz metal	8.41	0.304
Casting alloys			Forging brass	8.44	0.305
242.0	2.81	0.102	Architectural bronze	8.47	0.306
295.0	2.81	0.102	Inhibited admiralty	8.53	0.308
356.0	2.68	0.097	Naval brass	8.41	0.304
380.0	2.76	0.099	Leaded naval brass	8.44	0.305
413.0	2.66	0.096	Manganese bronze (A)	8.36	0.302
443.0	2.69	0.097	Phosphor bronze, 5% (A)	8.86	0.320
514.0	2.65	0.096	Phosphor bronze, 8% (C)	8.80	0.318
520.0	2.57	0.093	Phosphor bronze, 10% (D)	8.78	0.317

(continued)

Table 8.1 (continued)

Metal or alloy	Density g/cm³	lb/in.³	Metal or alloy	Density g/cm³	lb/in.³
Wrought alloys (continued)			**Casting alloys (continued)**		
Phosphor bronze, 1.25%............	8.89	0.321	20% Ni.......................	8.85	0.319
Free-cutting phosphor bronze.........	8.89	0.321	25% Ni.......................	8.8	0.32
Cupronickel, 30%.................	8.94	0.323	Silicon bronze.................	8.30	0.300
Cupronickel, 10%.................	8.94	0.323	Silicon brass..................	8.30	0.300
Nickel silver, 65-18...............	8.73	0.315	**Iron and iron alloys**		
Nickel silver, 55-18...............	8.70	0.314	Pure iron......................	7.874	0.2845
High-silicon bronze (A)............	8.53	0.308	Ingot iron.....................	7.866	0.2842
Low-silicon bronze (B).............	8.75	0.316	Wrought iron...................	7.7	0.28
Aluminum bronze, 5% Al...........	8.17	0.294	Gray cast iron..................	7.15(a)	0.258(a)
Aluminum bronze, (3)..............	7.78	0.281	Malleable iron..................	7.27(b)	0.262(b)
Aluminum-silicon bronze...........	7.69	0.278	0.06% C steel..................	7.871	0.2844
Aluminum bronze, (1)..............	7.58	0.274	0.23% C steel..................	7.859	0.2839
Aluminum bronze, (2)..............	7.58	0.274	0.435% C steel.................	7.844	0.2834
Beryllium copper..................	8.23	0.297	1.22% C steel..................	7.830	0.2829
Casting alloys			**Low-carbon chromium-molybdenum steels**		
Chromium copper (1% Cr).........	8.7	0.31	0.5% Mo steel..................	7.86	0.283
88Cu-10Sn-2Zn..................	8.7	0.31	1Cr-0.5Mo steel................	7.86	0.283
88Cu-8Sn-4Zn...................	8.8	0.32	1.25Cr-0.5Mo steel..............	7.86	0.283
89Cu-11Sn......................	8.78	0.317	2.25Cr-1.0Mo steel..............	7.86	0.283
88Cu-6Sn-1.5Pb-4.5Zn............	8.7	0.31	5Cr-0.5Mo steel................	7.78	0.278
87Cu-8Sn-1Pb-4Zn...............	8.8	0.32	7Cr-0.5Mo steel................	7.78	0.278
87Cu-10Sn-1Pb-2Zn..............	8.8	0.32	9Cr-1Mo steel..................	7.67	0.276
80Cu-10Sn-10Pb.................	8.95	0.323	**Medium-carbon alloy steels**		
83Cu-7Sn-7Pb-3Zn...............	8.93	0.322	1Cr-0.35Mo-0.25V steel..........	7.86	0.283
85Cu-5Sn-9Pb-1Zn...............	8.87	0.320	H11 die steel (5Cr-1.5Mo-0.4V)...	7.79	0.281
78Cu-7Sn-15Pb...................	9.25	0.334	**Other iron-base alloys**		
70Cu-5Sn-25Pb...................	9.30	0.336	A-286.........................	7.94	0.286
85Cu-5Sn-5Pb-5Zn...............	8.80	0.318	16-25-6 alloy...................	8.08	0.292
83Cu-4Sn-6Pb-7Zn...............	8.6	0.31	RA-330.......................	8.03	0.290
81Cu-3Sn-7Pb-9Zn...............	8.7	0.31	Incoloy.......................	8.02	0.290
76Cu-2.5Sn-6.5Pb-15Zn...........	8.77	0.317	Incoloy T.....................	7.98	0.288
72Cu-1Sn-3Pb-24Zn..............	8.50	0.307	Incoloy 901....................	8.23	0.297
67Cu-1Sn-3Pb-29Zn..............	8.45	0.305	T1 tool steel...................	8.67	0.313
61Cu-1Sn-1Pb-37Zn..............	8.40	0.304	M2 tool steel...................	8.16	0.295
Manganese bronze			H41 tool steel..................	7.88	0.285
60 ksi.........................	8.2	0.30	20W-4Cr-2V-12Co steel..........	8.89	0.321
65 ksi.........................	8.3	0.30	Invar (36% Ni).................	8.00	0.289
90 ksi.........................	7.9	0.29	Hipernik (50% Ni)..............	8.25	0.298
110 ksi........................	7.7	0.28	4% Si.........................	7.6	0.27
Aluminum bronze			10.27% Si.....................	6.97	0.252
Alloy 9A.......................	7.8	0.28			
Alloy 9B.......................	7.55	0.272	**Stainless steels and heat-resistant alloys**		
Alloy 9C.......................	7.5	0.27	**Corrosion-resistant steel castings**		
Alloy 9D.......................	7.7	0.28			
Nickel silver					
12% Ni........................	8.95	0.323	CA-15........................	7.612	0.2750
16% Ni........................	8.95	0.323			

(continued)

Table 8.1 (continued)

Metal or alloy	Density g/cm³	lb/in.³	Metal or alloy	Density g/cm³	lb/in.³
Corrosion-resistant steel castings (continued)			**Wrought stainless and heat-resistant steels (continued)**		
CA-40	7.612	0.2750	Type 416	7.7	0.28
CB-30	7.53	0.272	Type 420	7.7	0.28
CC-50	7.53	0.272	Type 430	7.7	0.28
CE-30	7.67	0.277	Type 430F	7.7	0.28
CF-8	7.75	0.280	Type 431	7.7	0.28
CF-20	7.75	0.280	Types 440A, 440B, 440C	7.7	0.28
CF-8M, CF-12M	7.75	0.280	Type 446	7.6	0.27
CF-8C	7.75	0.280	Type 501	7.7	0.28
CF-16F	7.75	0.280	Type 502	7.8	0.28
CH-20	7.72	0.279	19-9DL	7.97	0.29
CK-20	7.75	0.280	**Precipitation-hardening stainless steels**		
CN-7M	8.00	0.289	PH15-7Mo	7.804	0.2819
Heat-resistant alloy castings			17-4 PH	7.8	0.28
HA	7.72	0.279	17-7 PH	7.81	0.282
HC	7.53	0.272	**Nickel-base alloys**		
HD	7.58	0.274	D-979	8.27	0.299
HE	7.67	0.277	Nimonic 80A	8.25	0.298
HF	7.75	0.280	Nimonic 90	8.27	0.299
HH	7.72	0.279	M-252	8.27	0.298
HI	7.72	0.279	Inconel	8.51	0.307
HK	7.75	0.280	Inconel "X" 550	8.30	0.300
HL	7.72	0.279	Inconel 700	8.17	0.295
HN	7.83	0.283	Inconel "713C"	7.913	0.2859
HT	7.92	0.286	Waspaloy	8.23	0.296
HU	8.04	0.290	René 41	8.27	0.298
HW	8.14	0.294	Hastelloy alloy B	9.24	0.334
HX	8.14	0.294	Hastelloy alloy C	8.94	0.323
Wrought stainless and heat-resistant steels			Hastelloy alloy X	8.23	0.297
Type 301	7.9	0.29	Udimet 500	8.07	0.291
Type 302	7.9	0.29	GMR-235	8.03	0.290
Type 302B	8.0	0.29	**Cobalt-chromium-nickel-base alloys**		
Type 303	7.9	0.29	N-155 (HS-95)	8.23	0.296
Type 304	7.9	0.29	S-590	8.36	0.301
Type 305	8.0	0.29	**Cobalt-base alloys**		
Type 308	8.0	0.29	S-816	8.68	0.314
Type 309	7.9	0.29	V-36	8.60	0.311
Type 310	7.9	0.29	HS-25	9.13	0.330
Type 314	7.72	0.279	HS-36	9.04	0.327
Type 316	8.0	0.29	HS-31	8.61	0.311
Type 317	8.0	0.29	HS-21	8.30	0.300
Type 321	7.9	0.29			
Type 347	8.0	0.29	**Molybdenum-base alloy**		
Type 403	7.7	0.28			
Type 405	7.7	0.28	Mo-0.5Ti	10.2	0.368
Type 410	7.7	0.28			

(continued)

Table 8.1 (continued)

Metal or alloy	Density g/cm³	lb/in.³	Metal or alloy	Density g/cm³	lb/in.³
Lead and lead alloys			**Wrought alloys (continued)**		
Chemical lead (99.90+% Pb)	11.34	0.4097	AZ80A	1.80	0.065
Corroding lead (99.73+% Pb)	11.36	0.4104	ZK60A, B	1.83	0.066
Arsenical lead	11.34	0.4097	ZE10A	1.76	0.064
Calcium lead	11.34	0.4097	HM21A	1.78	0.064
5-95 solder	11.0	0.397	HM31A	1.81	0.065
20-80 solder	10.2	0.368	**Nickel and nickel alloys**		
50-50 solder	8.89	0.321	Nickel (99.95% Ni + Co)	8.902	0.322"
Antimonial lead alloys			"A" Nickel	8.885	0.321"
1% antimonial lead	11.27	0.407	"D" Nickel	8.78	0.317
Hard lead (96Pb-4Sb)	11.04	0.399	Duranickel	8.26	0.298
Hard lead (94Pb-6Sb)	10.88	0.393	Cast nickel	8.34	0.301
8% antimonial lead	10.74	0.388	Monel	8.84	0.319"
9% antimonial lead	10.66	0.385	"K" Monel	8.47	0.306
Lead-base babbitt alloys			Monel (cast)	8.63	0.312"
Lead-base babbitt			"H" Monel (cast)	8.5	0.31"
SAE 13	10.24	0.370	"S" Monel (cast)	8.36	0.302
SAE 14	9.73	0.352	Inconel	8.51	0.307
Alloy 8	10.04	0.363	Inconel (cast)	8.3	0.30
Arsenical lead			**Nickel-molybdenum-chromium-iron alloys**		
Babbitt (SAE 15)	10.1	0.365"	Hastelloy B	9.24	0.334
"G" Babbitt	10.1	0.365	Hastelloy C	8.94	0.323
Magnesium and magnesium alloys			Hastelloy D	7.8	0.282
Magnesium (99.8%)	1.738	0.06279	Hastelloy F	8.17	0.295
Casting alloys			Hastelloy N	8.79	0.317
AM100A	1.81	0.065	Hastelloy W	9.03	0.326
AZ63A	1.84	0.066	Hastelloy X	8.23	0.297
AZ81A	1.80	0.065	**Nickel-chromium-molybdenum-copper alloys**		
AZ91A, B, C	1.81	0.065	Illium G	8.58	0.310
AZ92A	1.82	0.066	Illium R	8.58	0.310
HK31A	1.79	0.065	**Electrical resistance alloys**		
HZ32A	1.83	0.066	80Ni-20Cr	8.4	0.30
ZH42, ZH62A	1.86	0.067	60Ni-24Fe-16Cr	8.247	0.298
ZK51A	1.81	0.065	35Ni-45Fe-20Cr	7.95	0.287
ZE41A	1.82	0.066	Constantan	8.9	0.32
EZ33A	1.83	0.066			
EK30A	1.79	0.065	**Tin and tin alloys**		
EK41A	1.81	0.065	Pure tin	7.3	0.264
			Soft solder (30% Pb)	8.32	0.301
Wrought alloys			Soft solder (37% Pb)	8.42	0.304
M1A	1.76	0.064	Tin babbitt		
A3A	1.77	0.064	Alloy 1	7.34	0.265
AZ31B	1.77	0.064	Alloy 2	7.39	0.267
PE	1.76	0.064	Alloy 3	7.46	0.269
AZ61A	1.80	0.065			

(continued)

Table 8.1 (continued)

Metal or alloy	Density g/cm³	lb/in.³	Metal or alloy	Density g/cm³	lb/in.³
Tin and tin alloys (continued)			**Precious metals (continued)**		
Alloy 4	7.53	0.272	Pt-5Ru	20.67	...
Alloy 5	7.75	0.280	Pt-10Ru	19.94	...
White metal	7.28	0.263	Palladium	12.02	0.4343
Pewter	7.28	0.263	60Pd-40Cu	10.6	0.383
Titanium and titanium alloys			95.5Pd-4.5Ru	12.07(a)	...
99.9% Ti	4.507	0.1628	95.5Pd-4.5Ru	11.62(b)	...
99.2% Ti	4.507	0.1628	**Permanent magnet materials**		
99.0% Ti	4.52	0.163	Cunico	8.30	0.300
Ti-6Al-4V	4.43	0.160	Cunife	8.61	0.311
Ti-5Al-2.5Sn	4.46	0.161	Comol	8.16	0.295
Ti-2Fe-2Cr-2Mo	4.65	0.168	Alnico I	6.89	0.249
Ti-8Mn	4.71	0.171	Alnico II	7.09	0.256
Ti-7Al-4Mo	4.48	0.162	Alnico III	6.89	0.249
Ti-4Al-4Mn	4.52	0.163	Alnico IV	7.00	0.253
Ti-4Al-3Mo-1V	4.507	0.1628	Alnico V	7.31	0.264
Ti-2.5Al-16V	4.65	0.168	Alnico VI	7.42	0.268
			Barium ferrite	4.7	0.17
Zinc and zinc alloys			Vectolite	3.13	0.113
Pure zinc	7.133	0.2577	**Pure metals**		
AG40A alloy	6.6	0.24	Antimony	6.62	0.239
AC41A alloy	6.7	0.24	Beryllium	1.848	0.067
Commercial rolled zinc			Bismuth	9.80	0.354
0.08% Pb	7.14	0.258	Cadmium	8.65	0.313
0.06 Pb, 0.06 Cd	7.14	0.258	Calcium	1.55	0.056
0.3 Pb, 0.3 Cd	7.14	0.258	Cesium	1.903	0.069
Copper-hardened, rolled zinc (1% Cu)	7.18	0.259	Chromium	7.19	0.260
Rolled zinc alloy (1 Cu, 0.010 Mg)	7.18	0.259	Cobalt	8.85	0.322
Zn-Cu-Ti alloy (0.8 Cu, 0.15 Ti)	7.18	0.259	Gallium	5.907	0.213
Precious metals			Germanium	5.323	0.192
Silver	10.49	0.379	Hafnium	13.1	0.473
Gold	19.32	0.698	Indium	7.31	0.264
70Au-30Pt	19.92	...	Iridium	22.5	0.813
Platinum	21.45	0.775	Lithium	0.534	0.019
Pt-3.5Rh	20.9	...	Manganese	7.43	0.270
Pt-5Rh	20.65	...	Mercury	13.546	0.489
Pt-10Rh	19.97	...	Molybdenum	10.22	0.369
Pt-20Rh	18.74	...	Niobium	8.57	0.310
Pt-30Rh	17.62	...	Osmium	22.583	0.816
Pt-40Rh	16.63	...	Plutonium	19.84	0.717
Pt-5Ir	21.49	...	Potassium	0.86	0.031
Pt-10Ir	21.53	...	Rhenium	21.04	0.756
Pt-15Ir	21.57	...	Rhodium	12.44	0.447
Pt-20Ir	21.61	...	Ruthenium	12.2	0.441
Pt-25Ir	21.66	...	Selenium	4.79	0.174
Pt-30Ir	21.70	...	Silicon	2.33	0.084
Pt-35Ir	21.79	...	Silver	10.49	0.379

(continued)

Table 8.1 (continued)

Metal or alloy	Density g/cm³	lb/in.³	Metal or alloy	Density g/cm³	lb/in.³
Pure metals (continued)			**Rare earth metals (continued)**		
Sodium	0.97	0.035			
Tantalum	16.6	0.600	Holmium	6.79(f)	...
Thallium	11.85	0.428	Lanthanum	6.19(d)	...
Thorium	11.72	0.423		6.18(c)	...
Tungsten	19.3	0.697		5.97(e)	...
Uranium	19.07	0.689	Lutetium	9.85(f)	...
Vanadium	6.1	0.22			
Zirconium	6.5	0.23	Neodymium	7.00(d)	...
				6.80(e)	...
Rare earth metals			Praseodymium	6.77(d)	...
Cerium	8.23(c)	...		6.64(e)	...
	6.66(d)	...	Samarium	7.49(g)	...
	6.77(e)	...	Scandium	2.99(f)	...
Dysprosium	8.55(f)	...	Terbium	8.25(f)	...
Erbium	9.15(f)	...	Thulium	9.31(f)	...
Europium	5.245(e)	...	Ytterbium	6.96(c)	...
Gadolinium	7.86(f)	...	Yttrium	4.47(f)	...

(a) 6.95 to 7.35 g/cm³ (0.251 to 0.265 lb/in.³). (b) 7.20 to 7.34 g/cm³ (0.260 to 0.265 lb/in.³). (c) Face-centered cubic. (d) Hexagonal. (e) Body-centered cubic. (f) Close-packed hexagonal. (g) Rhombohedral

Table 8.2 Linear thermal expansion of metals and alloys

Metal or alloy	Temperature, °C	Coefficient of expansion, μ in./in. · °C	Metal or alloy	Temperature, °C	Coefficient of expansion, μ in./in. · °C
Aluminum and aluminum alloys			**Wrought alloys (continued)**		
Aluminum (99.996%)	20–100	23.6	7075	20–100	23.2
Wrought alloys			7079, 7178	20–100	23.4
EC, 1060, 1100	20–100	23.6	**Casting alloys**		
2011, 2014	20–100	23.0	242.0	20–100	22.5
2024	20–100	22.8	295.0	20–100	22.9
2218	20–100	22.3	356.0	20–100	21.4
3003	20–100	23.2	380.0	20–100	21.2
4032	20–100	19.4	413.0	20–100	20.5
5005, 5050, 5052	20–100	23.8	443.0	20–100	22.1
5056	20–100	24.1	514.0	20–100	23.9
5083	20–100	23.4	520.0	20–100	25.2
5086	60–300	23.9	**Copper and copper alloys**		
5154	20–100	23.9	**Wrought coppers**		
5357	20–100	23.7	Pure copper	20	16.5
5456	20–100	23.9	Electrolytic tough pitch copper (ETP)	20–100	16.8
6061, 6063	20–100	23.4			
6101, 6151	20–100	23.0			

(continued)

(a) Longitudinal; 23.4 transverse. (b) Longitudinal; 21.1 transverse. (c) Longitudinal; 19.4 transverse

Physical Properties of Metals and Alloys 93

Table 8.2 (continued)

Metal or alloy	Temperature, °C	Coefficient of expansion, µ in./in. · °C	Metal or alloy	Temperature, °C	Coefficient of expansion, µ in./in. · °C
Wrought coppers (continued)			**Casting alloys (continued)**		
Deoxidized copper, high residual phosphorus (DHP)	20–300	17.7	89Cu-11Sn	20–300	18.4
			88Cu-6Sn-1.5Pb-4.5Zn	21–260	18.5
Oxygen-free copper	20–300	17.7	87Cu-8Sn-1Pb-4Zn	21–177	18.0
Free-machining copper, 0.5% Te or 1% Pb	20–300	17.7	87Cu-10Sn-1Pb-2Zn	21–177	18.0
			80Cu-10Sn-10Pb	21–204	18.5
Wrought alloys			78Cu-7Sn-15Pb	21–204	18.5
Gilding, 95%	20–300	18.1	85Cu-5Sn-5Pb-5Zn	21–204	18.1
Commercial bronze, 90%	20–300	18.4	72Cu-1Sn-3Pb-24Zn	21–93	20.7
Jewelry bronze, 87.5%	20–300	18.6	67Cu-1Sn-3Pb-29Zn	21–93	20.2
Red brass, 85%	20–300	18.7	61Cu-1Sn-1Pb-37Zn	21–260	21.6
Low brass, 80%	20–300	19.1	Manganese bronze		
Cartridge brass, 70%	20–300	19.9	60 ksi	21–204	20.5
Yellow brass	20–300	20.3	65 ksi	21–93	21.6
Muntz metal	20–300	20.8	110 ksi	21–260	19.8
Leaded commercial bronze	20–300	18.4	Aluminum bronze		
Low-leaded brass	20–300	20.2	Alloy 9A	...	17
Medium-leaded brass	20–300	20.3	Alloy 9B	20–250	17
High-leaded brass	20–300	20.3	Alloys 9C, 9D	...	16.2
Extra-high-leaded brass	20–300	20.5	**Iron and iron alloys**		
Free-cutting brass	20–300	20.5	Pure iron	20	11.7
Leaded Muntz metal	20–300	20.8	Fe-C alloys		
Forging brass	20–300	20.7	0.06% C	20–100	11.7
Architectural bronze	20–300	20.9	0.22% C	20–100	11.7
Inhibited admiralty	20–300	20.2	0.40 C	20–100	11.3
Naval brass	20–300	21.2	0.56 C	20–100	11.0
Leaded naval brass	20–300	21.2	1.08% C	20–100	10.8
Manganese bronze (A)	20–300	21.2	1.45% C	20–100	10.1
Phosphor bronze, 5% (A)	20–300	17.8	Invar (36% Ni)	20	0–2
Phosphor bronze, 8% (C)	20–300	18.2	13Mn-1.2C	20	18.0
Phosphor bronze, 10% (D)	20–300	18.4	13Cr-0.35C	20–100	10.0
Phosphor bronze, 1.25%	20–300	17.8	12.3Cr-0.4Ni-0.09C	20–100	9.8
Free-cutting phosphor bronze	20–300	17.3	17.7Cr-9.6Ni-0.06C	20–100	16.5
Cupro-nickel, 305	20–300	16.2	18W-4Cr-1V	0–100	11.2
Cupro-nickel, 10%	20–300	17.1	Gray cast iron	0–100	10.5
Nickel silver, 65-18	20–300	16.2	Malleable iron (pearlitic)	20–400	12
Nickel silver, 55-18	20–300	16.7	**Lead and lead alloys**		
Nickel silver, 65-12	20–300	16.2			
High-silicon bronze(a)	20–300	18.0	Corroding lead (99.73 + % Pb)	17–100	29.3
Low-silicon bronze(b)	20–300	17.9	5-95 solder	15–110	28.7
Aluminum bronze (3)	20–300	16.4	20-80 solder	15–110	26.5
Aluminum-silicon bronze	20–300	18.0	50-50 solder	15–110	23.4
Aluminum bronze (1)	20–300	16.8	1% antimonial lead	20–100	28.8
Beryllium copper	20–300	17.8	Hard lead (96Pb-6Sb)	20–100	28.8
			Hard lead (94Pb-6Sb)	20–100	27.2
Casting alloys			8% antimonial lead	20–100	26.7
88Cu-8Sn-4Zn	21–177	18.0	9% antimonial lead	20–100	26.4

(continued)

(a) Longitudinal; 23.4 transverse. (b) Longitudinal; 21.1 transverse. (c) Longitudinal; 19.4 transverse

Table 8.2 (continued)

Metal or alloy	Temperature, °C	Coefficient of expansion, µ in./in. · °C	Metal or alloy	Temperature, °C	Coefficient of expansion, µ in./in. · °C
Lead and lead alloys (continued)			**Nickel and nickel alloys (continued)**		
Lead-base babbitt			Constantan	20–1000	18.8
SAE 14	20–100	19.6	**Tin and tin alloys**		
Alloy 8	20–100	24.0	Pure tin	0–100	23
Magnesium and magnesium alloys			Solder (70Sn-30Pb)	15–110	21.6
Magnesium (99.8%)	20	25.2	Solder (63Sn-37Pb)	15–110	24.7
Casting alloys			**Titanium and titanium alloys**		
AM100A	18–100	25.2	99.9% Ti	20	8.41
AZ63A	20–100	26.1	99.0% Ti	93	8.55
AZ91A, B, C	20–100	26	Ti-5Al-2.5Sn	93	9.36
AZ92A	18–100	25.2	Ti-8Mn	93	8.64
HZ32A	20–200	26.7	**Zinc and zinc alloys**		
ZH42	20–200	27	Pure zinc	20–250	39.7
ZH62A	20–200	27.1	AG40A alloy	20–100	27.4
ZK51A	20	26.1	AC41A alloy	20–100	27.4
EZ33A	20–100	26.1	**Commercial rolled zinc**		
EK30A, EK41A	20–100	26.1	0.08 Pb	20–40	32.5
Wrought alloys			0.3 Pb, 0.3 Cd	20–98	33.9(a)
M1A, A3A	20–100	26	Rolled zinc alloy		
AZ31B, PE	20–100	26	(1 Cu, 0.010 Mg)	20–100	34.8(b)
AZ61A, AZ80A	20–100	26	Zn-Cu-Ti alloy (0.8 Cu, 0.15 Ti)	20–100	24.9(c)
ZK60A, B	20–100	26	**Pure metals**		
HM31A	20–93	26.1	Beryllium	25–100	11.6
Nickel and nickel alloys			Cadmium	20	29.8
Nickel (99.95% Ni + Co)	0–100	13.3	Calcium	0–400	22.3
Duranickel	0–100	13.0	Chromium	20	6.2
Monel	0–100	14.0	Cobalt	20	13.8
Monel (cast)	25–100	12.9	Gold	20	14.2
Inconel	20–100	11.5	Iridium	20	6.8
Ni-o-nel	27–93	12.9	Lithium	20	56
Hastelloy B	0–100	10.0	Manganese	0–100	22
Hastelloy C	0–100	11.3	Palladium	20	11.76
Hastelloy D	0–100	11.0	Platinum	20	8.9
Hastelloy F	20–100	14.2	Rhenium	20–500	6.7
Hastelloy N	21–204	10.4	Rhodium	20–100	8.3
Hastelloy W	23–100	11.3	Ruthenium	20	9.1
Hastelloy X	26–100	13.8	Silicon	0–1400	5
Illium G	0–100	12.19	Silver	0–100	19.68
Illium R	0–100	12.02	Tungsten	27	4.6
80Ni-20Cr	20–1000	17.3	Vanadium	23–100	8.3
60Ni-24Fe-16Cr	20–1000	17.0	Zirconium	...	5.85
35Ni-45Fe-20Cr	20–500	15.8			

(a) Longitudinal; 23.4 transverse. (b) Longitudinal; 21.1 transverse. (c) Longitudinal; 19.4 transverse

Table 8.3 Thermal conductivity of metals and alloys

Metal or alloy	Thermal conductivity near room temperature, cal/cm$^2 \cdot$ cm $\cdot$ s $\cdot$ °C	Metal or alloy	Thermal conductivity near room temperature, cal/cm$^2 \cdot$ cm $\cdot$ s $\cdot$ °C
Aluminum and aluminum alloys		**Wrought coppers (continued)**	
Wrought alloys		Free-machining copper (0.5% Te)	0.88
EC (O)	0.57	Free-machining copper (1% Pb)	0.92
1060 (O)	0.56	**Wrought alloys**	
1100	0.53	Gilding, 95%	0.56
2011 (T3)	0.34	Commercial bronze, 90%	0.45
2014 (O)	0.46	Jewelry bronze, 87.5%	0.41
2024 (L)	0.45	Red brass, 85%	0.38
2218 (T72)	0.37	Low brass, 80%	0.33
3003 (O)	0.46	Cartridge brass, 70%	0.29
4032 (O)	0.37	Yellow brass	0.28
5005	0.48	Muntz metal	0.29
5050 (O)	0.46	Leaded-commercial bronze	0.43
5052 (O)	0.33	Low-leaded brass (tube)	0.28
5056 (O)	0.28	Medium-leaded brass	0.28
5083	0.28	High-leaded brass (tube)	0.28
5086	0.30	High-leaded brass	0.28
5154	0.30	Extra-high-leaded brass	0.28
5357	0.40	Leaded Muntz metal	0.29
5456	0.28	Forging brass	0.28
6061 (O)	0.41	Architectural bronze	0.29
6063 (O)	0.52	Inhibited admiralty	0.26
6101 (T6)	0.52	Naval brass	0.28
6151 (O)	0.49	Leaded naval brass	0.28
7075 (T6)	0.29	Manganese bronze (A)	0.26
7079 (T6)	0.29	Phosphor bronze, 5% (A)	0.17
7178	0.29	Phosphor bronze, 8% (C)	0.15
Casting alloys		Phosphor bronze, 10% (D)	0.12
242.0 (T77, sand)	0.36	Phosphor bronze, 1.25%	0.49
295.0 (T4, sand)	0.33	Free-cutting phosphor bronze	0.18
356.0 (T51, sand)	0.40	Cupro-nickel, 30%	0.07
380.0 (F, die)	0.26	Cupro-nickel, 10%	0.095
413.0 (F, die)	0.37	Nickel silver, 65-18	0.08
443.0 (F, sand)	0.35	Nickel silver, 55-18	0.07
514.0 (F, sand)	0.33	Nickel silver, 65-12	0.10
520.0 (T4, sand)	0.21	High-silicon bronze (A)	0.09
		Low-silicon bronze (B)	0.14
Copper and copper alloys		Aluminum bronze, 5% Al	0.198
Wrought coppers		Aluminum bronze, (3)	0.18
Pure copper	0.941	Aluminum-silicon bronze	0.108
Electrolytic tough pitch copper (ETP)	0.934	Aluminum bronze, (1)	0.144
Deoxidized copper, high residual phosphorus (DHP)	0.81	Aluminum bronze, (2)	0.091
		Beryllium copper	0.20(a)

(continued)

(a) Depends on processing. (b) 18% of Cu. (c) 12% of Cu. (d) 9.05% of Cu. (f) 16% of Cu. (g) 11% of Cu. (h) 7% of Cu. (i) 6% of Cu. (j) 6.5% of Cu

Table 8.3 (continued)

Metal or alloy	Thermal conductivity near room temperature, cal/cm$^2 \cdot$ cm $\cdot$ s $\cdot$ °C	Metal or alloy	Thermal conductivity near room temperature, cal/cm$^2 \cdot$ cm $\cdot$ s $\cdot$ °C
Casting alloys		**Magnesium and magnesium alloys**	
Chromium copper (1% Cr)	0.4(a)	Magnesium (99.8%)	0.367
89Cu-11Sn	0.121	**Casting alloys**	
88Cu-6Sn-1.5Pb-4.5Zn	(b)	AM100A	0.17
87Cu-8Sn-1Pb-4Zn	(c)	AZ63A	0.18
87Cu-10Sn-1Pb-2Zn	(c)	AZ81A (T4)	0.12
80Cu-10Sn-10Pb	(c)	AZ91A, B, C	0.17
Manganese bronze, 110 ksi	(d)	AZ92A	0.17
Aluminum bronze		HK31A (T6, sand cast)	0.22
Alloy 9A	(e)	HZ32A	0.26
Alloy 9B	(f)	ZH42	0.27
Alloy 9C	(b)	ZH62A	0.26
Alloy 9D	(c)	ZK51A	0.26
Propeller bronze	(g)	ZE41A (T5)	0.27
Nickel silver		EZ33A	0.24
12% Ni	(h)	EK30A	0.26
16% Ni	(h)	EK41A (T5)	0.24
20% Ni	(i)	**Wrought alloys**	
25% Ni	(j)	M1A	0.33
Silicon bronze	(h)	AZ31B	0.23
Iron and iron alloys		AZ61A	0.19
Pure iron	0.178	AZ80A	0.18
Cast iron (3.16 C, 1.54 Si, 0.57 Mn)	0.112	ZK60A, B (F)	0.28
Carbon steel (0.23 C, 0.64 Mn)	0.124	ZE10A (O)	0.33
Carbon steel (1.22 C, 0.35 Mn)	0.108	HM21A (O)	0.33
Alloy steel (0.34 C, 0.55 Mn, 0.78 Cr, 3.53 Ni, 0.39 Mo, 0.05 Cu)	0.079	HM31A	0.25
Type 410	0.057		
Type 304	0.036	**Nickel and nickel alloys**	
T1 tool steel	0.058	Nickel (99.95% Ni + Co)	0.22
Lead and lead alloys		"A" nickel	0.145
Corroding lead (99.73 + % Pb)	0.083	"D" nickel	0.115
5-95 solder	0.085	Monel	0.062
20-80 solder	0.089	"K" Monel	0.045
50-50 solder	0.111	Inconel	0.036
1% antimonial lead	0.080	Hastelloy B	0.027
Hard lead (96Pb-4Sb)	0.073	Hastelloy C	0.03
Hard lead (94Pb-6Sb)	0.069	Hastelloy D	0.05
8% antimonial lead	0.065	Illium G	0.029
9% antimonial lead	0.064	Illium R	0.031
Lead-base babbitt (SAE 14)	0.057	60Ni-24Fe-16Cr	0.032
Lead-base babbitt (alloy 8)	0.058	35Ni-45Fe-20Cr	0.031

(continued)

(a) Depends on processing. (b) 18% of Cu. (c) 12% of Cu. (d) 9.05% of Cu. (f) 16% of Cu. (g) 11% of Cu. (h) 7% of Cu. (i) 6% of Cu. (j) 6.5% of Cu

Table 8.3 (continued)

Metal or alloy	Thermal conductivity near room temperature, cal/cm$^2 \cdot$ cm $\cdot$ s $\cdot$ °C	Metal or alloy	Thermal conductivity near room temperature, cal/cm$^2 \cdot$ cm $\cdot$ s $\cdot$ °C
Nickel and nickel alloys (continued)		**Pure metals (continued)**	
Constantan	0.051	Chromium	0.16
Tin and tin alloys		Cobalt	0.165
Pure tin	0.15	Germanium	0.14
Soft solder (63Sn-37Pb)	0.12	Gold	0.71
Tin foil (92Sn-8Zn)	0.14	Indium	0.057
Titanium and titanium alloys		Iridium	0.14
Titanium (99.0%)	0.043	Lithium	0.17
Ti-5Al-2.5Sn	0.019	Molybdenum	0.34
Ti-2Fe-2Cr-2Mo	0.028	Niobium	0.13
Ti-8Mn	0.026	Palladium	0.168
Zinc and zinc alloys		Platinum	0.165
Pure zinc	0.27	Plutonium	0.020
AG40A alloy	0.27	Rhenium	0.17
AC41A alloy	0.26	Rhodium	0.21
Commercial rolled zinc		Silicon	0.20
0.08 Pb	0.257	Silver	1.0
0.06 Pb, 0.06 Cd	0.257	Sodium	0.32
Rolled zinc alloy (1 Cu, 0.010 Mg)	0.25	Tantalum	0.130
Zn-Cu-Ti alloy (0.8 Cu, 0.15 Ti)	0.25	Thallium	0.093
		Thorium	0.090
		Tungsten	0.397
Pure metals		Uranium	0.071
Beryllium	0.35	Vanadium	0.074
Cadmium	0.22	Yttrium	0.035

(a) Depends on processing. (b) 18% of Cu. (c) 12% of Cu. (d) 9.05% of Cu. (f) 16% of Cu. (g) 11% of Cu. (h) 7% of Cu. (i) 6% of Cu. (j) 6.5% of Cu

Table 8.4 Electrical conductivity and resistivity of metals and alloys

Metal or alloy	Conductivity, % IACS	Resistivity, $\mu\Omega \cdot$ cm	Metal or alloy	Conductivity, % IACS	Resistivity, $\mu\Omega \cdot$ cm
Aluminum and aluminum alloys			**Aluminum and aluminum alloys (continued)**		
1100 (O)	59	2.9	5056 (O)	29	5.9
2024 (O)	50	3.4	6061 (T6)	43	4.0
3003 (O)	50	3.4	6101 (T6)	57	3.0
4032 (O)	40	4.3	7075 (O)	45	3.8
5052 (all)	35	4.9	7075 (T6)	33	5.2

(continued)

(a) Precipitation hardened; depends on processing. (b) A heat-treatable alloy. (c) Annealed and quenched. (d) At low field strength and high electrical resistance. (e) At higher field strength; annealed for optimal magnetic properties

Table 8.4 (continued)

Metal or alloy	Conductivity, % IACS	Resistivity, µΩ·cm	Metal or alloy	Conductivity, % IACS	Resistivity, µΩ·cm
Copper and copper alloys			**Silver and silver alloys (continued)**		
Wrought copper			97Ag-3Pd	58	2.9
Pure copper	103.06	1.67	90Ag-10Pd	27	5.3
Electrolytic tough pitch			90Ag-10Au	40	4.2
copper (ETP)	101	1.71	60Ag-40Pd	8	23
Oxygen-free copper (OF)	101	1.71	70Ag-30Pd	12	14.3
Free-machining copper			**Platinum and platinum alloys**		
0.5% Te	95	1.82	Platinum	16	10.6
1.0% Pb	98	1.76	95Pt-5Ir	9	19
Wrought alloys			90Pt-10Ir	7	25
Cartridge brass, 70%	28	6.2	85Pt-15Ir	6	28.5
Yellow brass	27	6.4	80Pt-20Ir	5.6	31
Leaded commercial bronze	42	4.1	75Pt-25Ir	5.5	33
Phosphor bronze, 1.25%	48	3.6	70Pt-30Ir	5	35
Nickel silver, 55-18	5.5	31	65Pt-35Ir	5	36
Low-silicon bronze (B)	12	14.3	95Pt-5Ru	5.5	31.5
Beryllium copper	22–30(a)	5.7–7.8(a)	90Pt-10Ru	4	43
			89Pt-11Ru	4	43
Casting alloys			86Pt-14Ru	3.5	46
Chromium copper (1% Cr)	80–90(a)	2.10	96Pt-4W	5	36
88Cu-8Sn-4Zn	11	15			
87Cu-10Sn-1Pb-2Zn	11	15	**Palladium and palladium alloys**		
Electrical contact materials			Palladium	16	10.8
Copper alloys			95.5Pd-4.5Ru	7	24.2
0.04 oxide	100	1.72	90Pd-10Ru	6.5	27
1.25 Sn + P	48	3.6	70Pd-30Ag	4.3	40
5 Sn + P	18	11	60Pd-40Ag	4.0	43
8 Sn + P	13	13	50Pd-50Ag	5.5	31.5
15 Zn	37	4.7	72Pd-26Ag-1.71-2Ni	4	43
20 Zn	32	5.4	60Pd-40Cu	5	35(c)
35 Zn	27	6.4	45Pd-30Ag-20Au-5Pt	4.5	39
2 Be + Ni or Co(b)	17–21	9.6–11.5	35Pd-30Ag-14Cu-10Pt-10Au-1Zn	5	35
			Gold and gold alloys		
			Gold	75	2.35
			90Au-10Cu	16	10.8
Silver and silver alloys			75Au-25Ag	16	10.8
Fine silver	104	1.59	72.5Au-14Cu-8.5Pt-4Ag-1Zn	10	17
92.5Ag-7.5Cu	88	2	69Au-25Ag-6Pt	11	15
90Ag-10Cu	85	2	41.7Au-32.5Cu-18.8Ni-7Zn	4.5	39
72Ag-28Cu	87	2	**Electrical heating alloys**		
72Ag-26Cu-2Ni	60	2.9	**Ni-Cr and Ni-Cr-Fe alloys**		
85Ag-15Cd	35	4.93	78.5Ni-20Cr-1.5Si (80-20)	1.6	108.05
97Ag-3Pt	45	3.5			

(continued)

(a) Precipitation hardened; depends on processing. (b) A heat-treatable alloy. (c) Annealed and quenched. (d) At low field strength and high electrical resistance. (e) At higher field strength; annealed for optimal magnetic properties

Table 8.4 (continued)

Metal or alloy	Conductivity, % IACS	Resistivity, $\mu\Omega \cdot cm$	Metal or alloy	Conductivity, % IACS	Resistivity, $\mu\Omega \cdot cm$
Ni-Cr and Ni-Cr-Fe alloys (continued)			**Fe-Cr-Al alloy**		
73.5Ni-20Cr-5Al-1.5Si	1.2	137.97	72Fe-23Cr-5Al-0.5Co	1.3	135.48
68Ni-20Cr-8.5Fe-2Si	1.5	116.36	**Pure metals**		
60Ni-16Cr-22.5Fe-1.5Si	1.5	112.20	Iron (99.99%)	17.75	9.71
35Ni-20Cr-43.5Fe-1.5Si	1.7	101.4	**Thermostat metals**		
Fe-Cr-Al alloys			75Fe-22Ni-3Cr	3	78.13
72Fe-23Cr-5Al	1.3	138.8	72Mn-18Cu-10Ni	1.5	112.2
55Fe-37.5Cr-7.5Al	1.2	166.23	67Ni-30Cu-1.4Fe-1Mn	3.5	56.52
Pure metals			75Fe-22Ni-3Cr	12	15.79
Molybdenum	34	5.2	66.5Fe-22Ni-8.5Cr	3.3	58.18
Platinum	16	10.64	**Permanent magnet materials**		
Tantalum	13.9	12.45	Carbon steel (0.65% C)	9.5	18
Tungsten	30	5.65	Carbon steel (1% C)	8	20
Nonmetallic heating element materials			Chromium steel (3.5% Cr)	6.1	29
Silicon carbide, SiC	1–1.7	100–200	Tungsten steel (6% W)	6	30
Molybdenum disilicide (MoSi2)	4.5	37.24	Cobalt steel (17% Co)	6.3	28
Graphite	...	910.1	Cobalt steel (36% Co)	6.5	27
Instrument and control alloys			**Intermediate alloys**		
Cu-Ni alloys			Cunico	7.5	24
98Cu-2Ni	35	4.99	Cunife	9.5	18
94Cu-6Ni	17	9.93	Comol	3.6	45
89Cu-11Ni	11	14.96	**Alnico alloys**		
78Cu-22Ni	5.7	29.92	Alnico I	3.3	75
55Cu-45Ni (constantan)	3.5	49.87	Alnico II	3.3	65
Cu-Mn-Ni alloys			Alnico III	3.3	60
87Cu-13Mn (manganin)	3.5	48.21	Alnico IV	3.3	75
83Cu-13Mn-4Ni (manganin)	3.5	48.21	Alnico V	3.5	47
85Cu-10Mn-4Ni (shunt manganin)	4.5	38.23	Alnico VI	3.5	50
70Cu-20Ni-10Mn	3.6	48.88	**Magnetically soft materials**		
67Cu-5Ni-27Mn	1.8	99.74	**Electrical steel sheet**		
Ni-base alloys			M-50	9.5	18
99.8 Ni	23	7.98	M-43	6–9	20–28
71Ni-29Fe	9	19.95	M-36	5.5–7.5	24–33
80Ni-20Cr	1.5	112.2	M-27	3.5–5.5	32–47
75Ni-20Cr-3Al + Cu or Fe	1.3	132.98	M-22	3.5–5	41–52
76Ni-17Cr-4Si-3Mn	1.3	132.98	M-19	3.5–5	41–56
60Ni-16Cr-24Fe	1.5	112.2	M-17	3–3.5	45–58
35Ni-20Cr-45Fe	1.7	101.4	M-15	3–3.5	45–69

(continued)

(a) Precipitation hardened; depends on processing. (b) A heat-treatable alloy. (c) Annealed and quenched. (d) At low field strength and high electrical resistance. (e) At higher field strength; annealed for optimal magnetic properties

Table 8.4 (continued)

Metal or alloy	Conductivity, % IACS	Resistivity, μΩ·cm
Electrical steel sheet (continued)		
M-14	3–3.5	58–69
M-7	3–3.5	45–52
M-6	3–3.5	45–52
M-5	3–3.5	45–52
Moderately high-permeability materials(d)		
Thermenol	0.5	162
16 Alfenol	0.7	153
Sinimax	2	90
Monimax	2.5	80
Supermalloy	3	65
4-79 Moly Permalloy, Hymu 80	3	58
Mumetal	3	60
1040 alloy	3	56
High Permalloy 49, A-L 4750, Armco 48	3.6	48
45 Permalloy	3.6	45
High-permeability materials(e)		
Supermendur	4.5	40
2V Permendur	4.5	40
35% Co, 1% Cr	9	20
Ingot iron	17.5	10
0.5% Si steel	6	28
1.75% Si steel	4.6	37
3.0% Si steel	3.6	47
Grain-oriented 3.0% Si steel	3.5	50
Grain-oriented 50% Ni iron	3.6	45
50% Ni iron	3.5	50
Relay steels and alloys after annealing		
Low-carbon iron and steel		
Low-carbon iron	17.5	10

Metal or alloy	Conductivity, % IACS	Resistivity, μΩ·cm
Low-carbon iron and steel (continued)		
1010 steel	14.5	12
Silicon steels		
1% Si	7.5	23
2.5% Si	4	41
3% Si	3.5	48
3% Si, grain-oriented	3.5	48
4% Si	3	59
Stainless steels		
Type 410	3	57
Type 416	3	57
Type 430	3	60
Type 443	3	68
Type 446	3	61
Nickel irons		
50% Ni	3.5	48
78% Ni	11	16
77% Ni (Cu, Cr)	3	60
79% Ni (Mo)	3	58
Stainless and heat-resistant alloys		
Type 302	3	72
Type 309	2.5	78
Type 316	2.5	74
Type 317	2.5	74
Type 347	2.5	73
Type 403	3	57
Type 405	3	60
Type 501	4.5	40
HH	2.5	80
HK	2	90
HT	1.7	100

(a) Precipitation hardened; depends on processing. (b) A heat-treatable alloy. (c) Annealed and quenched. (d) At low field strength and high electrical resistance. (e) At higher field strength; annealed for optimal magnetic properties

Table 8.5 Approximate melting temperatures of metals and alloys

Metal or alloy	Temperature °C	Temperature °F	Metal or alloy	Temperature °C	Temperature °F
Aluminum and aluminum alloys			**Casting alloys (continued)**		
Wrought alloys			M bronze (8.5% Sn, 4% Zn)	1000	1832
1100	655	1215	83Cu-7Sn-7Pb-3Zn	980	1800
2017	640	1185	Nickel-tin bronze (A)	1025	1880
Alclad 2024	635	1180	Nickel-tin bronze (B)	1025	1880
3003	655	1210	Nickel-aluminum bronze	1045–1060	1910–1940
5052	650	1200	Copper nickel (10% Ni)	1150	2100
6061	650	1205	Copper-nickel (30% Ni)	1240	2260
7075	635	1175	Nickel-silver, 12% Ni	1000	1830
			Nickel-silver, 20% Ni	1145	2090
Casting alloys			Nickel-silver, 25% Ni	1045	1910
242.0	635	1175	**Low-melting-point metals and alloys**		
295.0	645	1190	55.5Bi-44.5Pb	125	255
336.0	565	1050	58Bi-42Sn	138	281
A380.0	595	1100	50Bi-26.7Pb-13.3Sn-10Cd	70	158
413.0	580	1080	44.7Bi-22.6Pb-19.1In-5.3Cd-8.3Sn	47	117
B443.0	630	1170	Lead (99.9% min Pb)	327	621
514.0	640	1185	Antimonial lead (10% Sb)	285	545
520.0	605	1120	Tellurium lead (0.04% The, 0.06% Cu)	327	621
Copper and copper alloys			50-50 lead-tin solder	216	421
Wrought copper and copper alloys			60-40 tin-lead solder	190	374
Pure copper	1080	1980	Tin (99.8% min Sn)	232	449
Beryllium copper (1.7% Be, 0.25% Co)	870–980	1600–1800	**Zinc and zinc alloys**		
Beryllium copper (1.9% Be, 0.25% Co)	870–980	1600–1800	Rolled zinc (0.08% Pb)	419	786
Red brass (15% Zn)	1025	1880	Rolled zinc (1% Cu, 0.01% Mg)	422	792
Cartridge brass (30% Zn)	955	1750	AG40A	387	728
Yellow brass	930	1710	AC41A	386	727
Muntz metal	905	1660	**Magnesium and magnesium alloys**		
Admiralty brass	940	1720			
Naval brass	900	1650	**Casting alloys**		
Phosphor bronze (5% Sn)	1050	1920	AZ91B	595	1105
Phosphor bronze (10% Sn)	1000	1830	AZ91C	595	1105
Aluminum bronze (7% Al, 2.5% Fe)	1045	1915	AZ92A	593	1100
High-silicon bronze (3.3% Si)	1025	1880	EZ33A	643	1189
Copper-nickel (10% Ni)	1150	2100	HZ32A	648	1198
Copper-nickel (21% Ni)	1150–1200	2100–2190	**Wrought alloys**		
Copper-nickel (30% Ni)	1240	2260	AZ31B	630	1170
Nickel-silver, 72-18	1150	2100	AZ80A	610	1130
Nickel-silver, 65-18	1110	2030	HK31A	651	1204
Nickel-silver, 55-18	1055	1930	ZK60A	635	1175
Casting alloys			**Titanium and titanium alloys**		
85Cu-5Zn-5Sn-5Pb	1005	1840			
G bronze (10% Sn, 2% Zn)	980	1800	Unalloyed titanium (ASTM grade 1)	1683	3063

(continued)

Table 8.5 (continued)

Metal or alloy	Temperature °C	Temperature °F	Metal or alloy	Temperature °C	Temperature °F
Titanium and titanium alloys (continued)			**High-temperature high-strength alloys**		
Unalloyed titanium (ASTM grade 2)	1704	3100	Incoloy alloy 800	1385	2525
Unalloyed titanium (ASTM grade 4)	1670	3038	Incoloy alloy 801	1385	2525
Ti-0.2Pd	1704	3100	Incoloy alloy 802	1370	2500
Ti-5Al-2.5Sn	1650	3002	Incoloy alloy 825	1400	2550
Ti-6Al-2Sn-4Zr-2Mo	1650	3000	Alloy 713C (casting)	1290	2350
Ti-6Al-6V-2Sn	1705	3100	Alloy 713LC	1320	2410
Ti-6Al-4V	1650	3000	A-286	1400	2550
Ti-6Al-2Sn-4Zr-6Mo	1675	3050	IN-100	1335	2435
			IN-102	1290	2350
Nickel and nickel alloys			IN-738	1230–1315	2250–2400
Low-alloy nickels and nickel-coppers			IN-939	1230–1340	2255–2444
Nickel 200, 201, and 205 (99.5% min Ni)	1445	2635	Alloy 25 (L-605)	1410	2570
Nickel 211 (95% Ni, 4.9% Mn, 0.1% C)	1425	2600	Alloy 188	1330	2425
Nickel 270 (99.98% min Ni)	1455	2650	MAR-M 246	1355	2475
Duranickel 301	1440	2620	M-252	1370	2500
Nickel-beryllium (2.7% Be)	1265	2310	René 41	1370	2500
Monel alloy 400	1350	2460	TD nickel	1455	2650
Monel alloy R-405	1350	2460	Udimet alloy 500	1395	2540
Monel alloy K-500	1350	2460	Udimet alloy 700	1400	2550
"S" Monel (cast)	1290	2350	Waspaloy	1355	2475
			B-1900	1300	2375
Nickel-chromium-iron alloys			Multimet alloy N-155	1355	2470
Inconel alloy 600	1415	2575	Discaloy	1465	2665
Inconel alloy 601	1370	2494			
Inconel alloy 617	1375	2510	**Irons and steels**		
Inconel alloy 625	1350	2460	**Irons**		
Inconel alloy 690	1375	2510	Ingot iron	1535	2795
Inconel alloy 718	1335	2437	Wrought iron	1510	2750
Inconel alloy X-750	1425	2600	Gray cast iron	1175	2150
Inconel alloy 751	1425	2600	Malleable iron	1230	2250
			Ductile iron	1175	2150
Nickel-chromium alloys			Ni-Resists (15.5–35% Ni)	1230	2250
Nimonic alloy 75	1380	2515	Duriron (14.5% Si)	1260	2300
Nimonic alloy 80A	1370	2500	**Steels**		
Nimonic alloy 90	1370	2500	Carbon steel (SAE 1020)	1515	2760
Nimonic alloy 115	1315	2400	4340 steel	1505	2740
			9% Ni steel	1500	2730
Nickel-chromium-molybdenum alloys			Maraging steels (18Ni-200)	1455	2650
Hastelloy alloy C-276	1370	2500			
Hastelloy alloy G	1345	2450	**Wrought stainless steels**		
Hastelloy alloy N	1400	2550			
Hastelloy alloy S	1380	2516	Type 201	1400–1450	2550–2650
Hastelloy alloy W	1315	2400	Type 304	1400–1450	2550–2650
Hastelloy alloy X	1355	2470	Type 316	1375–1400	2500–2550
Chlorimet 2 and 3	1315	2400	Type 405	1480–1530	2700–2790

(continued)

Table 8.5 (continued)

Metal or alloy	Temperature °C	Temperature °F	Metal or alloy	Temperature °C	Temperature °F
Wrought stainless steels (continued)			**Heat-resistant alloys (continued)**		
Type 409	1480–1530	2700–2790	HK	1400	2550
Type 420	1450–1510	2650–2750	HL	1425	2600
Type 430	1425–1510	2600–2750	HN	1370	2500
Type 440C	1370–1480	2500–2700	HP, HT, and HU	1345	2450
Type 446	1425–1510	2600–2750	HW AND HX	1290	2350
17-4 PH	1400–1440	2560–2625	**Refractory metals and alloys**		
AM 350	1400	2550	Niobium	2470	4475
Cast stainless steels			Nb-1Zr	2400	4350
Corrosion-resistant alloys			Molybdenum	2610	4730
CA-15	1510	2750	Mo-0.5Ti	2610	4730
CA-40	1495	2725	Tantalum	2995	5425
CB-7Cu-1	1510	2750	Tungsten	3410	6170
CF-8	1425	2600	W-25Re	3100	5612
CF-3M	1400	2550	W-50Re	2550	4622
CF-20	1415	2575	**Precious metals and alloys**		
CH-20	1425	2600	Gold (99.995% min Au)	1063	1945
CN-7M	1455	2650	Silver (99.9% min Ag)	961	1761
Heat-resistant alloys			BAg-1 (brazing alloy)	635	1175
HA	1510	2750	Platinum	1791	3256
HC	1495	2725	Palladium	1552	2826
HD	1480	2700	Iridium	2443	4429
HE	1455	2650	Rhodium	1960	3560
HF	1400	2550	90Pt-10Ir	1782	3240
HH	1370	2500	60Pd-40Ag	1338	2440

9 Mechanical Properties of Metals and Alloys

Table 9.1 Mechanical properties of selected carbon and alloy steels in the hot-rolled, normalized, and annealed condition

Data were obtained from specimens 12.8 mm (0.505 in.) diameter that were machined from 25 mm (1 in.) rounds.

AISI No.(a)	Treatment	Austenitizing temperature °C	°F	Tensile strength MPa	ksi	Yield strength MPa	ksi	Elongation, %	Reduction in area, %	Hardness, HB
1015	As-rolled	...	...	420.6	61.0	313.7	45.5	39.0	61.0	126
	Normalized	925	1700	424.0	61.5	324.1	47.0	37.0	69.6	121
	Annealed	870	1600	386.1	56.0	284.4	41.3	37.0	69.7	111
1020	As-rolled	...	...	448.2	65.0	330.9	48.0	36.0	59.0	143
	Normalized	870	1600	441.3	64.0	346.5	50.3	35.8	67.9	131
	Annealed	870	1600	394.7	57.3	294.8	42.8	36.5	66.0	111
1022	As-rolled	...	...	503.3	73.0	358.5	52.0	35.0	67.0	149
	Normalized	925	1700	482.6	70.0	358.5	52.0	34.0	67.5	143
	Annealed	870	1600	429.2	62.3	317.2	46.0	35.0	63.6	137
1030	As-rolled	...	...	551.6	80.0	344.7	50.0	32.0	57.0	179
	Normalized	925	1700	520.6	75.5	344.7	50.0	32.0	60.8	149
	Annealed	845	1550	463.7	67.3	341.3	49.5	31.2	57.9	126
1040	As-rolled	...	...	620.5	90.0	413.7	60.0	25.0	50.0	201
	Normalized	900	1650	589.5	85.5	374.0	54.3	28.0	54.9	170
	Annealed	790	1450	518.8	75.3	353.4	51.3	30.2	57.2	149
1050	As-rolled	...	...	723.9	105.0	413.7	60.0	20.0	40.0	229
	Normalized	900	1650	748.1	108.5	427.5	62.0	20.0	39.4	217
	Annealed	790	1450	636.0	92.3	365.4	53.0	23.7	39.9	187
1060	As-rolled	...	...	813.6	118.0	482.6	70.0	17.0	34.0	241
	Normalized	900	1650	775.7	112.5	420.6	61.0	18.0	37.2	229
	Annealed	790	1450	625.7	90.8	372.3	54.0	22.5	38.2	179
1080	As-rolled	...	...	965.3	140.0	586.1	85.0	12.0	17.0	293
	Normalized	900	1650	1010.1	146.5	524.0	76.0	11.0	20.6	293
	Annealed	790	1450	615.4	89.3	375.8	54.5	24.7	45.0	174
1095	As-rolled	...	...	965.3	140.0	572.3	83.0	9.0	18.0	293
	Normalized	900	1650	1013.5	147.0	499.9	72.5	9.5	13.5	293
	Annealed	790	1450	656.7	95.3	379.2	55.0	13.0	20.6	192
1117	As-rolled	...	...	486.8	70.6	305.4	44.3	33.0	63.0	143
	Normalized	900	1650	467.1	67.8	303.4	44.0	33.5	63.8	137
	Annealed	855	1575	429.5	62.3	279.2	40.5	32.8	58.0	121
1118	As-rolled	...	...	521.2	75.6	316.5	45.9	32.0	70.0	149
	Normalized	925	1700	477.8	69.3	319.2	46.3	33.5	65.9	143
	Annealed	790	1450	450.2	65.3	284.8	41.3	34.5	66.8	131

(continued)

Table 9.1 (continued)

AISI No.(a)	Treatment	Austenitizing temperature °C	Austenitizing temperature °F	Tensile strength MPa	Tensile strength ksi	Yield strength MPa	Yield strength ksi	Elongation, %	Reduction in area, %	Hardness, HB
1137	As-rolled	...	...	627.4	91.0	379.2	55.0	28.0	61.0	192
	Normalized	900	1650	668.8	97.0	396.4	57.5	22.5	48.5	197
	Annealed	790	1450	584.7	84.8	344.7	50.0	26.8	53.9	174
1141	As-rolled	...	...	675.7	98.0	358.5	52.0	22.0	38.0	192
	Normalized	900	1650	706.7	102.5	405.4	58.8	22.7	55.5	201
	Annealed	815	1500	598.5	86.8	353.0	51.2	25.5	49.3	163
1144	As-rolled	...	...	703.3	102.0	420.6	61.0	21.0	41.0	212
	Normalized	900	1650	667.4	96.8	399.9	58.0	21.0	40.4	197
	Annealed	790	1450	584.7	84.8	346.8	50.3	24.8	41.3	167
1340	Normalized	870	1600	836.3	121.3	558.5	81.0	22.0	62.9	248
	Annealed	800	1475	703.3	102.0	436.4	63.3	25.5	57.3	207
3140	Normalized	870	1600	891.5	129.3	599.8	87.0	19.7	57.3	262
	Annealed	815	1500	689.5	100.0	422.6	61.3	24.5	50.8	197
4130	Normalized	870	1600	668.8	97.0	436.4	63.3	25.5	59.5	197
	Annealed	865	1585	560.5	81.3	360.6	52.3	28.2	55.6	156
4140	Normalized	870	1600	1020.4	148.0	655.0	95.0	17.7	46.8	302
	Annealed	815	1500	655.0	95.0	417.1	60.5	25.7	56.9	197
4150	Normalized	870	1600	1154.9	167.5	734.3	106.5	11.7	30.8	321
	Annealed	815	1500	729.5	105.8	379.2	55.0	20.2	40.2	197
4320	Normalized	895	1640	792.9	115.0	464.0	67.3	20.8	50.7	235
	Annealed	850	1560	579.2	84.0	609.5	61.6	29.0	58.4	163
4340	Normalized	870	1600	1279.0	185.5	861.8	125.0	12.2	36.3	363
	Annealed	810	1490	744.6	108.0	472.3	68.5	22.0	49.9	217
4620	Normalized	900	1650	574.3	83.3	366.1	53.1	29.0	66.7	174
	Annealed	855	1575	512.3	74.3	372.3	54.0	31.3	60.3	149
4820	Normalized	860	1580	75.0	109.5	484.7	70.3	24.0	59.2	229
	Annealed	815	1500	681.2	98.8	464.0	67.3	22.3	58.8	197
5140	Normalized	870	1600	792.9	115.0	472.3	68.5	22.7	59.2	229
	Annealed	830	1525	572.3	83.0	293.0	42.5	28.6	57.3	167
5150	Normalized	870	1600	870.8	126.3	529.5	76.8	20.7	58.7	255
	Annealed	825	1520	675.7	98.0	357.1	51.8	22.0	43.7	197
5160	Normalized	855	1575	957.0	138.8	530.9	77.0	17.5	44.8	269
	Annealed	815	1495	722.6	104.8	275.8	40.0	17.2	30.6	197
6150	Normalized	870	1600	939.8	136.3	615.7	89.3	21.8	61.0	269
	Annealed	815	1500	667.4	96.8	412.3	59.8	23.0	48.4	197
8620	Normalized	915	1675	632.9	91.8	357.1	51.8	26.3	59.7	183
	Annealed	870	1600	536.4	77.8	385.4	55.9	31.3	62.1	149
8630	Normalized	870	1600	650.2	94.3	429.5	62.3	23.5	53.5	187
	Annealed	845	1550	564.0	81.8	372.3	54.0	29.0	58.9	156
8650	Normalized	870	1600	1023.9	148.5	688.1	99.8	14.0	40.4	302
	Annealed	795	1465	715.7	103.8	386.1	56.0	22.5	46.4	212
8740	Normalized	870	1600	929.4	134.8	606.7	88.0	16.0	47.9	269
	Annealed	815	1500	695.0	100.8	415.8	60.3	22.2	46.4	201
9255	Normalized	900	1650	932.9	135.3	579.2	84.0	19.7	43.4	269
	Annealed	845	1550	774.3	112.3	486.1	70.5	21.7	41.1	229
9310	Normalized	890	1630	906.7	131.5	570.9	82.8	18.8	58.1	269
	Annealed	845	1550	820.5	119.0	439.9	63.8	17.3	42.1	241

(a) All grades are fine-grained except for those in the 1100 series, which are coarse-grained. Heat-treated specimens were oil quenched unless otherwise indicated.

Table 9.2 Mechanical properties of selected carbon and alloy steels in the quenched-and-tempered condition

Data were obtained from specimens 12.8 mm (0.505 in.) in diameter that were machined from 25 mm (1 in.) rounds.

AISI No.(a)	Tempering temperature °C	Tempering temperature °F	Tensile strength MPa	Tensile strength ksi	Yield strength MPa	Yield strength ksi	Elongation, %	Reduction in area, %	Hardness, HB
1030(b)	205	400	848	123	648	94	17	47	495
	315	600	800	116	621	90	19	53	401
	425	800	731	106	579	84	23	60	302
	540	1000	669	97	517	75	28	65	255
	650	1200	586	85	441	64	32	70	207
1040(b)	205	400	896	130	662	96	16	45	514
	315	600	889	129	648	94	18	52	444
	425	800	841	122	634	92	21	57	352
	540	1000	779	113	593	86	23	61	269
	650	1200	669	97	496	72	28	68	201
1040	205	400	779	113	593	86	19	48	262
	315	600	779	113	593	86	20	53	255
	425	800	758	110	552	80	21	54	241
	540	1000	717	104	490	71	26	57	212
	650	1200	634	92	434	63	29	65	192
1050(b)	205	400	1124	163	807	117	9	27	514
	315	600	1089	158	793	115	13	36	444
	425	800	1000	145	758	110	19	48	375
	540	1000	862	125	655	95	23	58	293
	650	1200	717	104	538	78	28	65	235
1050	205	400	...	...	...	...	...	...	...
	315	600	979	142	724	105	14	47	321
	425	800	938	136	655	95	20	50	277
	540	1000	876	127	579	84	23	53	262
	650	1200	738	107	469	68	29	60	223
1060	205	400	1103	160	779	113	13	40	321
	315	600	1103	160	779	113	13	40	321
	425	800	1076	156	765	111	14	41	311
	540	1000	965	140	669	97	17	45	277
	650	1200	800	116	524	76	23	54	229
1080	205	400	1310	190	979	142	12	35	388
	315	600	1303	189	979	142	12	35	388
	425	800	1289	187	951	138	13	36	375
	540	1000	1131	164	807	117	16	40	321
	650	1200	889	129	600	87	21	50	255
1095(b)	205	400	1489	216	1048	152	10	31	601
	315	600	1462	212	1034	150	11	33	534
	425	800	1372	199	958	139	13	35	388
	540	1000	1138	165	758	110	15	40	293
	650	1200	841	122	586	85	20	47	235
1095	205	400	1289	187	827	120	10	30	401
	315	600	1262	183	813	118	10	30	375
	425	800	1213	176	772	112	12	32	363
	540	1000	1089	158	676	98	15	37	321
	650	1200	896	130	552	80	21	47	269
1137	205	400	1082	157	938	136	5	22	352
	315	600	986	143	841	122	10	33	285
	425	800	876	127	731	106	15	48	262
	540	1000	758	110	607	88	24	62	229
	650	1200	655	95	483	70	28	69	197
1137(b)	205	400	1496	217	1165	169	5	17	415
	315	600	1372	199	1124	163	9	25	375
	425	800	1103	160	986	143	14	40	311
	540	1000	827	120	724	105	19	60	262
	650	1200	648	94	531	77	25	69	187
1141	205	400	1634	237	1213	176	6	17	461

(continued)

Table 9.2 (continued)

AISI No.(a)	Tempering temperature °C	Tempering temperature °F	Tensile strength MPa	Tensile strength ksi	Yield strength MPa	Yield strength ksi	Elongation, %	Reduction in area, %	Hardness, HB
1141	315	600	1462	212	1282	186	9	32	415
(continued)	425	800	1165	169	1034	150	12	47	331
	540	1000	896	130	765	111	18	57	262
	650	1200	710	103	593	86	23	62	217
1144	205	400	876	127	627	91	17	36	277
	315	600	869	126	621	90	17	40	262
	425	800	848	123	607	88	18	42	248
	540	1000	807	117	572	83	20	46	235
	650	1200	724	105	503	73	23	55	217
1330(b)	205	400	1600	232	1455	211	9	39	459
	315	600	1427	207	1282	186	9	44	402
	425	800	1158	168	1034	150	15	53	335
	540	1000	876	127	772	112	18	60	263
	650	1200	731	106	572	83	23	63	216
1340	205	400	1806	262	1593	231	11	35	505
	315	600	1586	230	1420	206	12	43	453
	425	800	1262	183	1151	167	14	51	375
	540	1000	965	140	827	120	17	58	295
	650	1200	800	116	621	90	22	66	252
4037	205	400	1027	149	758	110	6	38	310
	315	600	951	138	765	111	14	53	295
	425	800	876	127	731	106	20	60	270
	540	1000	793	115	655	95	23	63	247
	650	1200	696	101	421	61	29	60	220
4042	205	400	1800	261	1662	241	12	37	516
	315	600	1613	234	1455	211	13	42	455
	425	800	1289	187	1172	170	15	51	380
	540	1000	986	143	883	128	20	59	300
	650	1200	793	115	689	100	28	66	238
4130(b)	205	400	1627	236	1462	212	10	41	467
	315	600	1496	217	1379	200	11	43	435
	425	800	1282	186	1193	173	13	49	380
	540	1000	1034	150	910	132	17	57	315
	650	1200	814	118	703	102	22	64	245
4140	205	400	1772	257	1641	238	8	38	510
	315	600	1551	225	1434	208	9	43	445
	425	800	1248	181	1138	165	13	49	370
	540	1000	951	138	834	121	18	58	285
	650	1200	758	110	655	95	22	63	230
4150	205	400	1931	280	1724	250	10	39	530
	315	600	1765	256	1593	231	10	40	495
	425	800	1517	220	1379	200	12	45	440
	540	1000	1207	175	1103	160	15	52	370
	650	1200	958	139	841	122	19	60	290
4340	205	400	1875	272	1675	243	10	38	520
	315	600	1724	250	1586	230	10	40	486
	425	800	1469	213	1365	198	10	44	430
	540	1000	1172	170	1076	156	13	51	360
	650	1200	965	140	855	124	19	60	280
5046	205	400	1744	253	1407	204	9	25	482
	315	600	1413	205	1158	168	10	37	401
	425	800	1138	165	931	135	13	50	336
	540	1000	938	136	765	111	18	61	282
	650	1200	786	114	655	95	24	66	235
50B46	205	400	...	...	...	...	...	...	560
	315	600	1779	258	1620	235	10	37	505
	425	800	1393	202	1248	181	13	47	405
	540	1000	1082	157	979	142	17	51	322
	650	1200	883	128	793	115	22	60	273

(continued)

Table 9.2 (continued)

AISI No.(a)	Tempering temperature °C	Tempering temperature °F	Tensile strength MPa	Tensile strength ksi	Yield strength MPa	Yield strength ksi	Elongation, %	Reduction in area, %	Hardness, HB
50B60	205	400	...	...	...	...	...	...	600
	315	600	1882	273	1772	257	8	32	525
	425	800	1510	219	1386	201	11	34	435
	540	1000	1124	163	1000	145	15	38	350
	650	1200	896	130	779	113	19	50	290
5130	205	400	1613	234	1517	220	10	40	475
	315	600	1496	217	1407	204	10	46	440
	425	800	1275	185	1207	175	12	51	379
	540	1000	1034	150	938	136	15	56	305
	650	1200	793	115	689	100	20	63	245
5140	205	400	1793	260	1641	238	9	38	490
	315	600	1579	229	1448	210	10	43	450
	425	800	1310	190	1172	170	13	50	365
	540	1000	1000	145	862	125	17	58	280
	650	1200	758	110	662	96	25	66	235
5150	205	400	1944	282	1731	251	5	37	525
	315	600	1737	252	1586	230	6	40	475
	425	800	1448	210	1310	190	9	47	410
	540	1000	1124	163	1034	150	15	54	340
	650	1200	807	117	814	118	20	60	270
5160	205	400	2220	322	1793	260	4	10	627
	315	600	1999	290	1772	257	9	30	555
	425	800	1606	233	1462	212	10	37	461
	540	1000	1165	169	1041	151	12	47	341
	650	1200	896	130	800	116	20	56	269
51B60	205	400	...	...	...	...	...	...	600
	315	600	...	...	...	...	...	...	540
	425	800	1634	237	1489	216	11	36	460
	540	1000	1207	175	1103	160	15	44	355
	650	1200	965	140	869	126	20	47	290
6150	205	400	1931	280	1689	245	8	38	538
	315	600	1724	250	1572	228	8	39	483
	425	800	1434	208	1331	193	10	43	420
	540	1000	1158	168	1069	155	13	50	345
	650	1200	945	137	841	122	17	58	282
81B45	205	400	2034	295	1724	250	10	33	550
	315	600	1765	256	1572	228	8	42	475
	425	800	1407	204	1310	190	11	48	405
	540	1000	1103	160	1027	149	16	53	338
	650	1200	896	130	793	115	20	55	280
8630	205	400	1641	238	1503	218	9	38	465
	315	600	1482	215	1392	202	10	42	430
	425	800	1276	185	1172	170	13	47	375
	540	1000	1034	150	896	130	17	54	310
	650	1200	772	112	689	100	23	63	240
8640	205	400	1862	270	1669	242	10	40	505
	315	600	1655	240	1517	220	10	41	460
	425	800	1379	200	1296	188	12	45	400
	540	1000	1103	160	1034	150	16	54	340
	650	1200	896	130	800	116	20	62	280
86B45	205	400	1979	287	1641	238	9	31	525
	315	600	1696	246	1551	225	9	40	475
	425	800	1379	200	1317	191	11	41	395
	540	1000	1103	160	1034	150	15	49	335
	650	1200	903	131	876	127	19	58	280
8650	205	400	1937	281	1675	243	10	38	525
	315	600	1724	250	1551	225	10	40	490
	425	800	1448	210	1324	192	12	45	420
	540	1000	1172	170	1055	153	15	51	340

(continued)

Table 9.2 (continued)

AISI No.(a)	Tempering temperature °C	Tempering temperature °F	Tensile strength MPa	Tensile strength ksi	Yield strength MPa	Yield strength ksi	Elongation, %	Reduction in area, %	Hardness, HB
8650 (con't)	650	1200	965	140	827	120	20	58	280
8660	205	400	...	...	...	...	...	...	580
	315	600	...	...	...	...	...	...	535
	425	800	1634	237	1551	225	13	37	460
	540	1000	1310	190	1213	176	17	46	370
	650	1200	1068	155	951	138	20	53	315
8740	205	400	1999	290	1655	240	10	41	578
	315	600	1717	249	1551	225	11	46	495
	425	800	1434	208	1358	197	13	50	415
	540	1000	1207	175	1138	165	15	55	363
	650	1200	986	143	903	131	20	60	302
9255	205	400	2103	305	2048	297	1	3	601
	315	600	1937	281	1793	260	4	10	578
	425	800	1606	233	1489	216	8	22	477
	540	1000	1255	182	1103	160	15	32	352
	650	1200	993	144	814	118	20	42	285
9260	205	400	...	...	...	...	...	...	600
	315	600	...	...	...	...	...	...	540
	425	800	1758	255	1503	218	8	24	470
	540	1000	1324	192	1131	164	12	30	390
	650	1200	979	142	814	118	20	43	295
94B30	205	400	1724	250	1551	225	12	46	475
	315	600	1600	232	1420	206	12	49	445
	425	800	1344	195	1207	175	13	57	382
	540	1000	1000	145	931	135	16	65	307
	650	1200	827	120	724	105	21	69	250

(a) All grades are fine-grained except for those in the 1100 series, which are coarse-grained. Heat-treated specimens were oil quenched unless otherwise indicated. (b) Water quenched

Table 9.3 Mechanical property data for stainless steels

Steel UNS No.	Common designation	Tensile strength MPa	Tensile strength ksi	Yield strength(a) MPa	Yield strength(a) ksi	Elongation, %	Reduction in area, %	Hardness (max), HRB
Annealed austenitic stainless steels									
S30100......	301	515	75	205	30	40	...	88
S30200......	302	515	75	205	30	40	...	88
S30215......	302B	515	75	205	30	...	...	...
S30430......	302Cu	450–585	65–85	...	...	...	...	...
S30300......	303	585(b)	85(b)	240(b)	35(b)	50(b)	55(b)	...
S30323......	303Se	585(b)	85(b)	240(b)	35(b)	50(b)	55(b)	...
S30400......	304	515	75	205	30	40	...	88
S30403......	304L	480	70	170	25	40	...	88
S30451......	304N	550	80	240	35	30	...	...
S31651......	316N	550	80	240	35	30	...	...
S30500......	305	480	70	170	25	40	...	88
S30800......	308	515	75	205	30	40	...	88
S32100......	321	515	75	205	30	40	...	88
S34700......	347	515	75	205	30	40	...	88
S34800......	348	515	75	205	30	40	...	88
S30900......	309	515	75	205	30	40	...	95

(continued)

Table 9.3 (continued)

Steel UNS No.	Common designation	Tensile strength MPa	Tensile strength ksi	Yield strength(a) MPa	Yield strength(a) ksi	Elongation, %	Reduction, in area, %	Hardness (max), HRB
Annealed austenitic stainless steels (continued)								
S30908	309S	515	75	205	30	40	...	95
S31000	310	515	75	205	30	40	...	95
S31008	310S	515	75	205	30	40	...	95
S31400	314	515	75	205	30	30	40	...
S31600	316	515	75	205	30	40	...	95
S31620	316F	585(b)	85(b)	240(b)	35(b)	40(b)	55(b)	...
S31700	317	515	75	205	30	35	...	95
S31703	317L	515	75	205	30	35	...	95
N08330	330	480	70	210	30	30	...	...
S38400	384	415–550	60–80	...	...	...	...	...
S38500	385	415–550	60–80	...	...	...	...	...
N08904	904L	490	71	220	31	35	...	95
N08366	AL-6X	515	75	205	30	30	...	...
S38100	18-18-2	515	75	205	30	40	...	96
N08700	JS-700	550	80	205	30	30	40	...
N08020	20Cb-3	585	85	275	40	30	...	95
...	304LN	515	75	205	30	...	...	...
...	308L	550(b)	80(b)	207(b)	30(b)	60(b)	70(b)	...
...	312	655	95	...	...	20	...	...
...	316LN	515(b)	75(b)	205(b)	30(b)	60(b)	70(b)	...
...	317LM	515	75	205	30	35	50	95
...	332	550(b)	80(b)	240(b)	35(b)	45(b)	70(b)	...
...	Crutemp 25	615(b)	89(b)	275(b)	40(b)	40(b)	...	...
...	JS-777	550	80	240	35	30	40	95
High-nitrogen austenitic stainless steels								
S20100	201	655	95	310	45	40	...	...
S20200	202	655	95	310	45	40	...	...
S20500	205	830(b)	120(b)	475(b)	69(b)	58(b)	62(b)	98(b)
S21600	216	690	100	415	60	40	...	100
S30451	304N	550	80	240	35	30	...	88
S30452	304HN	620	90	345	50	30	...	100
S31651	316N	550	80	240	35	30	...	95
S24100	Nitronic 32	690	100	380	55	30	50	...
S24000	Nitronic 33	690	100	415	60	40	...	...
S21900	Nitronic 40	690	100	415	60	40	...	...
S20910	Nitronic 50	825	120	515	75	30	...	...
S21800	Nitronic 60	655	95	345	50	35	55	...
S28200	18-18 Plus	760	110	415	60	35	55	...
S21400	Tenelon	860	125	485	70	40	...	...
Annealed ferritic stainless steels								
S40500	405	415	60	170	25	20	...	88
S40900	409	415	60	205	30	22(c)	...	80
S42900	429	450	65	205	30	22(c)	...	88
S43000	430	450	65	205	30	22(c)	...	88
S43020	430F	585–860	85–125	...	...	...	...	...
S43400	434	530(b)	77(b)	365(b)	53(b)	23(b)	...	83
S43600	436	530(b)	77(b)	365(b)	53(b)	23(b)	...	83
S44200	442	515	75	275	40	20	...	95
S44400	444	415	60	275	40	20	...	95
S44600	446	515	75	275	40	20	...	95
S44625	E-Brite 26-1	450	65	275	40	22(c)	...	90
S44660	Sea-cure/SC-1	550	80	380	55	20	...	100

(continued)

Table 9.3 (continued)

Steel UNS No.	Common designation	Tensile strength MPa	ksi	Yield strength(a) MPa	ksi	Elongation, %	Reduction, in area, %	Hardness (max), HRB
Annealed ferritic stainless steels (continued)								
S44700	29-4	550	80	415	60	20	...	88
S44800	29-4-2	550	80	415	60	20	...	98
...	18SR	620(b)	90(c)	450(c)	65(c)	25(e)	...	90 min(b)
Annealed duplex stainless steels								
S31200	...	690	100	450	65	25	...	...
S31500	...	630	92	440	64	30	...	30.5 HRC(b)
S31803	...	620	90	450	65	25	...	30.5 HRC(b)
S32304	...	600	87	400	58	25	...	30.5 HRC(b)
S32550	...	760	110	550	80	15	...	31.5 HRC(b)
S32750	...	800	116	550	80	15	...	32 HRC(b)
S32760	...	750	109	550	80	25	...	...
S32900	...	620	90	485	70	20	...	28 HRC(b)
S32950	...	690	100	480	70	20	...	30.5 HRC(b)
Martensitic stainless steels								
S40300	403	485	70	205	30	25(c)	...	88
S41000	410	450	65	205	30	22(c)	...	95
S41008	410S	415	60	205	30	22	...	95
S41040	410Cb	485	70	275	40	12	35	...
S41400	414	795	115	620	90	15	45	...
S41800	418(d)	1450(b)	210(b)	1210(b)	175(b)	18(b)	52(b)	...
S42000	420(e)	1720	250	1480(b)	215(b)	8(b)	25(b)	52 HRC(b)
S42200	422(f)	965	140	760	110	13	30	...
S43100	431(d)	1370(b)	198(d)	1030(b)	149(b)	16(b)	55(b)	...
S44002	440A	725(b)	105(d)	415(b)	60(b)	20(b)	...	95
S44003	440B	740(b)	107(d)	425(b)	62(b)	18(b)	...	96
S44004	440C	760(b)	110(d)	450(b)	65(b)	14(b)	...	97
...	414L	795(b)	115(d)	550(b)	80(b)	20(b)	60(b)	...
...	416 Plus X	515(b)	75	275(b)	40(b)	30(b)	60(b)	...
Precipitation-hardening stainless steels								
S13800	PH13-8Mo(g)	1520	220	1410	205	6–10	...	45 HRC (min)
S15500	15-5PH(h)	1310	190	1170	170	6–10	...	40 HRC (min)
S17400	17-4PH(h)	1310	190	1170	170	5–10	...	40 HRC (min)
S45000	Custom 450(h)	1240	180	1170	170	3–5	...	40 HRC (min)
S45500	Custom 455(g)	1530	222	1410	205	≤4	...	44 HRC (min)
S15700	PH15-7Mo(h)	1650	240	1590	230	1	...	46 HRC (min)
S17700	17-7PH(g)	1450	210	1310	190	1–6	...	43 HRC (min)
S35000	AM-350(i)	1140	165	1000	145	2–8	...	36 HRC (min)
S35500	AM-355(i)	1170	170	1030	150	12	...	37 HRC (min)
S66286	A-286(j)	896–965	125–140	655	95	4–15	...	24 HRC (min)

(a) At 0.2% offset. (b) Typical values. (c) 20% elongation for thicknesses of 1.3 mm (0.050 in.) or less. (d) Tempered at 260 °C (500 °F). (e) Tempered at 205 °C (400 °F). (f) Intermediate and hard tempers. (g) Aged at 510 °C (950 °F). (h) Aged at 480 °C (900 °F). (i) Aged at 535 °C (1000 °F). (j) Aged at 730 °C (1350 °F)

Table 9.4 Typical mechanical properties of commonly used wrought aluminum alloys

Alloy and temper	Ultimate tensile strength MPa	ksi	Yield strength MPa	ksi	Elongation in 50 mm (2 in.), % 1.6 mm (1/16 in.) thick specimen	13 mm (1/2 in.) diameter specimen
1100-O	90	13	34	5	35	45
1100-H14	124	18	117	17	9	20
1350-O	83	12	28	4	...	...
1350-H19	186	27	165	24	...	...
2011-T3	379	55	296	43	...	15
2011-T8	407	59	310	45	...	12
2014-O	186	27	97	14	...	18
2014-T4, T451	427	62	290	42	...	20
2014-T6, T651	483	70	414	60	...	13
2017-T4, T451	427	62	276	40	...	22
2024-O	186	27	76	11	20	22
2024-T3	483	70	345	50	18	...
2024-T4, T351	469	68	324	47	20	19
Alclad:						
2024-O	179	26	76	11	20	...
2024-T3	448	65	310	45	18	...
2024-T4, T351	441	64	290	42	19	...
2219-O	172	25	76	11	18	...
2219-T87	476	69	393	57	10	...
3003-O	110	16	41	6	30	40
3003-H12	131	19	124	18	10	20
3003-H14	152	22	145	21	8	16
3003-H16	179	26	172	25	5	14
3004-O	179	26	69	10	20	25
3004-H38	283	41	248	36	5	6
3105-O	117	17	55	8	24	...
3105-H14	172	25	152	22	5	...
3105-H18	214	31	193	28	3	...
3105-H25	179	26	159	23	...	...
5005-H34	159	23	138	20	8	...
5052-O	193	28	90	13	25	30
5052-H112	...	...	...	...	...	...
5052-H32	228	33	193	28	12	18
5052-H34	262	38	214	31	10	14
5056-O	290	42	152	22	...	35
5056-H18	434	63	407	59	...	10
5083-O	290	42	145	21	...	22
5083-H321	317	46	228	33	...	16
5086-H32	290	42	207	30	12	...
5086-H34	324	47	255	37	10	...
5086-H112	269	39	131	19	14	...
5454-O	248	36	117	17	22	...
5454-H32	276	40	207	30	10	...
5454-H34	303	44	241	35	10	...
5454-H112	248	36	124	18	18	...
5456-H321 and -H116	352	51	255	37	...	16
5657-H25	159	23	138	20	12	...
6061-O	124	18	55	8	25	30
6061-T4	241	35	145	21	22	25
6061-T6 and -T651	310	45	276	40	12	17
6063-T5	186	27	145	21	12	...
6063-T6	241	35	214	31	12	...
7050-T7651	552	80	490	71	...	11
7075-T6 and -T651	572	83	503	73	11	11
Alclad:						
7075-O	221	32	97	14	17	...
7075-T6 and -T651	524	76	462	67	11	...

Table 9.5 Typical tensile properties for separately cast test bars of common aluminum casting alloys

Alloy	Product(a)	Temper	Tensile strength MPa	ksi	Yield strength(b) MPa	ksi	Elongation(c), %
201.0.	S	T4	365	53	215	31	20
	S	T6	485	70	435	63	7
	S	T7	460	67	415	60	4.5
206.0, A206.0.	S	T7	435	63	345	50	11.7
208.0.	S	F	145	21	97	14	2.5
242.0.	S	T21	185	27	125	18	1.0
	S	T571	220	32	205	30	0.5
	S	T77	205	30	160	23	2.0
	P	T571	275	40	235	34	1.0
	P	T61	325	47	290	42	0.5
295.0.	S	T4	220	32	110	16	8.5
	S	T6	250	36	165	24	5.0
	S	T62	285	41	220	32	2.0
296.0.	P	T4	255	37	130	19	9.0
	P	T6	275	40	180	26	5.0
	P	T7	270	39	140	20	4.5
308.0.	P	F	195	28	110	16	2.0
319.0.	S	F	185	27	125	18	2.0
	S	T6	250	36	165	24	2.0
	P	F	235	34	130	19	2.5
	P	T6	280	40	185	27	3.0
336.0.	P	T351	250	36	195	28	0.5
	P	T65	325	47	295	43	0.5
354.0.	P	T61	380	55	285	41	6.0
355.0.	S	T51	195	28	160	23	1.5
	S	T6	240	35	175	25	3.0
	S	T61	270	39	240	35	1.0
	S	T7	265	38	250	36	0.5
	S	T71	175	35	200	29	1.5
	P	T51	210	30	165	24	2.0
	P	T6	290	42	190	27	4.0
	P	T62	310	45	280	40	1.5
	P	T7	280	40	210	30	2.0
	P	T71	250	36	215	31	3.0
356.0.	S	T51	175	25	140	20	2.0
	S	T6	230	33	165	24	3.5
	S	T7	235	34	210	30	2.0
	S	T71	195	28	145	21	3.5
	P	T6	265	38	185	27	5.0
	P	T7	220	32	165	24	6.0
357.0, A357.0.	S	T62	360	52	290	42	8.0
359.0.	P	T61	330	48	255	37	6.0
		T62	345	50	290	42	5.5
360.0.	D	F	325	47	170	25	3.0
A360.0.	D	F	320	46	165	24	5.0
380.	D	F	330	48	165	24	3.0
383.0.	D	F	310	45	150	22	3.5
384.0, A384.0.	D	F	330	48	165	24	2.5
390.0.	D	F	280	41	240	35	1.0
	D	T5	300	43	260	38	1.0
A390.0.	S	F, T5	180	26	180	26	<1.0
	S	T6	280	40	280	40	<1.0
	S	T7	250	36	250	36	<1.0
	P	F, T5	200	29	200	29	1.0

(continued)

Table 9.5 (continued)

Alloy	Product(a)	Temper	Tensile strength MPa	ksi	Yield strength(b) MPa	ksi	Elongation(c), %
A390.0 (continued)							
	P	T6	310	45	310	45	<1.0
	P	T7	260	38	260	38	<1.0
413.0	D	F	300	43	140	21	2.5
A413.0	D	F	290	42	130	19	3.5
443.0	S	F	130	19	55	8	8.0
B443.0	P	F	159	23	62	9	10.0
C443.0	D	F	228	33	110	16	9.0
514.0	S	F	170	25	85	12	9.0
518.0	D	F	310	45	190	28	5.0–8.0
520.0	S	T4	330	48	180	26	16
535.0	S	F	275	40	140	20	13
712.0	S	F	240	35	170	25	5.0
713.0	S	T5	210	30	150	22	3.0
	P	T5	220	32	150	22	4.0
771.0	S	T6	345	50	275	40	9.0
850.0	P	T5	160	23	75	11	10.0

(a) S, sand casting; P, permanent mold casting; D, die casting. (b) 0.2% offset. (c) 12.7 mm (½ in.) diam specimen

Table 9.6 Mechanical properties of wrought copper and copper alloys

Alloy No. (and name)	Tensile strength MPa	ksi	Yield strength MPa	ksi	Elongation in 50 mm (2 in.)(a), %
Coppers					
C10100 (oxygen-free electronic copper)	221–455	32–66	69–365	10–53	55–4
C10200 (oxygen-free copper)	221–455	32–66	69–365	10–53	55–4
C10300 (oxygen-free extra-low-phosphorus copper)	221–379	32–55	69–345	10–50	50–6
C10400, C10500, C10700 (oxygen-free silver-bearing copper)	221–455	32–66	69–365	10–53	55-4
C10800 (oxygen-free low-phosphorus copper)	221–379	32–55	69–345	10–50	50–4
C11000 (electrolytic tough pitch copper)	221–455	32–66	69–365	10–53	55–4
C11100 (electrolytic tough pitch anneal-resistant copper)	455	66	...	...	1.5 in 1500 mm (60 in.)
C11300, C11400, C11500, C11600 (silver-bearing tough pitch copper)	221–455	32–66	69–365	10–53	55–4
C12000, C12100	221–393	32–57	69–365	10–53	55–4
C12200 (phosphorus-deoxidized copper, high residual phosphorus)	221–379	32–55	69–345	10–50	45–8
C12500, C12700, C12800, C12900, C13000 (fire-refined tough pitch with silver)	221–462	32–67	69–365	10–53	55–4
C14200 (phosphorus-deoxidized arsenical copper)	221–379	32–55	69–345	10–50	45–8
C14300	221–400	32–58	76–386	11–56	42–1
C14310	221–400	32–58	76–386	11–56	42–1
C14500 (phosphorus-deoxidized tellurium-bearing copper)	221–386	32–56	69–352	10–51	50–3
C14700 (sulfur-bearing copper)	221–393	32–57	69–379	10–55	52–8
C15000 (zirconium-copper)	200–524	29–76	41–496	6–72	54–1.5
C15100	262–469	38–68	69–455	10–66	36–2

(continued)

Table 9.6 (continued)

Alloy No. (and name)	Tensile strength MPa	Tensile strength ksi	Yield strength MPa	Yield strength ksi	Elongation in 50 mm (2 in.)(a), %
Coppers (continued)					
C15500	276–552	40–80	124–496	18–72	40–3
C15710	324–724	47–105	268–689	39–100	20–10
C15720	462–614	67–89	365–586	53–85	20–3.5
C15735	483–586	70–85	414–565	60–82	16–10
C15760	483–648	70–94	386–552	56–80	20–8
High-copper alloys					
C16200 (cadmium-copper)	241–689	35–100	48–476	7–69	57–1
C16500	276–655	40–95	97–490	14–71	53–1.5
C17000 (beryllium-copper)	483–1310	70–190	221–1172	32–170	45–3
C17200 (beryllium-copper)	469–1462	68–212	172–1344	25–195	48–1
C17300 (beryllium-copper)	469–1479	68–200	172–1255	25–182	48–3
C17400	620–793	90–115	172–758	25–110	12–4
C17500 (copper-cobalt-beryllium alloy)	310–793	45–115	172–758	25–110	28–5
C18200, C18400, C18500 (chromium-copper)	234–593	34–86	97–531	14–77	40–5
C18700 (leaded copper)	221–379	32–55	69–345	10–50	45–8
C18900	262–655	38–95	62–359	9–52	48–14
C19000 (copper-nickel-phosphorus alloy)	262–793	38–115	138–552	20–80	50–2
C19100 (copper-nickel-phosphorus-tellurium alloy)	248–717	36–104	69–634	10–92	27–6
C19200	255–531	37–77	76–510	11–74	40–2
C19400	310–524	45–76	165–503	24–73	32–2
C19500	552–669	80–97	448–655	65–95	15–2
C19700	344–517	50–75	165–503	24–73	32–2
Brasses					
C21000 (gilding, 95%)	234–441	34–64	69–400	10–58	45–4
C22000 (commercial bronze, 90%)	255–496	37–72	69–427	10–62	50–3
C22600 (jewelry bronze, 87.5%)	269–669	39–97	76–427	11–62	46–3
C23000 (red brass, 85%)	269–724	39–105	69–434	10–63	55–3
C24000 (low brass, 80%)	290–862	42–125	83–448	12–65	55–3
C26000 (cartridge brass, 70%)	303–896	44–130	76–448	11–65	66–3
C26800, C27000 (yellow brass)	317–883	46–128	97–427	14–62	65–3
C28000 (Muntz metal)	372–510	54–74	145–379	21–55	52–10
Leaded brasses					
C31400 (leaded commercial bronze)	255–414	37–60	83–379	12–55	45–10
C31600 (leaded commercial bronze, nickel-bearing)	255–462	37–67	83–407	12–59	45–12
C33000 (low-leaded brass tube)	324–517	47–75	103–414	15–60	60–7
C33200 (high-leaded brass tube)	359–517	52–75	138–414	20–60	50–7
C33500 (low-leaded brass)	317–510	46–74	97–414	14–60	65–8
C34000 (medium-leaded brass)	324–607	47–88	103–414	15–60	60–7
C34200 (high-leaded brass)	338–586	49–85	117–427	17–62	52–5
C34900	365–469	53–68	110–379	16–55	72–18
C35000 (medium-leaded brass)	310–655	45–95	90–483	13–70	66–1
C35300 (high-leaded brass)	338–586	49–85	117–427	17–62	52–5
C35600 (extra-high-leaded brass)	338–510	49–74	117–414	17–60	50–7
C36000 (free-cutting brass)	338–469	49–68	124–310	18–45	53–18
C36500 to C36800 (leaded Muntz metal)(b)	372	54	138	20	45
C37000 (free-cutting Muntz metal)	372–552	54–80	138–414	20–60	40–6
C37700 (forging brass)(c)	359	52	138	20	45
C38500 (architectural bronze)(c)	414	60	138	20	30
Tin brasses					
C40500	269–538	39–78	83–483	12–70	49–3

(continued)

Table 9.6 (continued)

Alloy No. (and name)	Tensile strength MPa	Tensile strength ksi	Yield strength MPa	Yield strength ksi	Elongation in 50 mm (2 in.)(a), %
Tin brasses (continued)					
C40800	290–545	42–79	90–517	13–75	43–3
C41100	269–731	39–106	76–496	11–72	13–2
C41300	283–724	41–105	83–565	12–82	45–2
C41500	317–558	46–81	117–517	17–75	44–2
C42200	296–607	43–88	103–517	15–75	46–2
C42500	310–634	45–92	124–524	18–76	49–2
C43000	317–648	46–94	124–503	18–73	55–3
C43400	310–607	45–88	103–517	15–75	49–3
C43500	317–552	46–80	110–469	16–68	46–7
C44300, C44400, C44500 (inhibited admiralty)	331–379	48–55	124–152	18–22	65–60
C46400 to C46700 (naval brass)	379–607	55–88	172–455	25–66	50–17
C48200 (naval brass, medium-leaded)	386–517	56–75	172–365	25–53	43–15
C48500 (leaded naval brass)	379–531	55–77	172–365	25–53	40–15
Phosphor bronzes					
C50500 (phosphor bronze, 1.25% E)	276–545	40–79	97–345	14–50	48–4
C51000 (phosphor bronze, 5% A)	324–965	47–140	131–552	19–80	64–2
C51100	317–710	46–103	345–552	50–80	48–2
C52100 (phosphor bronze, 8% C)	379–965	55–140	165–552	24–80	70–2
C52400 (phosphor bronze, 10% D)	455–1014	66–147	193 (Annealed)	28	70-3
Leaded phosphor bronzes					
C54400 (free-cutting phosphor bronze)	303–517	44–75	131–434	19–63	50–16
Aluminum bronzes					
C60800 (aluminum bronze, 5%)	414	60	186	27	55
C61000	483–552	70–80	207–379	30–55	65–25
C61300	483–586	70–85	207–400	30–58	42–35
C61400 (aluminum bronze, D)	524–614	76–89	228–414	33–60	45–32
C61500	483–1000	70–145	152–965	22–140	55–1
C61800	552–586	80–85	269–293	39–42.5	28–23
C61900	634–1048	92–152	338–1000	49–145	30–1
C62300	517–676	75–98	241–359	35–52	35–22
C62400	621–724	90–105	276–359	40–52	18–14
C62500(k)	689	100	379	55	1
C63000	621–814	90–118	345–517	50–75	20–15
C63200	621–724	90–105	310–365	45–53	25–20
C63600	414–579	60–84	...	...	64–29
C63800	565–896	82–130	372–786	54–114	36–4
C64200	517–703	75–102	241–469	35–68	32–22
Silicon bronzes					
C65100 (low-silicon bronze, B)	276–655	40–95	103–476	15–69	55–11
C65400	276–793	40–115	130–744	20–108	40–3
C65500 (high-silicon bronze, A)	386–1000	56–145	145–483	21–70	63–3
Other copper-zinc alloys					
C66700 (manganese brass)	315–689	45.8–100	83–638	12–92.5	60–2
C67400	483–634	70–92	234–379	34–55	28–20
C67500 (manganese bronze, A)	448–579	65–84	207–414	30–60	33–19
C68700 (aluminum brass, arsenical)	414	60	186	27	55
C68800	565–889	82–129	379–786	55–114	36–2
C69000	496–896	72–130	345–807	50–117	40–2
C69400 (silicon red brass)	552–689	80–100	276–393	40–57	25–20

(continued)

Table 9.6 (continued)

Alloy No. (and name)	Tensile strength MPa	Tensile strength ksi	Yield strength MPa	Yield strength ksi	Elongation in 50 mm (2 in.)(a), %
Copper nickels					
C70250	586–758	85–110	552–784	80–105	40–3
C70400	262–531	38–77	276–524	40–76	46–2
C70600 (copper-nickel, 10%)	303–414	44–60	110–393	16–57	42–10
C71000 (copper-nickel, 20%)	338–655	49–95	90–586	13–85	40–3
C71300	338–655	49–95	90–586	13–85	40–3
C71500 (copper-nickel, 30%)	372–517	54–75	138–483	20–70	45–15
C71700	483–1379	70–200	207–1241	30–180	40–4
C72500	379–827	55–120	152–745	22–108	35–1
Nickel-silvers					
C73500	345–758	50–110	103–579	15–84	37–1
C74500 (nickel silver, 65-10)	338–896	49–130	124–524	18–76	50–1
C75200 (nickel silver, 65-18)	386–710	56–103	172–621	25–90	45–3
C75400 (nickel silver, 65-15)	365–634	53–92	124–545	18–79	43–2
C75700 (nickel silver, 65-12)	359–641	52–93	124–545	18–79	48–2
C76200	393–841	57–122	145–758	21–110	50–1
C77000 (nickel silver, 55-18)	414–1000	60–145	186–621	27–90	40–2
C72200	317–483	46–70	124–455	18–66	46–6
C78200 (leaded nickel silver, 65-8-2)	365–627	53–91	159–524	23–76	40–3

(a) Ranges are from softest to hardest commercial forms. The strength of the standard copper alloys depends on the temper (annealed grain size or degree of cold work) and the section thickness of the mill product. Ranges cover standard tempers for each alloy. (b) Values are for hot-rolled material. (c) Values are for as-extruded material. Source: Copper Development Association Inc.

Table 9.7 Mechanical properties of cast copper and copper alloys

UNS designation	Tensile strength MPa	Tensile strength ksi	Yield strength MPa	Yield strength ksi	Elongation in 50 mm (2 in.), %	Hardness Rockwell	Brinell 500 kg	Brinell 3000 kg
Coppers								
C80100	172	25	62	9	40	...	44	...
C80300	172	25	62	9	40	...	44	...
C80500	172	25	62	9	40	...	44	...
C80700	172	25	62	9	40	...	44	...
C80900	172	25	62	9	40	...	44	...
C81100	172	25	62	9	40	...	44	...
High-copper alloys								
C81300	(365)	(53)	(248)	(36)	(11)	...	(39)	...
C81400	(365)	(53)	(248)	(36)	(11)	(B 69)	...	...
C81500	(352)	(51)	(276)	(40)	(17)	...	(105)	...
C81700	(634)	(92)	(469)	(68)	(8)	...	...	(217)
C81800	345	50	172	25	20	B 55	...	...
	(703)	(102)	(517)	(75)	(8)	(B 96)		
C82000	345	50	138	20	20	B 55	...	(195)
	(689)	(100)	(517)	(75)	(8)	(B 95)		
C82100	(634)	(92)	(469)	(68)	(8)	...	...	(217)
C82200	393	57	207	30	20	B 60	...	...
	(655)	(95)	(517)	(75)	(8)	(B 96)		

(continued)

Table 9.7 (continued)

UNS designation	Tensile strength MPa	Tensile strength ksi	Yield strength MPa	Yield strength ksi	Elongation in 50 mm (2 in.), %	Rockwell	Brinell 500 kg	Brinell 3000 kg
High-copper alloys (continued)								
C82400	496 (1034)	72 (150)	255 (965)	37 (140)	20 (1)	B 78 (C 38)	...	...
C82500	552 (1103)	80 (160)	310	45	20 (1)	B 82 (C 40)	...	...
C82600	565 (1138)	82 (165)	324 (1069)	47 (155)	20 (1)	B 83 (C 43)	...	...
C82700	(1069)	(155)	(896)	(130)	(0)	(C 39)	...	...
C82800	669 (1138)	97 (165)	379 (1000)	55 (145)	20 (1)	B 85 (C 45)	...	...
Red brasses and leaded red brasses								
C83300	221	32	69	10	35	...	35	...
C83400	241	35	69	10	30	F 50	...	...
C83600	255	37	117	17	30	...	60	...
C83800	241	35	110	16	25	...	60	...
Semired brasses and leaded semired brasses								
C84200	193	28	103	15	27	...	60	...
C84400	234	34	103	15	26	...	55	...
C84500	241	35	97	14	28	...	55	...
C84800	248	36	97	14	30	...	55	...
Yellow brasses and leaded yellow brasses								
C85200	262	38	90	13	35	...	45	...
C85400	234	34	83	12	35	...	50	...
C85500	414	60	159	23	40	B 55	85	...
C85700	345	50	124	18	40	...	75	...
C85800	379	55	207	30	15	B 55	...	...
Manganese and leaded manganese bronze alloys								
C86100	655	95	345	50	20	...	...	180
C86200	655	95	331	48	20	...	...	180
C86300	793	115	572	83	15	...	...	225
C86400	448	65	172	25	20	...	90	105
C86500	490	71	193	28	30	...	100	130
C86700	586	85	290	42	20	B 80	...	155
C86800	565	82	262	38	22	...	...	80
Silicon bronzes and silicon brasses								
C87200	379	55	172	25	30	...	85	...
C87400	379	55	165	24	30	...	70	100
C87500	462	67	207	30	21	...	115	134
C87600	455	66	221	32	20	B 76	110	135
C87800	586	85	345	50	25	B 85	...	...
C87900	483	70	241	35	25	B 70	...	...
Tin bronzes								
C90200	262	38	110	16	30	...	70	...
C90300	310	45	145	21	30	...	70	...
C90500	310	45	152	22	25	...	75	...
C90700	303 (379)	44 (55)	152 (207)	22 (30)	20 (16)	...	80 (102)	...
C90800								
C90900	276	40	138	20	15	...	90	...
C91000	221	32	172	25	2	...	105	...
C91100	241	35	172	25	2	...	...	135
C91300	241	35	207	30	0.5	...	...	170
C91600	303 (414)	44 (60)	152 (221)	22 (32)	16 (16)	...	85 (106)	...
C91700	303 (414)	44 (60)	152 (221)	22 (32)	16 (16)	...	85 (106)	...

(continued)

Table 9.7 (continued)

UNS designation	Tensile strength MPa	Tensile strength ksi	Yield strength MPa	Yield strength ksi	Elongation in 50 mm (2 in.), %	Hardness Rockwell	Hardness Brinell 500 kg	Hardness Brinell 3000 kg
Leaded tin bronzes								
C92200	276	40	138	20	30	...	65	...
C92300	276	40	138	20	25	...	70	...
C92500	303	44	138	20	20	...	80	...
C92600	303	44	138	20	30	F 78	70	...
C92700	290	42	145	21	20	...	77	...
C92800	276	40	207	30	1	B80	...	...
C92900	324	47	179	26	20	...	80	...
	(324)	(47)	(179)	(26)	(20)		(80)	
High-leaded tin bronzes								
C93200	241	35	124	18	20	...	65	...
C93400	221	32	110	16	20	...	60	...
C93500	221	32	110	16	20	...	60	...
C93700	241	35	124	18	20	...	60	...
C93800	207	30	110	16	18	...	55	...
C93900	221	32	152	22	7	...	63	...
C94300	186	27	90	13	15	...	48	...
C94400	221	32	110	16	18	...	55	...
C94500	172	25	83	12	12	...	50	...
Nickel-tin bronzes								
C94700	345	50	159	23	35	...	85	(180)
	(586)	(85)	(414)	(60)	(10)			
C94800	310	45	159	23	35	...	80	...
	(414)	(60)	(207)	(30)	(8)		(120)	
Aluminum bronzes								
C95200	552	80	186	27	35	...	...	125
C95300	517	75	186	27	25	...	...	140
	(586)	(85)	(290)	(42)	(15)			(174)
C95400	586	85	241	35	18	...	...	170
	(724)	(105)	(372)	(54)	(8)			(195)
C95410								
C95500	689	100	303	44	12	...	...	192
	(827)	(120)	(469)	(68)	(10)			(230)
C95600	517	75	234	34	18	...	...	140
C95700	655	95	310	45	26	...	...	180
C95800	655	95	262	38	25	...	...	159
Copper-nickels								
C96200	310	45	172	25	20	...	...	...
C96300	517	75	379	55	10	...	150	...
C96400	469	68	255	37	28	...	...	140
C96600	(758)	(110)	(482)	(70)	(7)	...	...	(230)
C96700	(1207)	(175)	(552)	(80)	(10)	C26	...	...
Nickel silvers								
C97300	241	35	117	17	20	...	55	...
C97400	262	38	117	17	20	...	70	...
C97600	310	45	165	24	20	...	80	...
C97800	379	55	207	30	15	...	...	130
Special alloys								
C99300	655	95	379	55	2	...	200	20
C99400	455	66	234	34	25	...	...	125
	(545)	(79)	(372)	(54)				(170)
C99500	483	70	276	40	12	...	145	50
C99600	558	81	248	36	34	B 72	...	130
	(558)	(81)	(303)	(44)	(27)			

(continued)

Table 9.7 (continued)

UNS designation	Typical mechanical properties, as-cast (heat treated)(a)							
	Tensile strength		Yield strength		Elongation in 50 mm (2 in.), %	Hardness		
						Rockwell	Brinell	
	MPa	ksi	MPa	ksi			500 kg	3000 kg
Special alloys (continued)								
C99700	379	55	172	25	25	...	...	110
C99750	448	65	221	32	30	B77	110	...
	(517)	(75)	(276)	(40)	(20)	(B82)	(119)	

(a) Values for C82700, C84200, C96200, and C96300 are minimum, not typical. As-cast values are for sand casting except C93900, continuous cast; and C85800, C87800, C87900, die cast. Heat-treated values, in parentheses, indicate that the alloy responds to heat treatment. If heat-treated values are not shown, the copper or copper alloy does not respond. Source: Copper Development Association Inc.

Table 9.8 Typcial mechanical properties of zinc alloy die castings

Alloy	UNS No.	Tensile strength		Yield strength		Elongation in 50 mm (2 in.), %	Hardness, HB
		MPa	ksi	MPa	ksi		
AG40A/No. 3	Z33520	283	41.0	...	...	10	82
AG40B/No. 7	Z33523	283	41.0	...	...	14	76
AC41A/No. 5	Z33531	329	47.7	...	...	7	91
AC43A/No. 2	Z33541	359	52.0	...	...	7	100
ZA-8	Z35635	372	54	290	42	6–10	95–110
ZA-12	Z35630	400	58	317	46	4–7	95–115
ZA-27	Z35840	421	61	365	53	1–3	105–125

Table 9.9 Minimum mechanical properties for magnesium alloys

Alloy-temper	Tensile strength		Yield strength		Elongation in 50 mm (2 in.), %	Hardness, HB
	MPa	ksi	MPa	ksi		
Sand and permanent mold castings						
AM100A-T6	241	35	117	17	...	69
AZ63A-T6	234	34	110	16	3	73
AZ81A-T4	234	34	76	11	7	55
AZ91C-T6	234	34	110	16	3	70
AZ91E-T6	234	34	110	16	3	70
AZ92A-T6	34	34	124	18	1	81
EQ21A-T6	234	34	172	25	2	78
EZ33A-T5	138	20	96	14	2	50
HK31A-T6	186	27	89	13	4	66
HZ32A-T5	186	27	89	13	4	55
K1A-F	165	24	41	6	14	...
QE22A-T6	241	35	172	25	2	78
QH21A-T6	241	35	186	27	2	...
WE43A-T6	221	32	172	25	2	85
WE54A-T6	255	37	179	26	2	85
ZC63A-T6	193	28	125	18	2	60
ZE41A-T5	200	29	133	19.5	2.5	62
ZE63A-T6	276	40	186	27	5	...
ZH62A-T5	241	35	152	22	5	70

(continued)

Table 9.9 (continued)

Alloy-temper	Tensile strength MPa	ksi	Yield strength MPa	ksi	Elongation in 50 mm (2 in.), %	Hardness, HB
Sand and permanent mold castings (continued)						
ZK51A-T5	234	34	138	20	5	65
ZK61A-T6	276	40	179	26	5	70
Die castings						
AM50A	200	29	110	16	10	...
AM60A and B	220	32	130	19	8	...
AS41A and B	210	31	140	20	6	...
AZ91A, B, and D	230	34	160	32	3	...
Extruded bars, rods, and shapes						
AZ31B-F	220–240	32–35	140–150	20–22	7	...
AZ61A-F	260–275	38–40	145–165	21–24	7–9	...
AZ80A-F	290–295	42–43	185–195	27–28	4–9	...
AZ80A-T5	310–325	45–47	205–230	30–33	2–4	...
M1A-F	200–205	29–32	...	...	2–3	...
ZK40A-T5	275	40	255	37	4	...
ZK60A-F	295	43	215	31	4–5	...
ZK60A-T5	295–310	43–45	215–250	31–36	4–6	...
Forgings						
AZ31B-F	234	34	131	19	6	...
AZ61A-F	262	38	152	22	6	...
AZ80A-F	290	42	179	26	5	...
AZ80A-T5	290	42	193	28	2	...
HM21A-T5	228	33	172	25	3	...
ZK60A-T5	290	42	179	26	7	...
ZK60A-T6	296	43	221	32	4	...
Sheet and plate						
AZ31B-O	221	32	...	...	9–12	...
AZ31A-H26	241–269	35–39	145–186	21–27	6	...
HK31A-O	200–207	29–30	97–124	14–18	12	...
HK31A-H24	228–234	33–34	172–179	25–26	4	...
HM21A-T81	234	34	172	25	4	...
LA141A-T7	124–131	18–19	103	15	10	...
ZE10A-O	200–207	29–30	83–124	12–18	12–15	...
ZE10A-H24	214–248	31–36	138–172	20–25	6	...

Note: Property ranges are dependent on specified thicknesses or diameters of test specimens.

Table 9.10 Tensile properties of common titanium and titanium alloys

Designation	Tensile strength (min) MPa	ksi	0.2% yield strength, min MPa	ksi
Unalloyed grades				
ASTM grade 1	240	35	170	25
ASTM grade 2	340	50	280	40
ASTM grade 3	450	65	380	55
ASTM grade 4	550	80	480	70
ASTM grade 7	340	50	280	40
ASTM grade 11	240	35	170	25
Alpha and near-alpha alloys				
Ti-0.3Mo-0.8Ni	480	70	380	55

(continued)

Table 9.10 (continued)

Designation	Tensile strength (min) MPa	ksi	0.2% yield strength, min MPa	ksi
Alpha and near-alpha alloys (continued)				
Ti-5Al-2.5Sn	790	115	760	110
Ti-5Al-2.5Sn-ELI	690	100	620	90
Ti-8Al-1Mo-1V	900	130	830	120
Ti-6Al-2Sn-4Zr-2Mo	900	130	830	120
Ti-6Al-2Nb-1Ta-0.8Mo	790	115	690	100
Ti-2.25Al-11Sn-5Zr-1Mo	1000	145	900	130
Ti-5.8Al-4Sn-3.5Zr-0.7Nb-0.5Mo-0.35Si	1030	149	910	132
Alpha-beta alloys				
Ti-6Al-4V(a)	900	130	830	120
Ti-6Al-4V-ELI(a)	830	120	760	110
Ti-6Al-6V-2Sn(a)	1030	150	970	140
Ti-6Al-2Sn-4Zr-6Mo(b)	1170	170	1100	160
Ti-5Al-2Sn-2Zr-4Mo-4Cr(b)	1125	163	1055	153
Ti-6Al-2Sn-2Zr-2Mo-2Cr(a)	1030	150	970	140
Ti-3Al-2.5V	620	90	520	75
Ti-4Al-4Mo-2Sn-0.5Si	1100	160	960	139
Beta alloys				
Ti-10V-2Fe-3Al	1170	170	1100	160
Ti-3Al-8V-6Cr-4Mo-4Zr	900	130	830	120
Ti-15V-3Cr-3Al-3Sn	1000(b)	145(b)	965(b)	140(b)
	1241(c)	180(c)	1172(c)	170(c)
Ti-15Mo-3Al-2.7Nb-0.2Si	862	125	793	115

(a) Mechanical properties given in the annealed condition; may be solution treated and aged to increase strength. (b) Mechanical properties given in the solution treated and aged condition; alloy not normally applied in the annealed condition. (c) Also solution treated and aged using an alternative aging temperature (900 °C, or 1650 °F)

Table 9.11 Mechanical properties of selected nickel-base alloys
Properties are for annealed sheet unless otherwise indicated.

Alloy	Ultimate tensile strength MPa	ksi	Yield strength (0.2% offset) MPa	ksi	Elongation in 50 mm (2 in.), %	Elastic modulus (tension) GPa	10^6 psi	Hardness
Commercially pure and low-alloy nickels								
Nickel 200	462	67	148	21.5	47	204	29.6	109 HB
Nickel 201	403	58.5	103	15	50	207	30	129 HB
Nickel 205	345	50	90	13	45	...	...	...
Nickel 211	530	77	240	35	40	...	...	...
Nickel 212	483	70	...	...	...	...	...	...
Nickel 222	380	55	...	...	...	...	...	...
Nickel 270	345	50	110	16	50	...	...	30 HRB
Duranickel 301 (precipitation hardened)	1170	170	862	125	25	207	30	30–40 HRC
Nickel-copper alloys								
Alloy 400	550	80	240	35	40	180	26	110–150 HB
Alloy 401	440	64	134	19.5	51	...	...	...
Alloy R-405	550	80	240	35	40	180	26	110–140 HB

(continued)

Table 9.11 (continued)

Alloy	Ultimate tensile strength MPa	ksi	Yield strength (0.2% offset) MPa	ksi	Elongation in 50 mm (2 in.), %	Elastic modulus (tension) GPa	10^6 psi	Hardness
Nickel-copper alloys (continued)								
Alloy 450	385	56	165	24	46	...	...	...
Alloy K-500 (precipitation hardened)	1100	160	790	115	20	180	26	300 HB
Nickel-chromium and nickel-chromium-iron alloys								
Alloy 230(a)	860	125	390	57	47.7	211	30.6	92.5 HRB
Alloy 600	655	95	310	45	40	207	30	75 HRB
Alloy 601	620	90	275	40	45	207	30	65–80 HRB
Alloy 617 (solution annealed)	755	110	350	51	58	211	30.6	173 HB
Alloy 625	930	135	517	75	42.5	207	30	190 HB
Alloy 690	725	105	348	50.5	41	211	30.6	88 HRB
Alloy 718 (precipitation hardened)	1240	180	1036	150	12	211	30.6	36 HRC
Alloy X750 (precipitation hardened)	1137	165	690	100	20	207	30	330 HB
Alloy 751 (precipitation hardened)	1310	190	976	141.5	22.5	214	31	352 HB
Alloy MA 754	965	140	585	85	22	...	...	...
Alloy C-22	785	114	372	54	62	...	...	209 HB
Alloy C-276	790	115	355	52	61	205	29.8	90 HRB
Alloy G3	690	100	320	47	50	199	28.9	79 HRB
Alloy HX (solution annealed)	793	115	358	52	45.5	205	29.7	90 HRB
Alloy S (solution annealed)	835	121	445	64.5	49	212	30.8	52 HRA
Alloy W (solution annealed)	850	123	370	53.5	55	...	...	...
Alloy X (solution annealed)	785	114	360	52.5	43	196	28.5	89 HRB
Iron-nickel-chromium alloys								
Alloy 556	815	118.1	410	59.5	47.7	205	29.7	91 HB
Alloy 800	600	87	295	43	44	193	28	138 HB
Alloy 800HT	See Alloy 800							
Alloy 825	690	100	310	45	45	206	29.8	...
Alloy 925(b)	1210	176	815	118	24	...	...	36.5 HRC
Controlled-expansion alloys								
Alloy 902 (precipitation hardened)	1210	175	760	110	25	...	...	...
Alloy 903 (precipitation hardened)	1310	190	1100	160	14	...	...	...
Alloy 907	See Alloy 903							
Alloy 909 (precipitation hardened)	1275	185	1035	150	15	159	23	...

(a) Cold rolled and solution annealed at 1230 °C (2250 °F). Sheet thickness, 1.2 to 1.6 mm (0.048 to 0.063 in.). (b) Annealed at 980 °C (1800 °F) for 30 min, air cooled, and aged at 760 °C (1400 °F) for 8 h, furnace cooled at a rate of 55 °C (100 °F)/h, heated to 620 °C (1150 °F) for 8 h, air cooled

Table 9.12 Mechanical properties of selected cobalt-base alloys

Alloy	Tensile strength MPa	Tensile strength ksi	Yield strength MPa	Yield strength ksi	Elongation, %	Hardness, HRC
Stellite 1	618	89.6	...	...	<1	55
Stellite 6	896	130	541	78.5	1	40
Stellite 12	834	135.5	649	94.1	<1	48
Stellite 21	694	100	494	71.6	9	32
Alloy 6B	998	145	619	89.8	11	37
Tribaloy T-800	...	...	...	...	...	58
Alloy 25	970	141	445	64.5	62	...
Alloy 188	945	137	464	67.3	53	...
MAR-M 509	780	113	585	85	...	...

Table 9.13 Elastic constants for polycrystalline metals at 20 °C

Metal	Young's modulus (E), 10^6 psi	Bulk modulus (K), 10^6 psi	Shear modulus (G), 10^6 psi	Poisson's ratio, ν
Aluminum	10.2	10.9	3.80	0.345
Brass, 30 Zn	14.6	16.2	5.41	0.350
Chromium	40.5	23.2	16.7	0.210
Copper	18.8	20.0	7.01	0.343
Iron				
Soft	30.7	24.6	11.8	0.293
Cast	22.1	15.9	8.7	0.27
Lead	2.34	6.64	0.811	0.44
Magnesium	6.48	5.16	2.51	0.291
Molybdenum	47.1	37.9	18.2	0.293
Nickel	28.9	25.7	11.0	0.312
Soft	28.9	25.7	11.0	0.312
Hard	31.8	27.2	12.2	0.306
Nickel-silver, 55Cu-18Ni-27Zn	19.2	19.1	4.97	0.333
Niobium	15.2	24.7	5.44	0 397
Silver	12.0	15.0	4.39	0.367
Steel				
Mild	30.7	24.5	11.9	0.291
0.75 C	30.5	24.5	11.8	0.293
0.75 C, hardened	29.2	23.9	11.3	0.296
Tool	30.7	24.0	11.9	0.287
Tool, hardened	29.5	24.0	11.4	0.295
Stainless, 2Ni-18Cr	31.2	24.1	12.2	0.283
Tantalum	26.9	28.5	10.0	0.342
Tin	7.24	8.44	2.67	0.357
Titanium	17.4	15.7	6.61	0.361
Tungsten	59.6	45.1	23.3	0.280
Vanadium	18.5	22.9	6.77	0.365
Zinc	15.2	10.1	6.08	0.249

10 Mechanical Properties Charts for Steels

The charts and corresponding tables that are presented in this chapter are guides to the mechanical properties that may be expected from various low-alloy steels. The charts can be used to give the approximate tempering temperatures to produce a given hardness or tensile strength. The values in these charts and tables represent average analyses and are based on fine-grained steels according to the McQuaid-Ehn test (grain size 5 to 8).

After determining from a chart that a given tensile strength corresponds with a definite yield point, elongation, reduction of area, and Brinell hardness, it must not be assumed that these values can be taken as minima. The relation of tensile strength to the other mechanical properties given has been found to be relatively consistent provided the steel has been heat treated in such a way that all carbides are substantially in solution at the time of quenching, and that the steel is hardened to a martensitic, or fully quenched, structure. However, if these conditions are not met in heat treatment if will be found, in general, that the ductility will be somewhat lower than indicated by the charts. In addition, the steel will have less resistance to tempering at a given temperature than the tempering response charts show. These limitations must be kept in mind when using these data. They may be used with assurance if it is known that the hardenability of the steel selected and the heat-treating practice employed is such that the steel will harden to 90% or more martensite during quenching. Indicated on the composite charts (Fig. 10.1–10.3), for each grade of steel, is the maximum size round that can be expected to harden at its center to this extent when heat treated as recommended. This size is based on the average chemical analysis of the grade and may vary above or below this value depending on normal differences in chemical analysis from heat to heat. The tempering response for a given grade should be expected to vary about ⨦50 HB. Similar variations in other properties should also be expected.

Following the three composite charts are individual property charts representative of alloy grades containing 0.3%, 0.4%, and 0.5% C, respectively. Mass effect charts for 4340 steel tempered at two different temperatures are also included so that estimates of mechanical properties for sections larger than those covered by the composite and individual charts can be determined.

126 Concise Metals Engineering Data Book

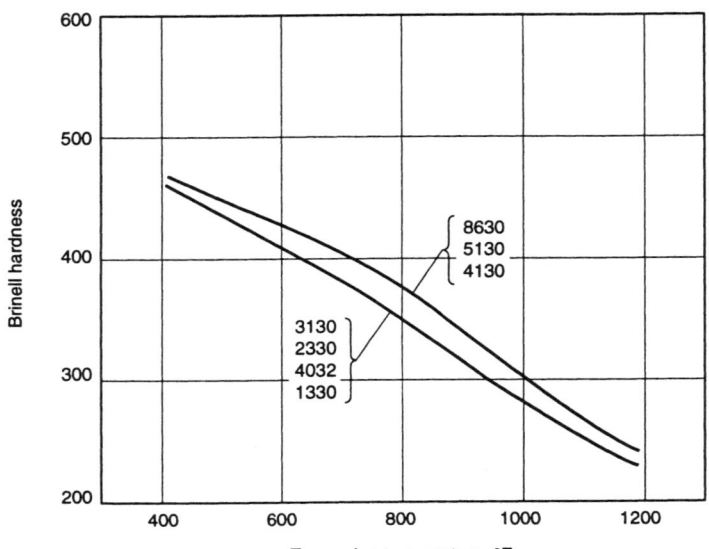

Grade	C	Mn	Si	Ni	Cr	Mo	Ac₁	Ac₂	Ar₁	Thermal treatment (forged)(a), °F	Annealed	Normalized	Maximum size(b), in.
1330	0.28–0.33	1.60–1.90	0.20–0.35	...	...	...	1325	1470	1170	2250	174	241	1.5
2330	0.28–0.33	0.60–0.80	0.20–0.35	3.25–3.75	...	...	1280	1375	1040	2200	183	241	1.4
3130	0.28–0.33	0.60–0.80	0.20–0.35	1.10–1.40	0.55–0.75	...	1345	1450	1230	2250	174	241	1.2
4032	0.30–0.35	0.70–0.90	0.20–0.35	...	...	0.20–0.30	1350	1500	1200	2250	174	228	1.0
4130	0.28–0.33	0.40–0.60	0.20–0.35	...	0.80–1.10	0.15–0.25	1395	1490	1280	2250	174	255	1.4
5130	0.28–0.33	0.70–0.90	0.20–0.35	...	0.80–1.10	...	1370	1490	1280	2250	170	241	1.25
8630	0.28–0.33	0.70–0.90	0.20–0.35	0.40–0.70	0.40–0.60	0.15–0.25	1355	1460	1220	2250	179	255	1.5

(a) After forging, all specimens were annealed at 1500 to 1700 °F and furnace cooled. (b) The maximum cross section in which the above tempering response can be expected when treated as recommended

Fig. 10.1 Brinell hardness versus tempering temperature for alloy steel grades containing nominally 0.3% C. Heat treatment for all grades: normalized at 1600 to 1650 °F; reheated to 1475 to 1550 °F; quenched in water; tempered 2 h

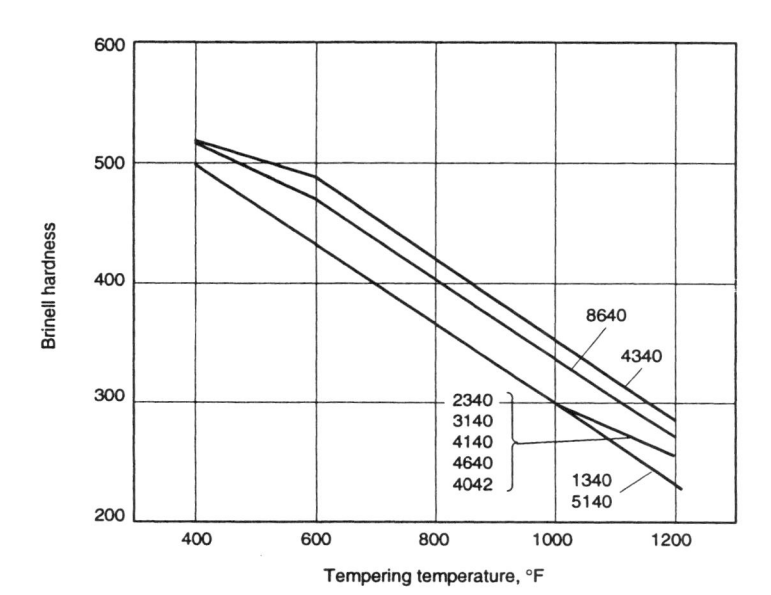

Mechanical Properties Charts for Steels 127

			Composition, wt%				Critical points, °F			Thermal treatment	Maximum Brinell hardness		Maximum
Grade	C	Mn	Si	Ni	Cr	Mo	Ac₁	Ac₂	Ar₁	(forged)(a), °F	Annealed	Normalized	size(b), in.
1340	0.38–0.43	1.60–1.90	0.20–0.35	...	...	...	1320	1430	1150	2250	183	269	1.25
2340	0.38–0.43	0.70–0.90	0.20–0.35	3.25–3.75	...	...	1275	1355	1050	2200	202	269	1.0
3140	0.38–0.43	0.70–0.90	0.20–0.35	1.10–1.40	0.35–0.75	...	1355	1410	1220	2250	187	302	1.1
4042	0.40–0.45	0.70–0.90	...	...	...	0.20–0.30	1340	1440	1210	2260	182	235	0.6
4140	0.38–0.43	0.75–1.00	0.20–0.35	...	0.80–1.10	0.15–0.25	1380	1450	1280	2250	197	300	2.0
4340	0.38–0.43	0.60–0.80	0.20–0.35	1.65–2.00	0.70–0.90	0.20–0.30	1335	1425	1210	2250	223	415	4.0
4640	0.38–0.43	0.60–0.80	0.20–0.35	1.65–2.00	...	0.20–0.30	1315	1430	1450	2250	197	285	1.1
5140	0.38–0.43	0.70–0.90	0.20–0.35	...	0.70–0.90	...	1360	1450	1280	2250	187	269	1.0
8640	0.38–0.43	0.75–1.00	0.20–0.35	0.40–0.70	0.40–0.60	0.15–0.25	1350	1435	1230	2250	197	302	1.1

(a) After forging, all specimens were annealed at 1500 to 1700 °F and furnace cooled. (b) The maximum cross section in which the above tempering response can be expected when treated as recommended

Fig. 10.2 Brinell hardness versus tempering temperature for alloy steel grades containing nominally 0.4% C. Heat treated for all grades: normalized at 1575 to 1600 °F; reheated to 1475 to 1550 °F; oil quenched; tempered 2 h

128 Concise Metals Engineering Data Book

[Graph: Brinell hardness vs Tempering temperature (°F), showing curves for grades 6150, 8650, 4150, 9850 (grouped) and 5150]

Grade	C	Mn	Si	Ni	Cr	Mo	V	Ac₁	Ac₃	Ar₁	Thermal treatment (forged)(a), °F	Annealed	Normalized	Maximum size(b), in.
4150	0.48–0.53	0.75–1.00	0.20–0.35	...	0.80–1.10	0.15–0.25	...	1370	1410	1280	2250	212	375	3.1
5150	0.48–0.53	0.70–0.90	0.20–0.35	...	0.70–0.90	...	...	1330	1420	1220	2250	202	293	1.2
6150	0.48–0.53	0.70–0.90	0.20–0.35	...	0.80–1.10	...	0.15 min	1380	1450	1280	2250	202	302	1.6
8650	0.48–0.53	0.75–1.00	0.20–0.35	0.40–0.70	0.40–0.60	0.15–0.25	...	1350	1420	1210	2250	212	355	2.0
9850	0.48–0.53	0.70–0.90	0.20–0.35	0.85–1.15	0.70–0.90	0.20–0.30	...	1330	1405	1210	2250	223	387	3.5

(a) After forging, all specimens were annealed at 1475 to 1725 °F and furnace cooled. (b) The maximum cross section in which the above tempering response can be expected when treated as recommended

Fig. 10.3 Brinell hardness versus tempering temperature for alloy steel grades containing nominally 0.5% C. Heat treatment for all grades: normalized at 1600 to 1700 °F; reheated to 1475 to 1550 °F; oil quenched; tempered 2 h

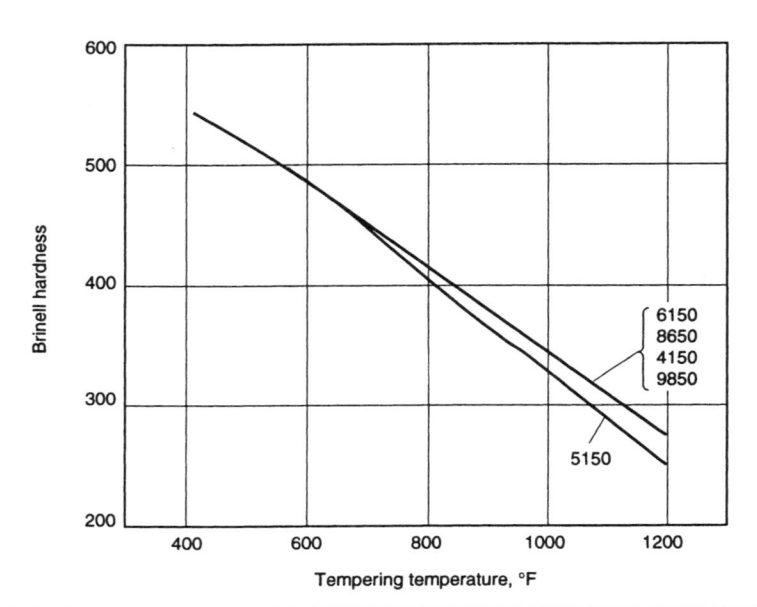

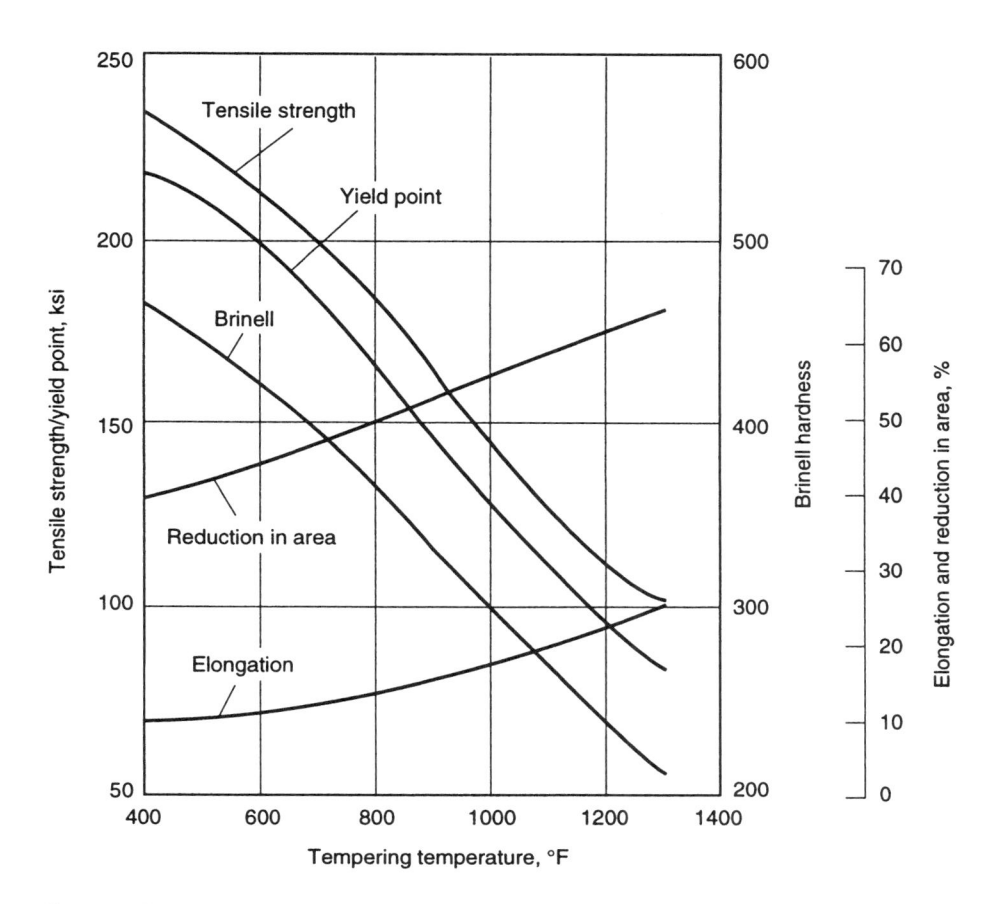

Fig. 10.4 Effect of tempering temperature on the properties of 4130 steel. Heat treatment: normalized at 1575 °F; reheated to 1550 °F; water quenched; tempered 400 to 1300 °F for 1 h. 1.0 in round heat treated; 0.505 in round tested. See also Fig. 10.1.

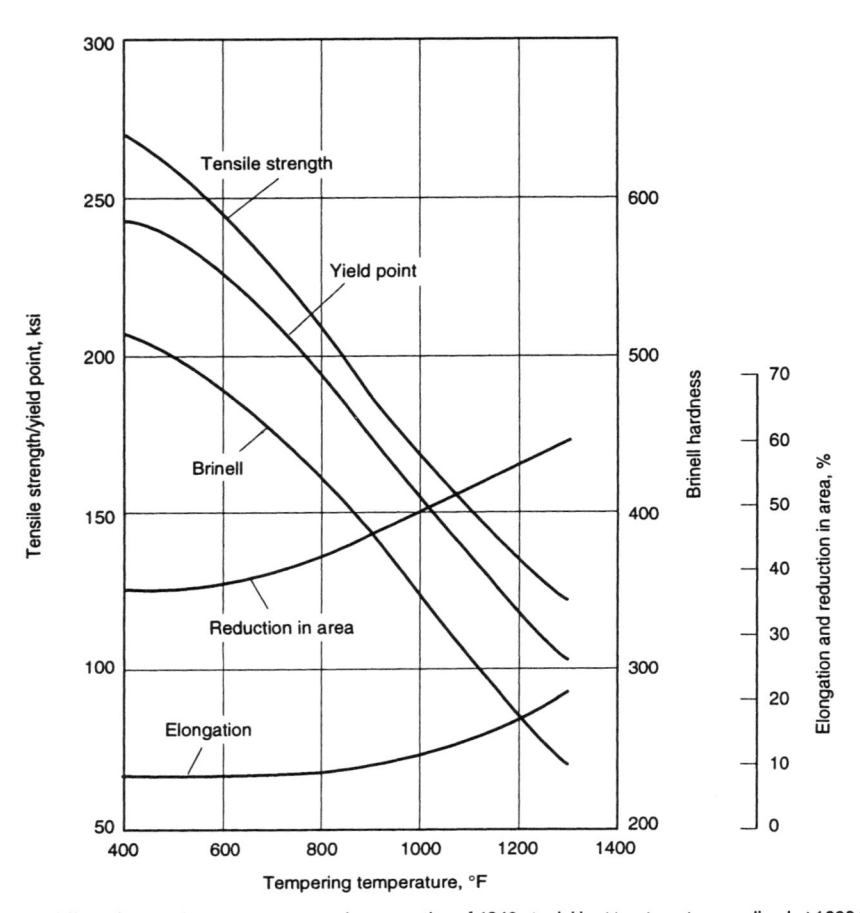

Fig. 10.5 Effect of tempering temperature on the properties of 4340 steel. Heat treatment: normalized at 1600 °F; reheated to 1525 °F; oil quenched; tempered 400 to 1300 °F for 1 h. 1.0 in round heat treated; 0.505 in round tested. See also Fig. 10.2.

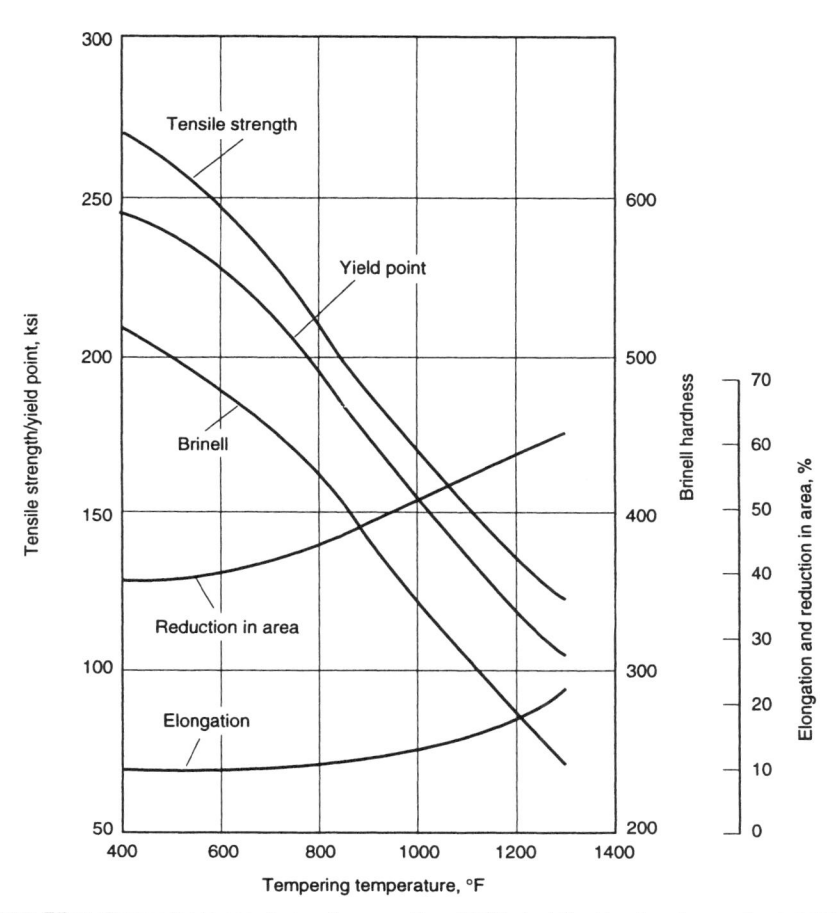

Fig. 10.6 Effect of tempering temperature on the properties of 8650 steel. Heat treatment: normalized at 1600 °F; reheated to 1500 °F; oil quenched; tempered 400 to 1300 °F for 1 h. 1.0 in round heat treated; 0.505 in round tested. See also Fig. 10.3.

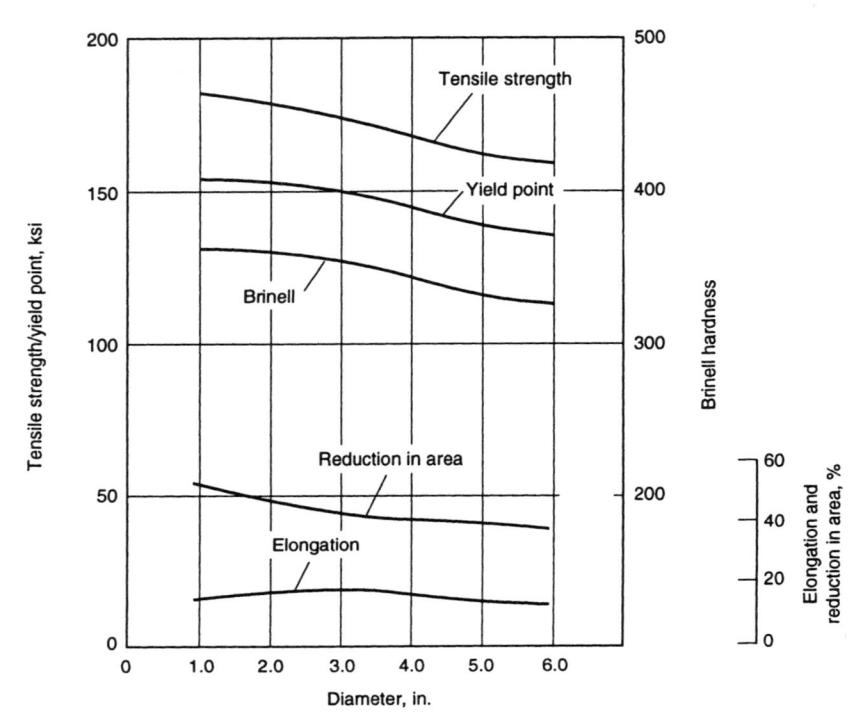

Fig. 10.7 Effect of mass on the properties of 4340 steel tempered at 1000 °F

Mechanical Properties Charts for Steels

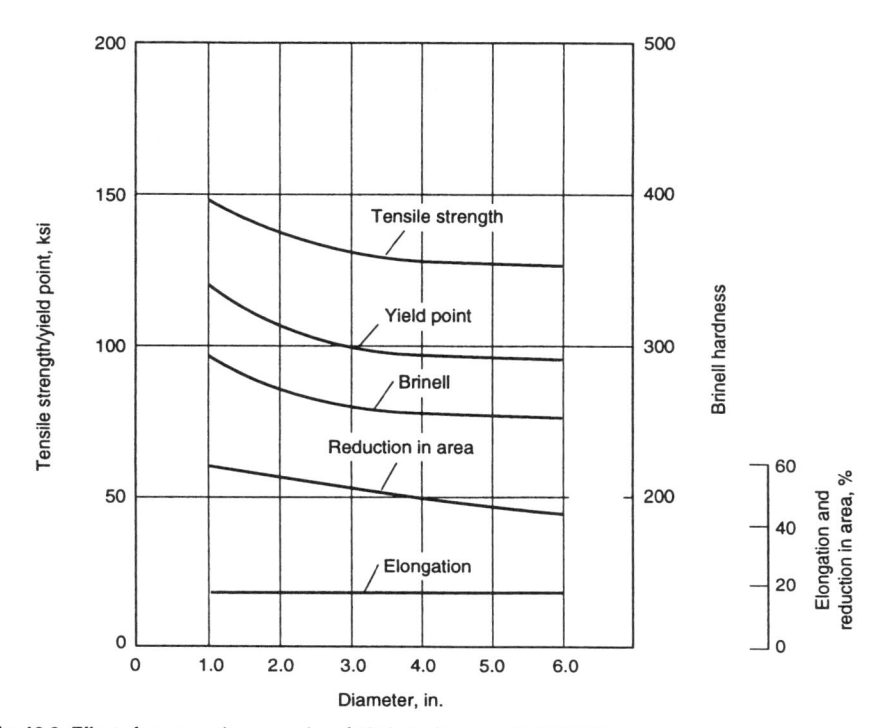

Fig. 10.8 Effect of mass on the properties of 4340 steel tempered at 1200 °F

11 Hardenability Data for Steels

Hardenability of steel is the property that determines the depth and distribution of harndess induced by quenching. Steels that exhibit deep hardness penetration are considered to have high hardenability, while those that exhibit shallow hardness penetration are of low hardenability. Because the primary objective in quenching is to obtain satisfactory hardening to some desired depth, it follows that hardenability is usually the most important single factor in the selection of steel for heat-treated parts.

Hardenability should not be confused with hardness as such or with maximum hardness. The maximum attainable hardness of any steel depends solely on carbon content. Also, the maximum hardness values that can be obtained with small test specimens under the fastest cooling rates of water quenching are nearly always higher than those developed under production heat-treating conditions, because hardenability limitations in quenching larger sizes may result in less than 100% martensite formation. Effects of carbon and martensite content on hardness are shown in Fig. 11.1.

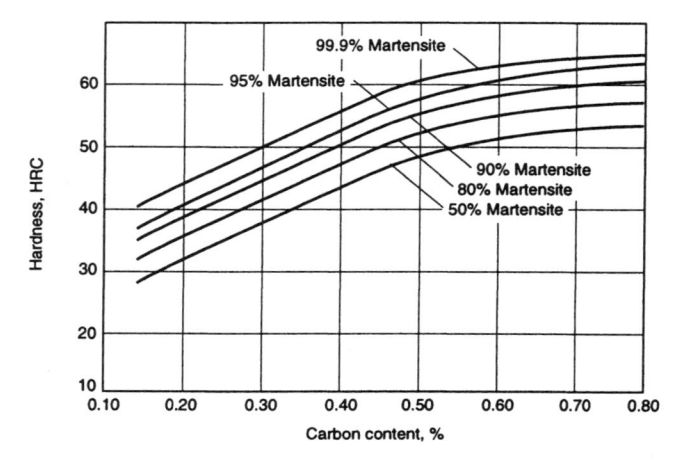

Fig. 11.1 Effect of carbon on the hardness of martensite structures

Hardenability is largely determined by the percentage of alloying elements in the steel. Austenitic grain size, time and temperature during austenitizing, and prior microstructure can also have significant effects.

Basically, the units of hardenability are those of cooling rate—for example, degrees per second. These cooling rates, as related to continuous-cooling-transformation behavior of the steel, determine the hardness and microstructural outcome of a quench. In practice, these cooling rates are often expressed as a distance, with other factors such as thermal conductivity of the steel and the rate of surface heat removal being held constant. Therefore, the terms Jominy distance, as described below, and ideal critical diameter (the diameter of a bar that can be quenched to 50% in the center when given a sufficiently severe quench that the heat-removal rate is controlled by the thermal diffusion of the metal and not by the surface heat-transfer rate) can be used. Both of these terms/concepts are described in greater detail in the article "Hardenability of Steels" in the *ASM Specialty Handbook: Carbon and Alloy Steels* published by ASM International.

End-Quench Hardenability Testing

The most commonly used method for determining the hardenability of steels is the end-quench (Jominy) test. In conducting this test, the test specimen, a 25 mm (1 in.) diam bar, 100 mm (4 in.) in length, is water quenched on one end face. The bar from which the specimen is made must be normalized before the test specimen is machined. The test involves heating the test specimen to

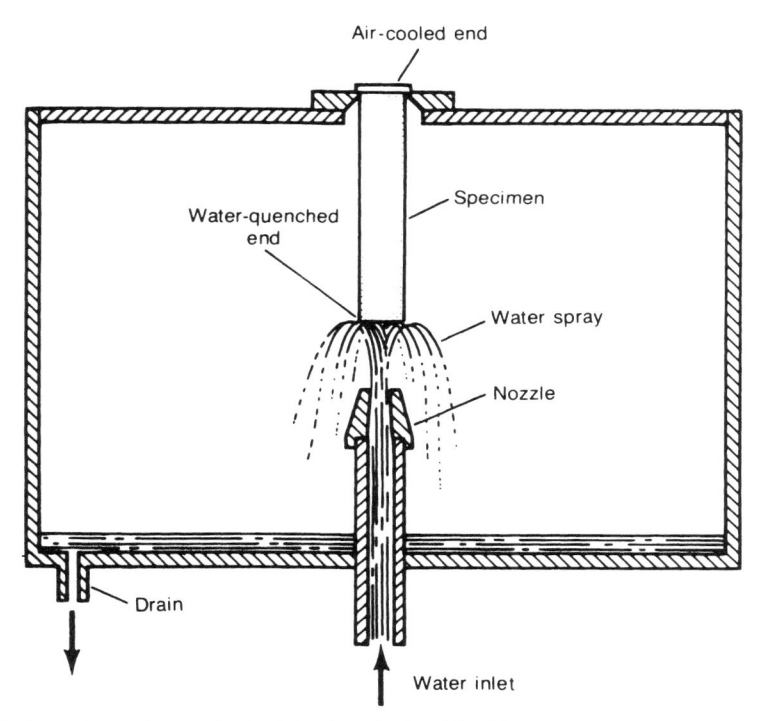

Fig. 11.2 Jominy end-quench apparatus for determining hardenability data

the proper austenitizing temperature and then transferring it to a quenching fixture so designed that the specimen is held vertically 13 mm (0.5 in.) above an opening through which a column of water can be directed against the bottom face of the specimen (Fig. 11.2). While the bottom end is being quenched by the column of water, the opposite end is cooling slowly in air, and intermediate positions along the specimen are cooling at intermediate rates.

After the specimen has been quenched, parallel flats 180° apart are ground 0.38 mm (0.015 in.) deep on the cylindrical surface. Rockwell C hardness is measured at intervals of $\frac{1}{16}$ in. (1.6 mm) for alloy steels and $\frac{1}{32}$ in. for carbon steels, starting from the water-quenched end. A typical plot of these hardness values and their positions on the test bar, as shown in Fig. 11.3, indicates the relation between hardness and cooling rate, which in effect is the hardenability of the steel. Figure 11.3 also shows the cooling rate for the designated test positions. Details of the standard test method are available in ASTM A 255 and SAE J406.

End-Quench Hardenability Curves. As shown in Fig. 11.3, a hardenability curve is plotted using Rockwell C readings as ordinates and distances from the quenched end as abscissas. Representative data have been accumulated for a variety of standard steel grades and are published by SAE International in SAE standards J1268 and J1868 as H-bands. These show graphically and in tabular form the high and low limits applicable to each grade. These H-bands/curves have also been reproduced in Volume 1 of the *ASM Handbook*. Steels specified to these limits are designated as H-grades. Chemical composition ranges for these steels can be found in Chapter 7, "Chemical Compositions of Metals and Alloys," in this Volume (see Tables 7.11 and 7.12).

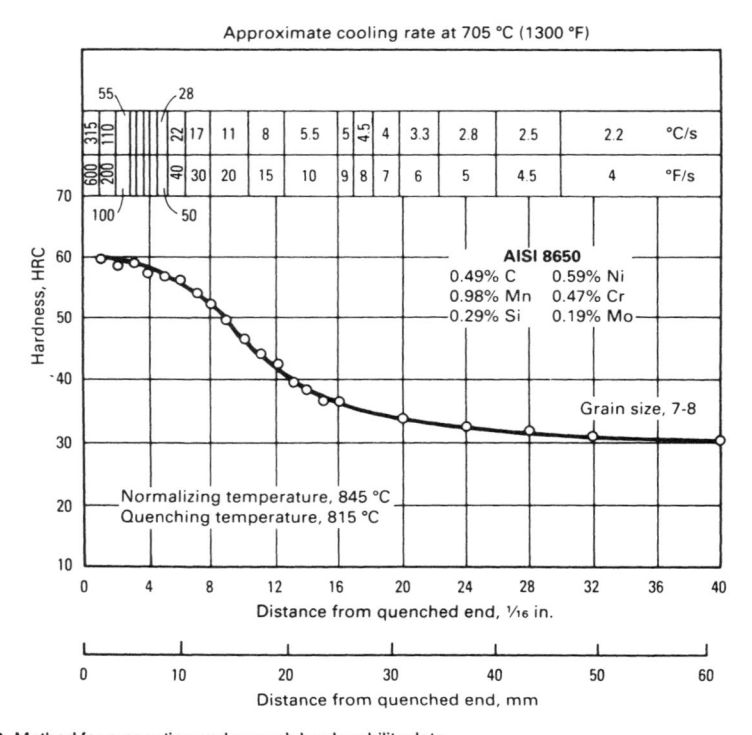

Fig. 11.3 Method for presenting end-quench hardenability data

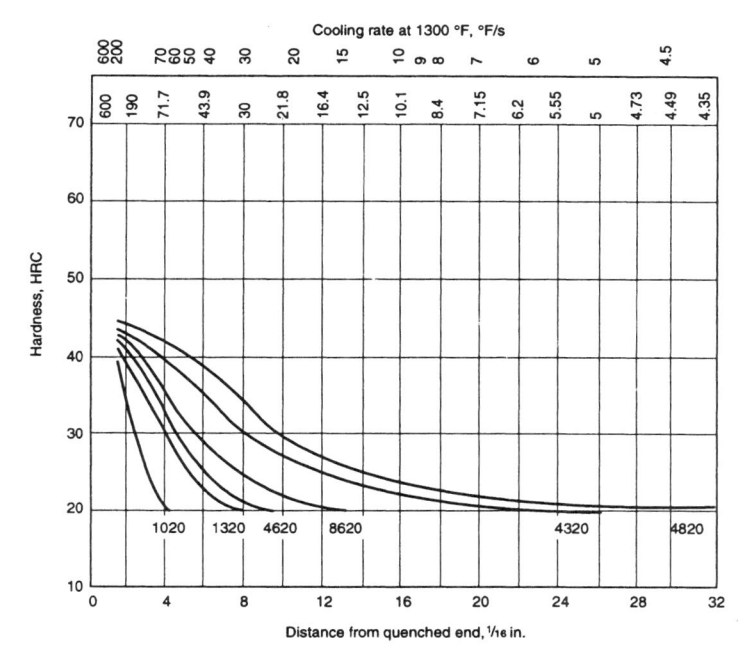

Fig. 11.4 Hardenability bands for carbon and alloy steels containing 0.20% C

The effect of common alloying elements and of their combinations upon the shape of these curves is illustrated in Fig. 11.4 and 11.5 by a series of curves characteristic of some common types of steel containing 0.20 and 0.40% C, respectively. The effect of variation in carbon content is shown in greater detail by reproducing a family of average curves for different carbon ranges of the 86xx SAE/AISI series of steel (Fig. 11.6). In general, the carbon content of steel largely determines the hardness at a distance of $\frac{1}{16}$ in. (1.6 mm) from the quenched end, while the hardness distribution further away from the end is governed by the nature and amount of alloying elements present. A shallow-hardening steel is characterized by a curve showing a rapid decrease in hardness at a short distance from the quenched end. A deep-hardening steel, on the other hand, will harden over the entire length of the end-quenched specimen, giving an almost horizontal curve.

Relation between End-Quench Hardenability and Cross-Section Hardness. The use of the end-quench method not only provides a simple and rapid means for determination of the hardenability of a steel, but also permits prediction of hardness of different shapes and sizes once the hardenability of the steel is known. This is based on the principle that different sections of the same steel have the same hardness if the sections are cooled at identical rates. In other words, the center hardness of a bar equals the hardness at that position on the end-quench specimen of the same steel that has been cooled at the same rate at the center of the bar.

To apply this principle to practice, the cooling rates for different positions within some standard shapes and sizes have been determined and published in the form of convenient diagrams. Two such diagrams for quenched round bars are shown in 11.7 and 11.8. Similar data for quenched plates and also for rounds and plates cooled in air are given in Fig. 11.9.

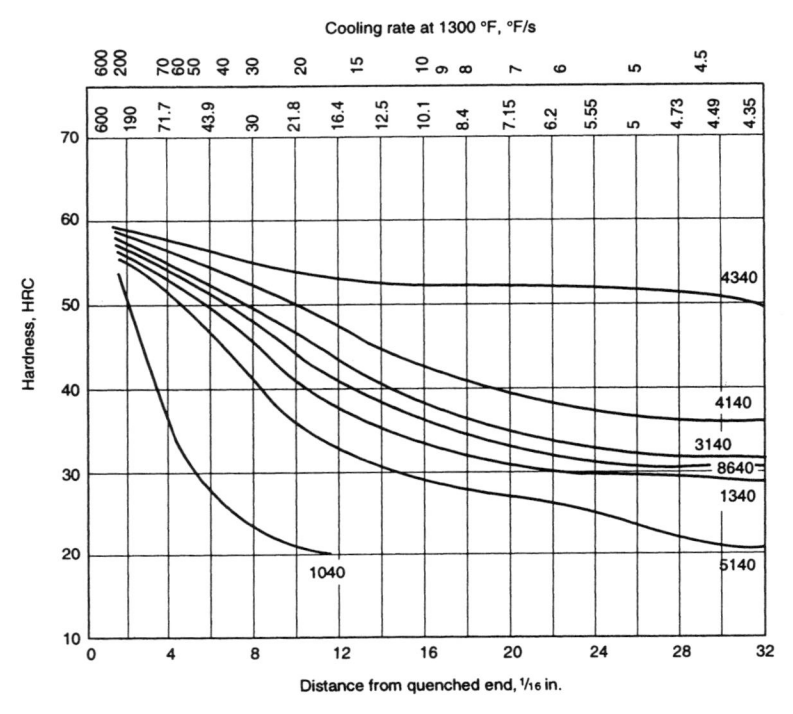

Fig. 11.5 Hardenability bands for carbon and alloy steels containing 0.40% C

Hardenability Data for Steels 139

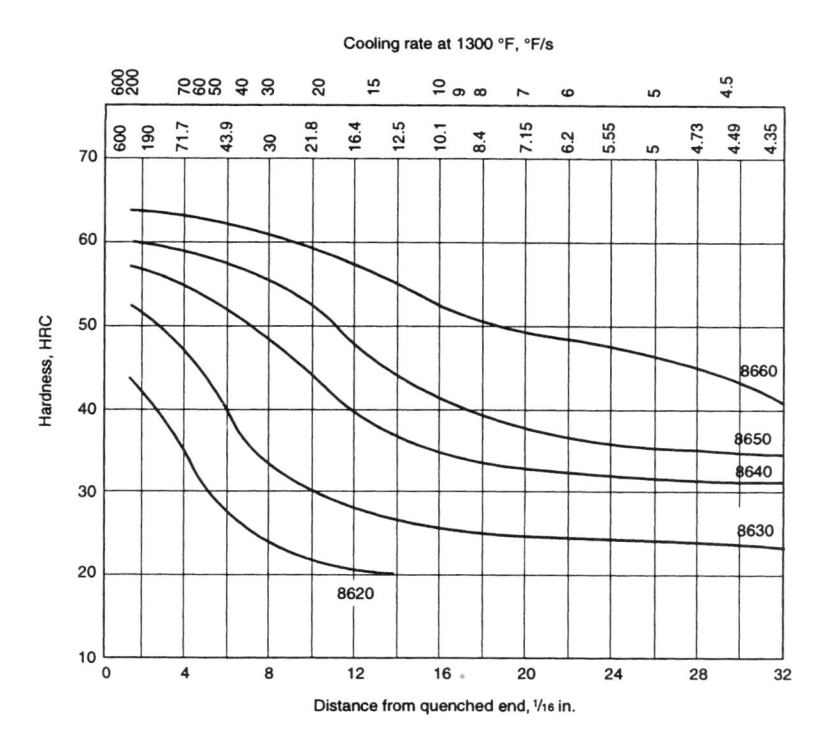

Fig. 11.6 Hardenability bands for 86xx series alloy steels containing 0.20 to 0.60% C

140 Concise Metals Engineering Data Book

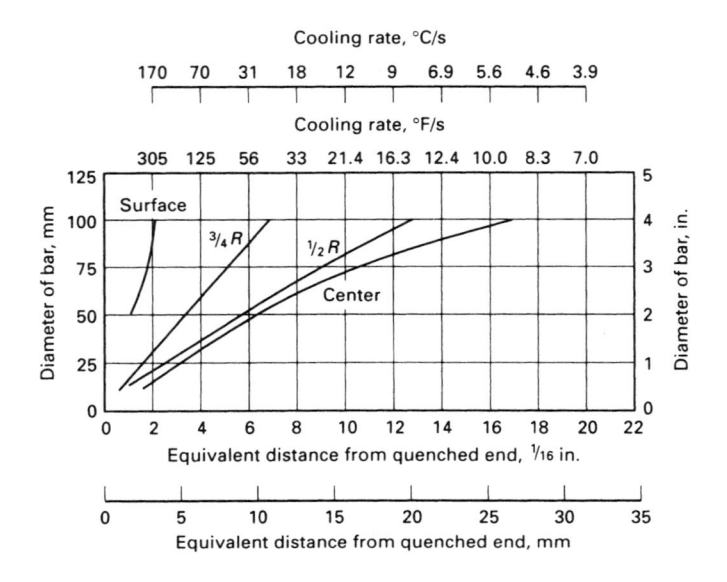

(a)

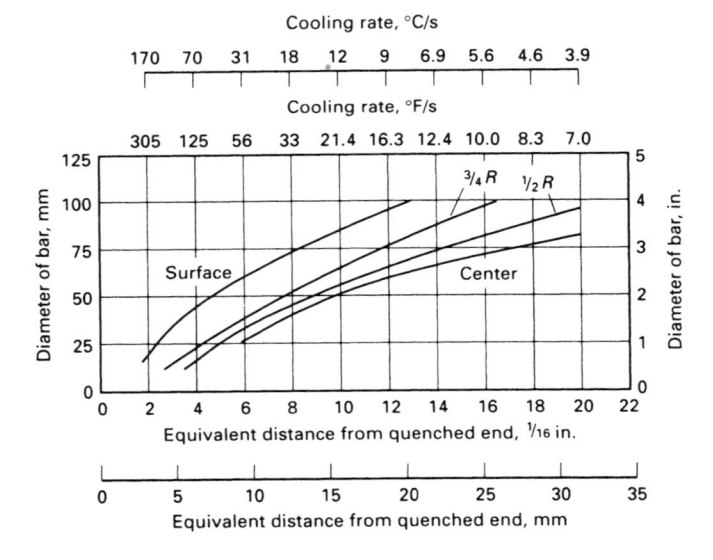

(b)

Fig. 11.7 Equivalent cooling rates at 705 °C (1300 °F) for round bars quenched in (a) mildly agitated water and (b) mildly agitated oil. These curves show the relationship between the distance on the Jominy bar and points at the surface, ¼ radius, ½ radius, and center of round bars up to 100 mm (4 in.) in diameter

Hardenability Data for Steels 141

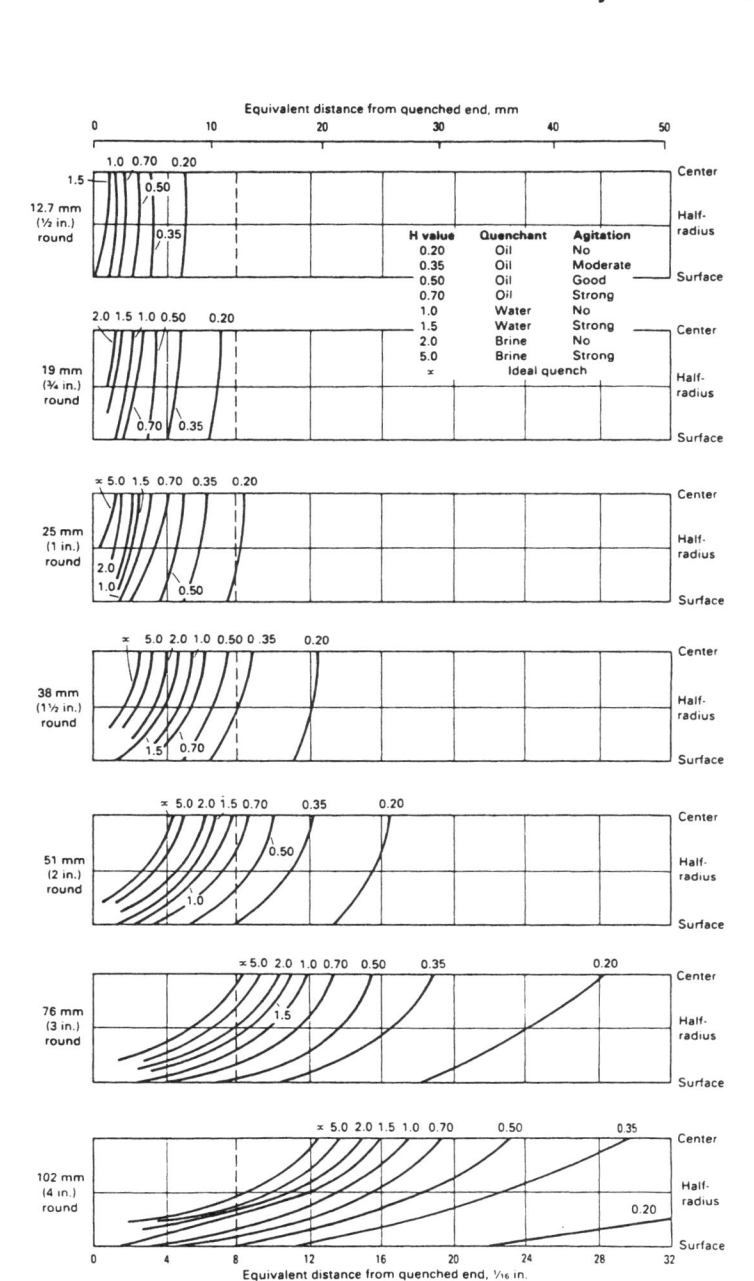

Fig. 11.8 Correlation of equivalent Jominy hardness positions in end-quenched hardenability specimen and various locations in round bars quenched in oil, water, and brine. The dashed line shows the various positions in $\frac{1}{2}$ to 4 in. diam rounds that are equivalent to the $\frac{1}{16}$ in. distance on the end-quenched bar. To determine cross-sectional hardnesses from results of end-quenched tests, pick out the end-quenched hardness at an appropriate point on the bottom line and extend the vertical line until it intersects the curved line that corresponds to the quenching severity needed to obtain that hardness for the given diameter of round.

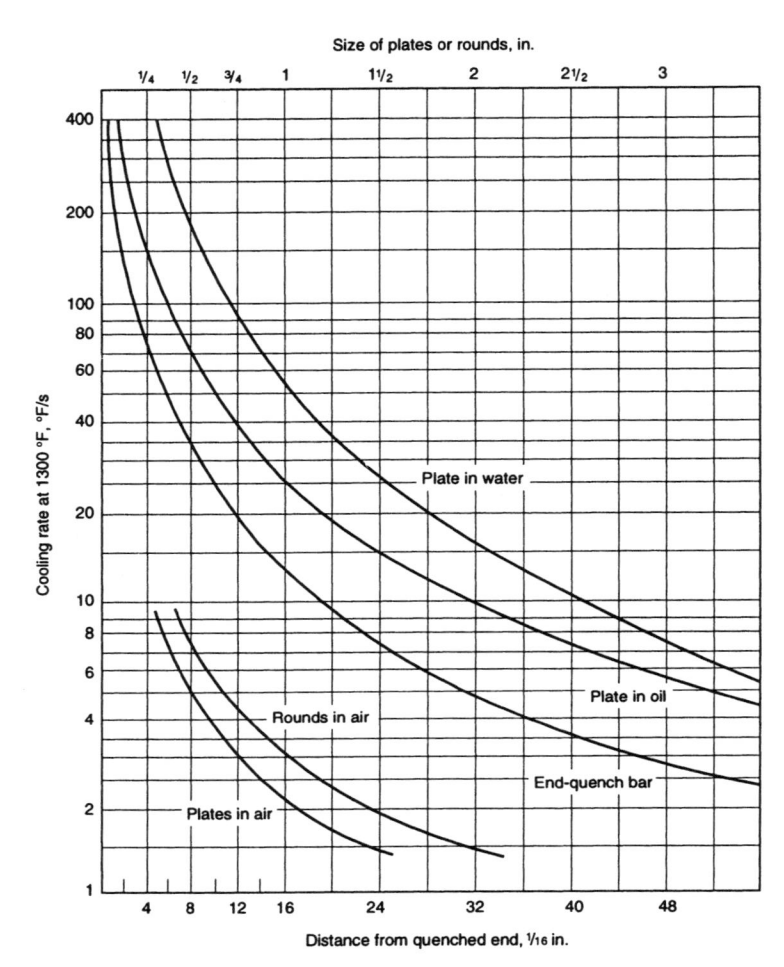

Fig. 11.9 Relation of end-quench hardenability to the center hardness of plates. To find the position on the end-quench curve corresponding to the center hardness, locate the size of the plate on the upper horizontal scale, and then draw a vertical line until the proper curve is reached. A horizontal line passing through this intersection represents the location of all the points cooled at the same rate, indicated by intersection of this horizontal line with the left-hand scale. For this reason, the intersection of this line with the end-quench curve indicates the distance (measured by bottom scale) on the end-quench specimen possessing the same hardness as the center of the plate.

12 Hardness Conversion Tables

From a practical standpoint, it is important to be able to convert the results of one type of hardness test into those of a different test. Because a hardness test does not measure a well-defined property of a material and because all the tests in common use are not based on the same type of measurements, it is not surprising that universal hardness-conversion relationships have not been developed.

Hardness conversions are empirical relationships. The oldest and most reliable hardness-conversion data exist for steels. Tables 12.1 and 12.2 on the following pages provide conversions among Rockwell, Brinell, Vickers, and Scleroscope hardness for steels. They are applicable to heat-treated carbon and alloy steels and to almost all alloy constructional steels and tool steels in the as-forged, annealed, normalized, and quenched-and-tempered conditions. However, different conversion tables are required for materials with greatly different elastic moduli, or with greater strain-hardening capacity. As a result, conversion tables have also been developed for austenitic (stainless) steels (Tables 12.3 to 12.5), abrasion-resistant high-alloy white irons (Table 12.6), wrought aluminum and aluminum alloy products (Table 12.7), wrought high-purity coppers (Table 12.8), cartridge brass (Table 12.9), and nickel and high-nickel alloys (Table 12.10). These tables are described in greater detail in ASTM E 140, "Standard Hardness Conversion Tables for Metals."

The indentation hardness for soft metals depends on the strain-hardening behavior of the material during the test, which in turn is dependent on the previous degree of strain hardening of the material before the test. The modulus of elasticity also has been shown to influence conversions at high hardness levels. At low hardness levels, conversions between hardness scales measuring depth and those measuring diameter are likewise influenced by differences in the modulus of elasticity.

Table 12.1 Approximate equivalent hardness numbers for nonaustenitic steels (Rockwell C hardness range)

For carbon and alloy steels in the annealed, normalized, and quenched-and-tempered conditions

Rockwell C hardness number 150 kgf, HRC	Vickers hardness number, HV	Brinell hardness No. 10 mm standard ball, 3000 kgf, HBS	Brinell hardness No. 10 mm carbide ball, 3000 kgf, HBW	Knoop hardness, number 500 gf and over, HK	Rockwell hardness No. A scale, 60 kgf, HRA	Rockwell hardness No. D scale, 100 kgf, HRD	Rockwell superficial hardness No. 15-N scale, 15 kgf, HR 15-N	Rockwell superficial hardness No. 30-N scale, 30 kgf, HR 30-N	Rockwell superficial hardness No. 45-N scale, 45 kgf, HR 45-N	Scleroscope hardness No.	Rockwell C hardness No. 150 kgf, HRC
68	940	...	...	920	85.6	76.9	93.2	84.4	75.4	97.3	68
67	900	...	...	895	85.0	76.1	92.9	83.6	74.2	95.0	67
66	865	...	...	870	84.5	75.4	92.5	82.8	73.3	92.7	66
65	832	...	(739)	846	83.9	74.5	92.2	81.9	72.0	90.6	65
64	800	...	(722)	822	83.4	73.8	91.8	81.1	71.0	88.5	64
63	772	...	(705)	799	82.8	73.0	91.4	80.1	69.9	86.5	63
62	746	...	(688)	776	82.3	72.2	91.1	79.3	68.8	84.5	62
61	720	...	(670)	754	81.8	71.5	90.7	78.4	67.7	82.6	61
60	697	...	(654)	732	81.2	70.7	90.2	77.5	66.6	80.8	60
59	674	...	(634)	710	80.7	69.9	89.8	76.6	65.5	79.0	59
58	653	...	615	690	80.1	69.2	89.3	75.7	64.3	77.3	58
57	633	...	595	670	79.6	68.5	88.9	74.8	63.2	75.6	57
56	613	...	577	650	79.0	67.7	88.3	73.9	62.0	74.0	56
55	595	...	560	630	78.5	66.9	87.9	73.0	60.9	72.4	55
54	577	...	543	612	78.0	66.1	87.4	72.0	59.8	70.9	54
53	560	...	525	594	77.4	65.4	86.9	71.2	58.6	69.4	53
52	544	(500)	512	576	76.8	64.6	86.4	70.2	57.4	67.9	52
51	528	(487)	496	558	76.3	63.8	85.9	69.4	56.1	66.5	51
50	513	(475)	481	542	75.9	63.1	85.5	68.5	55.0	65.1	50
49	498	(464)	469	526	75.2	62.1	85.0	67.6	53.8	63.7	49
48	484	451	455	510	74.7	61.4	84.5	66.7	52.5	62.4	48
47	471	442	443	495	74.1	60.8	83.9	65.8	51.4	61.1	47
46	458	432	432	480	73.6	60.0	83.5	64.8	50.3	59.8	46
45	446	421	421	466	73.1	59.2	83.0	64.0	49.0	58.5	45
44	434	409	409	452	72.5	58.5	82.5	63.1	47.8	57.3	44
43	423	400	400	438	72.0	57.7	82.0	62.2	46.7	56.1	43
42	412	390	390	426	71.5	56.9	81.5	61.3	45.5	54.9	42
41	402	381	381	414	70.9	56.2	80.9	60.4	44.3	53.7	41
40	392	371	371	402	70.4	55.4	80.4	59.5	43.1	52.6	40
39	382	362	362	391	69.9	54.6	79.9	58.6	41.9	51.5	39
38	372	353	353	380	69.4	53.8	79.4	57.7	40.8	50.4	38
37	363	344	344	370	68.9	53.1	78.8	56.8	39.6	49.3	37
36	354	336	336	360	68.4	52.3	78.3	55.9	38.4	48.2	36
35	345	327	327	351	67.9	51.5	77.7	55.0	37.2	47.1	35
34	336	319	319	342	67.4	50.8	77.2	54.2	36.1	46.1	34
33	327	311	311	334	66.8	50.0	76.6	53.3	34.9	45.1	33
32	318	301	301	326	66.3	49.2	76.1	52.1	33.7	44.1	32
31	310	294	294	318	65.8	48.4	75.6	51.3	32.5	43.1	31
30	302	286	286	311	65.3	47.7	75.0	50.4	31.3	42.2	30
29	294	279	279	304	64.8	47.0	74.5	49.5	30.1	41.3	29
28	286	271	271	297	64.3	46.1	73.9	48.6	28.9	40.4	28
27	279	264	264	290	63.8	45.2	73.3	47.7	27.8	39.5	27
26	272	258	258	284	63.3	44.6	72.8	46.8	26.7	38.7	26
25	266	253	253	278	62.8	43.8	72.2	45.9	25.5	37.8	25
24	260	247	247	272	62.4	43.1	71.6	45.0	24.3	37.0	24
23	254	243	243	266	62.0	42.1	71.0	44.0	23.1	36.3	23
22	248	237	237	261	61.5	41.6	70.5	43.2	22.0	35.5	22
21	243	231	231	256	61.0	40.9	69.9	42.3	20.7	34.8	21
20	238	226	226	251	60.5	40.1	69.4	41.5	19.6	34.2	20

Note: Values in parentheses are beyond the normal range and are presented for information only. Source: ASTM E 140

Table 12.2   Approximate equivalent hardness numbers for nonaustenitic steels (Rockwell B hardness range)

For carbon and alloy steels in the annealed, normalized, and quenched-and-tempered conditions

Rockwell B hardness No., 100 kgf, HRB	Vickers hardness No., HV	Brinell hardness No., 3000 kgf, HBS	Knoop hardness No., 500 gf, and over, HK	Rockwell A hardness No. 60 kgf, HRA	Rockwell F hardness No. 60 kgf, HRF	Rockwell superficial hardness No. 15-T scale, 15 kgf, HR 15-T	30-T scale, 30 kgf, HR 30-T	45-T scale, 45 kgf, HR 45-T	Rockwell B hardness No. 100 kgf, HRB
100	240	240	251	61.5	...	93.1	83.1	72.9	100
99	234	234	246	60.9	...	92.8	82.5	71.9	99
98	228	228	241	60.2	...	92.5	81.8	70.9	98
97	222	222	236	59.5	...	92.1	81.1	69.9	97
96	216	216	231	58.9	...	91.8	80.4	68.9	96
95	210	210	226	58.3	...	91.5	79.8	67.9	95
94	205	205	221	57.6	...	91.2	79.1	66.9	94
93	200	200	216	57.0	...	90.8	78.4	65.9	93
92	195	195	211	56.4	...	90.5	77.8	64.8	92
91	190	190	206	55.8	...	90.2	77.1	63.8	91
90	185	185	201	55.2	...	89.9	76.4	62.8	90
89	180	180	196	54.6	...	89.5	75.8	61.8	89
88	176	176	192	54.0	...	89.2	75.1	60.8	88
87	172	172	188	53.4	...	88.9	74.4	59.8	87
86	169	169	184	52.8	...	88.6	73.8	58.8	86
85	165	165	180	52.3	...	88.2	73.1	57.8	85
84	162	162	176	51.7	...	87.9	72.4	56.8	84
83	159	159	173	51.1	...	87.6	71.8	55.8	83
82	156	156	170	50.6	...	87.3	71.1	54.8	82
81	153	153	167	50.0	...	86.9	70.4	53.8	81
80	150	150	164	49.5	...	86.6	69.7	52.8	80
79	147	147	161	48.9	...	86.3	69.1	51.8	79
78	144	144	158	48.4	...	86.0	68.4	50.8	78
77	141	141	155	47.9	...	85.6	67.7	49.8	77
76	139	139	152	47.3	...	85.3	67.1	48.8	76
75	137	137	150	46.8	99.6	85.0	66.4	47.8	75
74	135	135	147	46.3	99.1	84.7	65.7	46.8	74
73	132	132	145	45.8	98.5	84.3	65.1	45.8	73
72	130	130	143	45.3	98.0	84.0	64.4	44.8	72
71	127	127	141	44.8	97.4	83.7	63.7	43.8	71
70	125	125	139	44.3	96.8	83.4	63.1	42.8	70
69	123	123	137	43.8	96.2	83.0	62.4	41.8	69
68	121	121	135	43.3	95.6	82.7	61.7	40.8	68
67	119	119	133	42.8	95.1	82.4	61.0	39.8	67
66	117	117	131	42.3	94.5	82.1	60.4	38.7	66
65	116	116	129	41.8	93.9	81.8	59.7	37.7	65
64	114	114	127	41.4	93.4	81.4	59.0	36.7	64
63	112	112	125	40.9	92.8	81.1	58.4	35.7	63
62	110	110	124	40.4	92.2	80.8	57.7	34.7	62
61	108	108	122	40.0	91.7	80.5	57.0	33.7	61
60	107	107	120	39.5	91.1	80.1	56.4	32.7	60
59	106	106	118	39.0	90.5	79.8	55.7	31.7	59
58	104	104	117	38.6	90.0	79.5	55.0	30.7	58
57	103	103	115	38.1	89.4	79.2	54.4	29.7	57
56	101	101	114	37.7	88.8	78.8	53.7	28.7	56
55	100	100	112	37.2	88.2	78.5	53.0	27.7	55
54	...	...	111	36.8	87.7	78.2	52.4	26.7	54
53	...	...	110	36.3	87.1	77.9	51.7	25.7	53
52	...	...	109	35.9	86.5	77.5	51.0	24.7	52
51	...	...	108	35.5	86.0	77.2	50.3	23.7	51
50	...	...	107	35.0	85.4	76.9	49.7	22.7	50

(continued)

Table 12.2 (continued)

Rockwell B hardness No., 100 kgf, HRB	Vickers hardness No., HV	Brinell hardness No., 3000 kgf, HBS	Knoop hardness No., 500 gf, and over, HK	Rockwell A hardness No. 60 kgf, HRA	Rockwell F hardness No. 60 kgf, HRF	Rockwell superficial hardness No. 15-T scale, 15 kgf, HR 15-T	Rockwell superficial hardness No. 30-T scale, 30 kgf, HR 30-T	Rockwell superficial hardness No. 45-T scale, 45 kgf, HR 45-T	Rockwell B hardness No., 100 kgf, HRB
49	...	...	106	34.6	84.8	76.6	49.0	21.7	49
48	...	...	105	34.1	84.3	76.2	48.3	20.7	48
47	...	...	104	33.7	83.7	75.9	47.7	19.7	47
46	...	...	103	33.3	83.1	75.6	47.0	18.7	46
45	...	...	102	32.9	82.6	75.3	46.3	17.7	45
44	...	...	101	32.4	82.0	74.9	45.7	16.7	44
43	...	...	100	32.0	81.4	74.6	45.0	15.7	43
42	...	...	99	31.6	80.8	74.3	44.3	14.7	42
41	...	...	98	31.2	80.3	74.0	43.7	13.6	41
40	...	...	97	30.7	79.7	73.6	43.0	12.6	40
39	...	...	96	30.3	79.1	73.3	42.3	11.6	39
38	...	...	95	29.9	78.6	73.0	41.6	10.6	38
37	...	...	94	29.5	78.0	72.7	41.0	9.6	37
36	...	...	93	29.1	77.4	72.3	40.3	8.6	36
35	...	...	92	28.7	76.9	72.0	39.6	7.6	35
34	...	...	91	28.2	76.3	71.7	39.0	6.6	34
33	...	...	90	27.8	75.7	71.4	38.3	5.6	33
32	...	...	89	27.4	75.2	71.0	37.6	4.6	32
31	...	...	88	27.0	74.6	70.7	37.0	3.6	31
30	...	...	87	26.6	74.0	70.4	36.3	2.6	30

Source: ASTM E 140

Table 12.3 Approximate equivalent hardness numbers for austenitic stainless steel sheet (Rockwell C hardness range)

For types 201, 202, 301, 302, 304, 304L, 305, 316, 316L, 321, and 347. Tempers range from annealed to extra hard for type 301, with a smaller range for the other types. Test coupon thickness: 0.1 to 0.050 in. (2.5 to 1.27 mm)

Rockwell hardness number		Rockwell superficial hardness No.		
C scale, 150 kgf diamond penetrator, HRC	A scale, 60 kgf, diamond penetrator, HRA	15-N scale, 15 kgf, superficial diamond penetrator, HR 15-N	30-N scale, 30 kgf, superficial diamond penetrator, HR 30-N	45-N scale, 45 kgf, superficial diamond penetrator, HR 45-N
48	74.4	84.1	66.2	52.1
47	73.9	83.6	65.3	50.9
46	73.4	83.1	64.5	49.8
45	72.9	82.6	63.6	48.7
44	72.4	82.1	62.7	47.5
43	71.9	81.6	61.8	46.4
42	71.4	81.0	61.0	45.2
41	70.9	80.5	60.1	44.1
40	70.4	80.0	59.2	43.0
39	69.9	79.5	58.4	41.8
38	69.3	79.0	57.5	40.7
37	68.8	78.5	56.6	39.6
36	68.3	78.0	55.7	38.4
35	67.8	77.5	54.9	37.3
34	67.3	77.0	54.0	36.1
33	66.8	76.5	53.1	35.0

(continued)

Table 12.3 (continued)

Rockwell hardness number		Rockwell superficial hardness No.		
C scale, 150 kgf diamond penetrator, HRC	A scale, 60 kgf, diamond penetrator, HRA	15-N scale, 15 kgf, superficial diamond penetrator, HR 15-N	30-N scale, 30 kgf, superficial diamond penetrator, HR 30-N	45-N scale, 45 kgf, superficial diamond penetrator, HR 45-N
32	66.3	75.9	52.3	33.9
31	65.8	75.4	51.4	32.7
30	65.3	74.9	50.5	31.6
29	64.8	74.4	49.6	30.4
28	64.3	73.9	48.8	29.3
27	63.8	73.4	47.9	28.2
26	63.3	72.9	47.0	27.0
25	62.8	72.4	46.2	25.9
24	62.3	71.9	45.3	24.8
23	61.8	71.3	44.4	23.6
22	61.3	70.8	43.5	22.5
21	60.8	70.3	42.7	21.3
20	60.3	69.8	41.8	20.2

Source: ASTM E 140

Table 12.4 Approximate equivalent hardness numbers for austenitic stainless steel sheet (Rockwell B hardness range)

For types 201, 202, 301, 302, 304, 304L, 305, 316, 316L, 321, and 347. Tempers range from annealed to extra hard for type 301, with a smaller range for the other types. Test coupon thickness: 0.1 to 0.050 in. (2.5 to 1.27 mm)

Rockwell hardness No.			Rockwell superficial hardness No.		
B scale, 100 kgf, 1/16 in. (1.588 mm) ball, HRB	A scale, 60 kgf, diamond penetrator, HRA	F scale, 60 kgf, 1/16 in. (1.588 mm) ball, HRF	15-T scale, 15 kgf, 1/16 in. (1.588 mm) ball, HR 15-T	30-T scale, 30 kgf, 1/16 in. (1.588 mm) ball, HR 30-T	45-T scale, 45 kgf, 1/16 in. (1.588 mm) ball, HR 45-T
100	61.5	(113.9)	91.5	80.4	70.2
99	60.9	(113.2)	91.2	79.7	69.2
98	60.3	(112.5)	90.8	79.0	68.2
97	59.7	(111.8)	90.4	78.3	67.2
96	59.1	(111.1)	90.1	77.7	66.1
95	58.5	(110.5)	89.7	77.0	65.1
94	58.0	(109.8)	89.3	76.3	64.1
93	57.4	(109.1)	88.9	75.6	63.1
92	56.8	(108.4)	88.6	74.9	62.1
91	56.2	(107.8)	88.2	74.2	61.1
90	55.6	(107.1)	87.8	73.5	60.1
89	55.0	(106.4)	87.5	72.8	59.0
88	54.5	(105.7)	87.1	72.1	58.0
87	53.9	(105.0)	86.7	71.4	57.0
86	53.3	(104.4)	86.4	70.7	56.0
85	52.7	(103.7)	86.0	70.0	55.0
84	52.1	(103.0)	85.6	69.3	54.0
83	51.5	(102.3)	85.2	68.6	52.9
82	50.9	(101.7)	84.9	67.9	51.9
81	50.4	(101.0)	84.5	67.2	50.9
80	49.8	(100.3)	84.1	66.5	49.9
79	49.2	99.6	83.8	65.8	48.9
78	48.6	99.0	83.4	65.1	47.9
77	48.0	98.3	83.0	64.4	46.8

(continued)

148 Concise Metals Engineering Data Book

Table 12.4 (continued)

Rockwell hardness No. B scale, 100 kgf, 1/16 in. (1.588 mm) ball, HRB	A scale, 60 kgf, diamond penetrator, HRA	F scale, 60 kgf, 1/16 in. (1.588 mm) ball, HRF	15-T scale, 15 kgf, 1/16 in. (1.588 mm) ball, HR 15-T	30-T scale, 30 kgf, 1/16 in. (1.588 mm) ball, HR 30-T	45-T scale, 45 kgf, 1/16 in. (1.588 mm) ball, HR 45-T
76	47.4	97.6	82.6	63.7	45.8
75	46.9	96.9	82.3	63.0	44.8
74	46.3	96.2	81.9	62.4	43.8
73	45.7	95.6	81.5	61.7	42.8
72	45.1	94.9	81.2	61.0	41.8
71	44.5	94.2	80.8	60.3	40.7
70	43.9	93.5	80.4	59.6	39.7
69	43.3	92.8	80.1	58.9	38.7
68	42.8	92.2	79.7	58.2	37.7
67	42.2	91.5	79.3	57.5	36.7
66	41.6	90.8	78.9	56.8	35.7
65	41.0	90.1	78.6	56.1	34.7
64	40.4	89.5	78.2	55.4	33.6
63	39.8	88.8	77.8	54.7	32.6
62	39.3	88.1	77.5	54.0	31.6
61	38.7	87.4	77.1	53.3	30.6
60	38.1	86.8	76.7	52.6	29.6

Note: Rockwell F numbers in parentheses are beyond the normal range and are presented for information only. Source: ASTM E 140

Table 12.5 Approximate Brinell-Rockwell B hardness numbers for equivalent austenitic stainless steel plate in the annealed condition

Rockwell hardness No. B scale (100 kgf, 1/16 in., 1.588 mm ball), HRB	Brinell hardness No. (3000 kgf, 10 mm steel ball), HBS	Rockwell hardness No. B scale (100 kgf, 1/16 in. 1.588 mm ball), HRB	Brinell hardness No. (3000 kgf, 10 mm steel ball), HBS
100	256	79	150
99	248	78	147
98	240	77	144
97	233	76	142
96	226	75	139
95	219	74	137
94	213	73	135
93	207	72	132
92	202	71	130
91	197	70	128
90	192	69	126
89	187	68	124
88	183	67	122
87	178	66	120
86	174	65	118
85	170	64	116
84	167	63	114
83	163	62	113
82	160	61	111
81	156	60	110
80	153		

Hardness Conversion Tables

Table 12.6 Approximate equivalent hardness numbers of alloyed white irons

Vickers hardness No., HV 50	Brinell hardness No.(a), HBW	Rockwell C hardness No., HRC	Vickers hardness No., HV 50	Brinell hardness No.(a), HBW	Rockwell C hardness No., HRC
1000	(903)	70	700	(639)	58
980	(886)	69	680	621	57
960	(868)	68	660	604	56
940	(850)	68	640	586	55
920	(833)	67	620	569	54
900	(815)	66	600	551	53
880	(798)	66	580	533	52
860	(780)	65	560	516	51
840	(762)	64	540	498	50
820	(745)	63	520	481	48
800	(727)	62	500	463	47
780	(710)	62	480	445	45
760	(692)	61	460	428	44
740	(674)	60	440	410	42
720	(657)	59	420	393	40

Note: Brinell hardness numbers in parentheses are beyond the normal range and are presented for information only. (a) 10 mm diam tungsten carbide ball; 3000 kgf load. Source: ASTM E 140

Table 12.7 Approximate equivalent hardness numbers for wrought aluminum products

Brinell hardness No. 500 kgf, 10 mm ball, HBS	Vickers hardness No. 15 kgf, HV	Rockwell hardness No. B scale 100 kgf, 1/16 in. ball, HRB	Rockwell hardness No. E scale 100 kgf, 1/8 in. ball, HRE	Rockwell hardness No. H scale 60 kgf, 1/8 in. ball, HRH	Rockwell superficial hardness No. 15-T scale 15 kgf, 1/16 in. ball, HR 15-T	Rockwell superficial hardness No. 30-T scale 30 kgf, 1/16 in. ball, HR 30-T	Rockwell superficial hardness No. 15-W scale 15 kgf, 1/8 in. ball, HR 15-W
160	189	91	...	...	89	77	95
155	183	90	...	...	89	76	95
150	177	89	...	...	89	75	94
145	171	87	...	...	88	74	94
140	165	86	...	...	88	73	94
135	159	84	...	...	87	71	93
130	153	81	...	...	87	70	93
125	147	79	...	...	86	68	92
120	141	76	101	...	86	67	92
115	135	72	100	...	86	65	91
110	129	69	99	...	85	63	91
105	123	65	98	...	84	61	91
100	117	60	...	...	83	59	90
95	111	56	96	...	82	57	90
90	105	51	94	108	81	54	89
85	98	46	91	107	80	52	89
80	92	40	88	106	78	50	88
75	86	34	84	104	76	47	87
70	80	28	80	102	74	44	86
65	74	...	75	100	72	...	85
60	68	...	70	97	70	...	83
55	62	...	65	94	67	...	82
50	56	...	59	91	64	...	80
45	50	...	53	87	62	...	79
40	44	...	46	83	59	...	77

Source: ASTM E 140

Table 12.8 Approximate equivalent hardness numbers for wrought coppers (>99% Cu, alloys C10200 through C14200)

Vickers hardness No. 1 kgf, HV	Knoop hardness No. 500 gf, HK	Rockwell superficial hardness No. 15-T scale, 15 kgf, 1/16 in. (1.588 mm) ball, HR 15-T 0.010 in. (0.25 mm) strip	Rockwell superficial hardness No. 15-T scale, 15 kgf, 1/16 in. (1.588 mm) ball, HR 15-T 0.020 in. (0.51 mm) strip	Rockwell superficial hardness No. 30-T scale, 30 kgf, 1/16 in. (1.588 mm) ball, HR 30-T 0.020 in. (0.51 mm) strip	Rockwell hardness No. B scale, 100 kgf, 1/16 in. (1.588 mm) ball, HRB	Rockwell hardness No. F scale, 60 kgf, 1/16 in. (1.588 mm) ball, HRF	Rockwell superficial hardness No. 15-T scale, 15 kgf, 1/16 in. (1.588 mm) ball, HR 15-T	Rockwell superficial hardness No. 30-T scale, 30 kgf, 1/16 in. (1.588 mm) ball, HR 30-T	Rockwell superficial hardness No. 45-T scale, 45 kgf, 1/16 in. (1.588 mm) ball, HR 45-T	Brinell hardness No. 500 kgf, 10 mm diameter ball, HBS 0.080 in. (2.03 mm) strip	Brinell hardness No. 20 kgf, 2 mm diameter ball, HBS 0.040 in. (1.02 mm) strip
								0.040 in. (1.02 mm) strip and greater			
130	133.8	...	85.0	...	67.0	99.0	...	69.5	49.0	...	119.0
128	132.1	83.0	84.5	...	66.0	98.0	87.0	68.5	48.0	...	117.5
126	130.4	...	84.0	...	65.0	97.0	...	67.5	46.5	120.0	115.0
124	128.7	82.5	83.5	...	64.0	96.0	86.0	66.5	45.0	117.5	113.0
122	127.0	...	83.0	...	62.5	95.5	85.5	66.0	44.0	115.0	111.0
120	125.2	82.0	82.5	...	61.0	95.0	...	65.0	42.5	112.0	109.0
118	123.5	81.5	...	...	59.5	94.0	85.0	64.0	41.0	110.0	107.5
116	121.7	...	82.0	...	58.5	93.0	...	63.0	40.0	107.5	105.5
114	119.9	81.0	81.5	...	57.0	92.5	84.5	62.0	38.5	105.0	103.5
112	118.1	80.5	81.0	...	55.0	91.5	...	61.0	37.0	102.0	102.0
110	116.3	80.0	...	...	53.5	91.0	84.0	60.0	36.0	99.5	100.0
108	114.5	...	80.5	...	52.0	90.5	83.5	59.0	34.5	97.0	98.0
106	112.6	79.5	80.0	...	50.0	89.5	...	58.0	33.0	94.5	96.0
104	110.1	79.0	79.5	...	48.0	88.5	83.0	57.0	32.0	92.0	94.0
102	108.0	78.5	79.0	...	46.5	87.5	82.5	56.0	30.0	89.5	92.0
100	106.0	78.0	78.0	...	44.5	87.0	82.0	55.0	28.5	87.0	90.0
98	104.0	77.5	77.5	...	42.0	85.5	81.0	53.5	26.5	84.5	88.0
96	102.1	77.0	77.0	...	40.0	84.5	80.5	52.0	25.5	82.0	86.5
94	100.0	76.5	76.5	...	38.0	83.0	80.0	51.0	23.0	79.5	85.0
92	98.0	76.0	75.5	...	35.5	82.0	79.0	49.0	21.0	77.0	83.0
90	96.0	75.5	75.0	...	33.0	81.0	78.0	47.5	19.0	74.5	81.0
88	94.0	75.0	74.5	...	30.5	79.5	77.0	46.0	16.5	...	79.0
86	92.0	74.5	73.5	...	28.0	78.0	76.0	44.0	14.0	...	77.0
84	90.0	74.0	73.0	...	25.5	76.5	75.0	43.0	12.0	...	75.0
82	87.9	73.5	72.0	...	23.0	74.5	74.5	41.0	9.5	...	73.0
80	86.0	72.5	71.0	...	20.0	73.0	73.5	39.5	7.0	...	71.5
78	84.0	72.0	70.0	...	17.0	71.0	72.5	37.5	5.0	...	69.5
76	81.9	71.5	69.5	...	14.5	69.0	71.5	36.0	2.0	...	67.5
74	79.9	71.0	68.5	...	11.5	67.5	70.0	34.0	...	...	66.0

(continued)

Table 12.8 (continued)

Vickers hardness No. 1 kgf, HV	100 gf, HV	Knoop hardness No. 1 kgf, HK	500 gf, HK	Rockwell superficial hardness No. 15-T scale, 15 kgf, 1/16 in. (1.588 mm) ball, HR 15-T 0.010 in. (0.25 mm) strip	15-T scale, 15 kgf, 1/16 in. (1.588 mm) ball, HR 15-T 0.020 in. (0.51 mm) strip	30-T scale, 30 kgf, 1/16 in. (1.588 mm) ball, HR 30-T	Rockwell hardness No. B scale, 100 kgf, 1/16 in. (1.588 mm) ball, HRB	F scale, 60 kgf, 1/16 in. (1.588 mm) ball, HRF 0.040 in. (1.02 mm) strip and greater	Rockwell superficial hardness No. 15-T scale, 15 kgf, 1/16 in. (1.588 mm) ball, HR 15-T	30-T scale, 30 kgf, 1/16 in. (1.588 mm) ball, HR 30-T	45-T scale, 45 kgf, 1/16 in. (1.588 mm) ball, HR 45-T	Brinell hardness No. 500 kgf, 10 mm diameter ball, HBS 0.080 in. (2.03 mm) strip	20 kgf, 2 mm diameter ball, HBS 0.040 in. (1.02 mm) strip
72	77.6	78.9	78.7	70.0	67.5	...	8.5	66.0	69.0	32.0	...	...	64.0
70	75.8	76.8	76.6	69.5	66.5	...	5.0	64.0	67.5	30.0	...	...	62.0
68	74.3	74.1	74.4	69.0	65.5	...	2.0	62.0	66.0	28.0	...	...	60.5
66	72.6	71.9	71.9	68.0	64.5	...	...	60.0	64.5	25.5	...	...	58.5
64	70.9	69.5	70.0	67.5	63.5	...	...	58.0	63.5	23.5	...	...	57.0
62	69.1	67.0	67.9	66.5	62.0	...	...	56.0	61.0	21.0	...	...	55.0
60	67.5	64.6	65.9	66.0	61.0	...	...	54.0	59.0	18.0	...	...	53.0
58	65.8	62.0	63.8	65.0	60.0	...	...	51.5	57.0	15.5	...	...	51.5
56	64.0	59.8	61.8	64.5	58.5	...	...	49.0	55.0	13.0	...	...	49.5
54	62.3	57.4	59.5	63.5	57.5	...	...	47.0	53.0	10.0	...	...	48.0
52	60.7	55.0	57.2	63.0	56.0	...	...	44.0	51.5	7.5	...	...	46.5
50	58.9	52.8	55.0	62.0	55.0	...	...	41.5	49.5	4.5	...	...	44.5
48	57.3	50.3	52.7	61.0	53.5	...	...	39.0	47.5	1.5	...	...	42.0
46	55.8	48.0	50.2	60.5	52.0	...	...	36.0	45.0	...	...	...	41.0
44	53.9	45.9	47.8	59.5	51.0	...	...	33.5	43.0	...	...	...	...
42	52.2	43.7	45.2	58.5	49.5	...	...	30.5	41.0	...	...	...	...
40	51.3	40.2	42.8	57.5	48.0	...	...	28.0	38.5	...	...	...	...

Source: ASTM E 140

Table 12.9 Approximate equivalent hardness numbers for cartridge brass (70% Cu, 30% Zn)

Vickers hardness No., HV	Rockwell hardness No. B scale, 100 kgf, 1/16 in. (1.588 mm) ball, HRB	Rockwell hardness No. F scale, 60 kgf 1/16 in. (1.588 mm) ball, HRF	Rockwell superficial hardness No. 15-T scale, 15 kgf, 1/16 in. (1.588 mm) ball, HR 15-T	Rockwell superficial hardness No. 30-T scale, 30 kgf, 1/16 in. (1.588 mm) ball, HR 30-T	Rockwell superficial hardness No. 45-T scale, 45 kgf, 1/16 in. (1.588 mm) ball, HR 45-T	Brinell hardness No. 500 kgf, 10 mm ball, HBS
196	93.5	110.0	90.0	77.5	66.0	169
194	...	109.5	...	...	65.5	167
192	93.0	...	...	77.0	65.0	166
190	92.5	109.0	...	76.5	64.5	164
188	92.0	...	89.5	...	64.0	162
186	91.5	108.5	...	76.0	63.5	161
184	91.0	...	...	75.5	63.0	159
182	90.5	108.0	89.0	...	62.5	157
180	90.0	107.5	...	75.0	62.0	156
178	89.0	...	...	74.5	61.5	154
176	88.5	107.0	...	...	61.0	152
174	88.0	...	88.5	74.0	60.5	150
172	87.5	106.5	...	73.5	60.0	149
170	87.0	...	...	...	59.5	147
168	86.0	106.0	88.0	73.0	59.0	146
166	85.5	...	...	72.5	58.5	144
164	85.0	105.5	...	72.0	58.0	142
162	84.0	105.0	87.5	...	57.5	141
160	83.5	...	...	71.5	56.5	139
158	83.0	104.5	...	71.0	56.0	138
156	82.0	104.0	87.0	70.5	55.5	136
154	81.5	103.5	...	70.0	54.5	135
152	80.5	103.0	...	...	54.0	133
150	80.0	...	86.5	69.5	53.5	131
148	79.0	102.5	...	69.0	53.0	129
146	78.0	102.0	...	68.5	52.5	128
144	77.5	101.5	86.0	68.0	51.5	126
142	77.0	101.0	...	67.5	51.0	124
140	76.0	100.5	85.5	67.0	50.0	122
138	75.0	100.0	...	66.5	49.0	121
136	74.5	99.5	85.0	66.0	48.0	120
134	73.5	99.0	...	65.5	47.5	118
132	73.0	98.5	84.5	65.0	46.5	116
130	72.0	98.0	84.0	64.5	45.5	114
128	71.0	97.5	...	63.5	45.0	113
126	70.0	97.0	83.5	63.0	44.0	112
124	69.0	96.5	...	62.5	43.0	110
122	68.0	96.0	83.0	62.0	42.0	108
120	67.0	95.5	...	61.0	41.0	106
118	66.0	95.0	82.5	60.5	40.0	105
116	65.0	94.5	82.0	60.0	39.0	103
114	64.0	94.0	81.5	59.5	38.0	101
112	63.0	93.0	81.0	58.5	37.0	99
110	62.0	92.6	80.5	58.0	35.5	97
108	61.0	92.0	...	57.0	34.5	95
106	59.5	91.2	80.0	56.0	33.0	94
104	58.0	90.5	79.5	55.0	32.0	92
102	57.0	89.8	79.0	54.5	30.5	90
100	56.0	89.0	78.5	53.5	29.5	88
98	54.0	88.0	78.0	52.5	28.0	86
96	53.0	87.2	77.5	51.5	26.5	85

(continued)

Table 12.9 (continued)

Vickers hardness No., HV	Rockwell hardness No.		Rockwell superficial hardness No.			Brinell hardness No. 500 kgf, 10 mm ball, HBS
	B scale, 100 kgf, 1/16 in. (1.588 mm) ball, HRB	F scale, 60 kgf 1/16 in. (1.588 mm) ball, HRF	15-T scale, 15 kgf, 1/16 in. (1.588 mm) ball, HR 15-T	30-T scale, 30 kgf, 1/16 in. (1.588 mm) ball, HR 30-T	45-T scale, 45 kgf, 1/16 in. (1.588 mm) ball, HR 45-T	
94	51.0	86.3	77.0	50.5	24.5	83
92	49.5	85.4	76.5	49.0	23.0	82
90	47.5	84.4	75.5	48.0	21.0	80
88	46.0	83.5	75.0	47.0	19.0	79
86	44.0	82.3	74.5	45.5	17.0	77
84	42.0	81.2	73.5	44.0	14.5	76
82	40.0	80.0	73.0	43.0	12.5	74
80	37.5	78.6	72.0	41.0	10.0	72
78	35.0	77.4	71.5	39.5	7.5	70
76	32.5	76.0	70.5	38.0	4.5	68
74	30.0	74.8	70.0	36.0	1.0	66
72	27.5	73.2	69.0	34.0	...	64
70	24.5	71.8	68.0	32.0	...	63
68	21.5	70.0	67.0	30.0	...	62
66	18.5	68.5	66.0	28.0	...	61
64	15.5	66.8	65.0	25.5	...	59
62	12.5	65.0	63.5	23.0	...	57
60	10.0	62.5	62.5	...	...	55
58	...	61.0	61.0	18.0	...	53
56	...	58.8	60.0	15.0	...	52
54	...	56.5	58.5	12.0	...	50
52	...	53.5	57.0	...	...	48
50	...	50.5	55.5	...	...	47
49	...	49.0	54.5	...	...	46
48	...	47.0	53.5	...	...	45
47	...	45.0	...	...	...	44
46	...	43.0	...	...	...	43
45	...	40.0	...	...	...	42

Source: ASTM E 140

154 Concise Metals Engineering Data Book

Table 12.10 Approximate equivalent hardness numbers for nickel and high-nickel alloys

Vickers hardness No. Vickers Indenter 1,5, 10, 30 kgf, HV	Knoop hardness No. Knoop Indenter 500 and 1000 gf, HK	Brinell hardness No. 10 mm standard ball, 3000 kgf, HBS	A scale: 60 kgf diamond penetrator, HRA	B scale: 100 kgf (⅟₁₆ in. mm) ball, HRB	C scale: 150 kgf diamond penetrator, HRC	D scale: 100 kgf diamond penetrator, HRD	E scale: 100 kgf ⅛ in. (3.175 mm) ball, HRE	F scale: 60 kgf ⅟₁₆ in. (1.588 mm) ball, HRF	G scale: 150 kgf ⅟₁₆ in. (1.588 mm) ball, HRG	K scale: 150 kgf ⅛ in. (3.175 mm) ball, HRK
513	...	(479)	75.5	...	50.0	63.0	...	...	...	...
481	...	450	74.5	...	48.0	61.5	...	...	...	...
452	...	425	73.5	...	46.0	60.0	...	...	...	...
427	...	403	72.5	...	44.0	58.5	...	...	...	...
404	...	382	71.5	...	42.0	57.0	...	...	...	...
382	436	363	70.5	...	40.0	55.5	...	...	...	...
362	413	346	69.5	...	38.0	54.0	...	...	...	...
344	392	329	68.5	...	36.0	52.5	...	...	...	...
326	372	313	67.5	...	34.0	50.5	...	...	...	...
309	352	298	66.5	...	32.0	49.5	...	...	94.0	...
285	325	275	64.5	(106)	28.5	46.5	...	(116.5)	91.0	...
266	304	258	63.0	(104)	25.5	44.5	...	(115.5)	87.5	...
248	283	241	61.5	(102)	22.5	42.0	...	(114.5)	84.5	...
234	267	228	60.5	100	20.0	40.0	...	(113.0)	81.5	...
220	251	215	59.0	98	(17.0)	38.0	(108.5)	(112.0)	78.5	100.0
209	239	204	57.5	96	(14.5)	36.0	(107.0)	(111.0)	75.5	98.0
198	226	194	56.5	94	(12.0)	34.0	(106.0)	(110.0)	72.0	96.5
188	215	184	55.0	92	(9.0)	32.0	(104.5)	(108.5)	69.0	94.5
179	204	176	53.5	90	(6.5)	30.0	(103.0)	(107.5)	65.5	93.0
171	195	168	52.5	88	(4.0)	28.0	(102.0)	(106.5)	62.5	91.0
164	187	161	51.5	86	(2.0)	26.5	(100.5)	(105.0)	59.5	89.0
157	179	155	50.0	84	...	24.5	...	(104.0)	56.5	87.5
151	173	149	49.0	82	...	22.5	...	(103.0)	53.0	85.5
145	166	144	47.5	80	...	21.0	99.5	(101.5)	50.0	83.5
140	160	139	46.5	78	...	(19.0)	98.5	(100.5)	47.0	82.0
135	154	134	45.5	76	...	(17.5)	98.0	99.5	43.5	80.0
130	149	129	44.0	74	...	(16.0)	97.0	98.5	40.5	78.0
126	144	125	43.0	72	...	(14.5)	95.5	97.0	37.5	76.5
122	140	121	42.0	70	...	(13.0)	94.5	96.0	34.5	74.5
119	136	118	41.0	68	...	(11.5)	93.0	95.0	31.0	72.5
115	...	114	40.0	66	...	(10.0)	91.5	93.5	...	71.0
								92.5		

(continued)

Table 12.10 (continued)

Vickers hardness No. Vickers indenter 1.5, 10, 30 kgf, HV	Knoop hardness No. Knoop indenter 500 and 1000 gf, HK	Brinell hardness No. 10 mm standard ball, 3000 kgf, HBS	Rockwell hardness No.						
			A scale: 60 kgf diamond penetrator, HRA	B scale: 100 kgf 1/16 in. mm) ball, HRB	C scale: 150 kgf diamond penetrator, HRC	D scale: 100 kgf diamond penetrator, HRD	E scale: 100 kgf (3.175 mm) ball, HRE	F scale: 60 kgf (1.588 mm) ball, HRF	G scale: 150 kgf 1/16 in. (1.588 mm) ball, HRG	K scale: 150 kgf 1/8 in. (3.175 mm) ball, HRK
112	...	111	39.0	62	...	(8.0)	90.5	91.5	...	69.0
108	...	108	...	60	...	...	89.0	90.0	...	67.5
106	...	106	...	58	...	...	88.0	89.0	...	65.5
103	...	103	...	56	...	...	86.5	88.0	...	63.5
100	...	100	...	54	...	...	85.5	87.0	...	62.0
98	...	98	...	52	...	...	84.0	85.5	...	60.0
95	...	95	...	50	...	...	83.0	84.5	...	58.0
93	...	93	...	48	...	...	81.5	83.5	...	56.5
91	...	91	...	46	...	...	80.5	82.0	...	54.5
89	...	89	...	44	...	...	79.0	81.0	...	52.5
87	...	87	...	42	...	...	78.0	80.0	...	51.0
85	...	85	...	40	...	...	76.5	79.0	...	49.0
83	...	83	...	38	...	...	75.0	77.5	...	47.0
81	...	81	...	36	...	...	74.0	76.5	...	45.5
79	...	79	...	34	...	...	72.5	75.5	...	43.5
78	...	78	...	32	...	...	71.5	74.0	...	42.0
77	...	77	...	30	...	...	70.0	73.0	...	40.0

Note: Values in parentheses are beyond the normal range and are presented for information only. Source: ASTM E 140

13 Corrosion Data

Table 13.1 Relationships among some of the units commonly used for corrosion rates
d is metal density in grams per cubic centimeter (g/cm³)

Unit	mdd	g/m²/d	μm/yr	mm/yr	mils/yr	in./yr
Milligrams per square decimeter per day (mdd)	1	0.1	36.5/*d*	0.0365/*d*	1.144/*d*	0.00144/*d*
Grams per square meter per day (g/m²/d)	10	1	365/*d*	0.365/*d*	14.4/*d*	0.0144/*d*
Microns per year (μm/yr)	0.0274*d*	0.00274*d*	1	0.001	0.0394	0.0000394
Millimeters per year (mm/yr)	27.4*d*	2.74*d*	1000	1	39.4	0.0394
Mils per year (mils/yr)	0.696*d*	0.0696*d*	25.4	0.0254	1	0.001
Inches per year (in./yr)	696*d*	69.6*d*	25,400	25.4	1000	1

Source: G. Wranglén, *An Introduction to Corrosion and Protection of Metals*, Chapman and Hall, 1985, p 238

Table 13.2 Corrosion rate calculation (from mass loss)

Corrosion rate $= \dfrac{(K \times W)}{(A \times T \times D)}$

where
  K = a constant (see below)
  T = time of exposure in hours to the nearest 0.01 h
  A = area in cm² to the nearest 0.01 cm²
  W = mass loss in g, to nearest 1 mg (corrected for any loss during cleaning)
  D = density in g/cm³

Many different units are used to express corrosion rates. Using the above units for T, A, W, and D, the corrosion rate can be calculated in a variety of units with the following appropriate value of K.

Corrosion rate units desired	Constant (K) in corrosion rate equation
Mils per year (mils/yr)	3.45×10^6
Inches per year (in./yr)	3.45×10^3
Inches per month (ipm)	2.87×10^2
Millimeters per year (mm/yr)	8.76×10^4
Micrometers per year (μm/yr)	8.76×10^7
Picometers per second (pm/s)	2.78×10^6
Grams per square meter per hour (g/m²h)	$1.00 \times 10^4 \times D$(a)
Milligrams per square decimeter per day (mdd)	$2.40 \times 10^6 \times D$(a)
Micrograms per square meter per second (μg/m²/s)	$2.78 \times 10^6 \times D$(a)

(a) Density is not needed to calculate the corrosion rate in these units. The density in the constant K cancels out the density in the corrosion rate equation. Source: *Corrosion Tests and Standards: Application and Interpretation*, ASTM, 1995, p 22

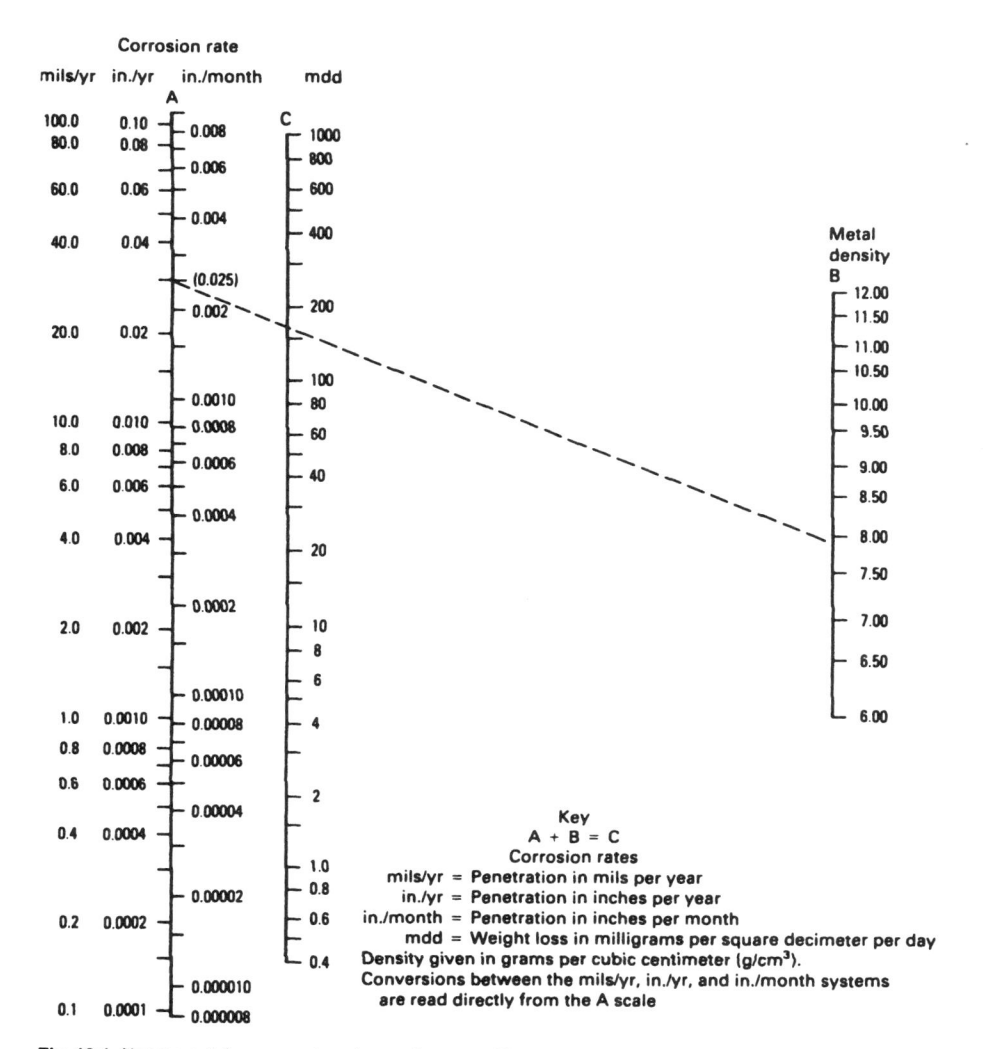

Fig. 13.1 Nomograph for conversion of corrosion rates. The example given is for type 304 stainless steel (density 7.87 g/cm3 and a corrosion rate of 30 mils/yr. Source: M.G. Fontana, Corrosion Engineering, 3rd ed., McGraw-Hill, 1986, p 217

Table 13.3 Reference potentials and conversion factors

Electrode	Potential (V) at 25 °C $E'(a)$	$E''(b)$	Thermal temperature coefficient(c), mV/°C
(Pt)/H$_2$ (α = 1)H$^+$ (α=1)(SHE)	0.000	...	+0.87
Ag/AgCl/1 M KCl	+0.235	...	+0.25
Ag/AgCl/0.6 M Cl$^-$ (seawater)	+0.25	...	...
Ag/AgCl/0.1 M Cl$^-$	+0.288	...	+0.22
Hg/Hg$_2$Cl$_2$/sat. KCl (SCE)	+0.241	+0.244	+0.22
Hg/Hg$_2$Cl$_2$/1 M KCl	+0.280	+0.283	+0.59
Hg/Hg$_2$Cl$_2$/0.1 M KCl	+0.334	+0.336	+0.79
Cu/CuSO$_4$ sat.	+0.30	...	+0.90
Hg/HgSO$_4$/H$_2$SO$_4$	+0.616	...	...

(a) E' is the standard potential for the half cell corrected for the concentration of the ions. (b) E'' also includes the liquid junction potentials for a saturated KCl salt bridge. To convert from one scale to another, add the value indicated. SHE = standard hydrogen electrode; SCE = saturated calomel electrode. (c) To convert from thermal to isothermal temperature coefficients, subtract 0.87 mV/°C. Thus the isothermal temperature coefficient for Ag-AgCl is –0.62 mV/°C.

From (E')	To SHE scale	To SCE scale (E')
H$_2$/H$^+$	...	–0.241
Ag/AgCl/1 M KCl	+0.235	–0.006
Ag/AgCl/0.6 M Cl (seawater)	+0.25	+0.009
Ag/AgCl/0.1 M Cl	+0.288	+0.047
Hg/Hg$_2$Cl$_2$/sat. KCl (SCE)	+0.241	...
Hg/Hg$_2$Cl$_2$, 1 M	+0.280	+0.039
Hg/Hg$_2$Cl$_2$, 0.1 M	+0.334	+0.093
Cu/CuSO$_4$ sat.	+0.30	+0.06
Hg/HgSO$_4$/H$_2$SO$_4$	+0.616	...

Example: An electrode potential of +1.000 V versus SCE would be (1.000 + 0.241) ± +1.241 V versus SHE. An electrode potential of –1.000 V versus SCE would give (–1.000 + 0.241) = –0.759 V versus SHE. Source: *Corrosion Tests and Standards: Application and Interpretation*, ASTM, 1995, p 23

Table 13.4 Electromotive force (emf) series

Electrode reaction	Standard potential at 25 °C (77 °F), volts versus SHE	Electrode reaction	Standard potential at 25 °C (77 °F), volts versus SHE
Au^{3+} + 3e^- → Au	1.50	Fe^{2+} + 2e^- → Fe	–0.440
Pd^{2+} + 2e^- → Pd	0.987	Ga^{3+} + 3e^- → Ga	–0.53
Hg^{2+} + 2e^- → Hg	0.854	Cr^{3+} + 3e^- → Cr	–0.74
Ag$^+$ + e^- → Ag	0.800	Zn^{2+} + 2e^- → Zn	–0.763
Hg$_2^{2+}$ + 2e^- → 2Hg	0.789	Mn^2 + 2e^- → Mn	–1.18
Cu$^+$ + e^- → Cu	0.521	Zr^{4+} _ 4e^- → Zr	–1.53
Cu^{2+} + 2e^- → Cu	0.337	Ti^{2+} + 2e^- → Ti	–1.63
2H$^+$ + 2e^- → H$_2$	(Reference) 0.000	Al^{3+} + 3e^- → Al	–1.66
		Hf^{4+} + 4e^- → Hf	–1.70
Pb^{2+} + 2e^- → Pb	–0.126	U^{3+} + 3e^- → U	–1.80
Sn$_2$ + 2e^- → Sn	–0.136	Be^{2+} + 2e^- → Be	–1.85
Ni^{2+} + 2e^- → Ni	–0.250	Mg^{2+} + 2e^- → Mg	–2.37
Co^{2+} + 2e^- → Ni	–0.277	Na$^+$ + e^- → Na	–2.71
Tl$^+$ + e^- → Tl	–0.336	Ca^{2+} + 2e^- → Ca	–2.87
In^{3+} + 3e^- → In	–0.342	K$^+$ + e^- → K	–2.93
Cd^{2+} + 2e^- → Cd	–0.403	Li$^+$ + e^- → Li	–3.05

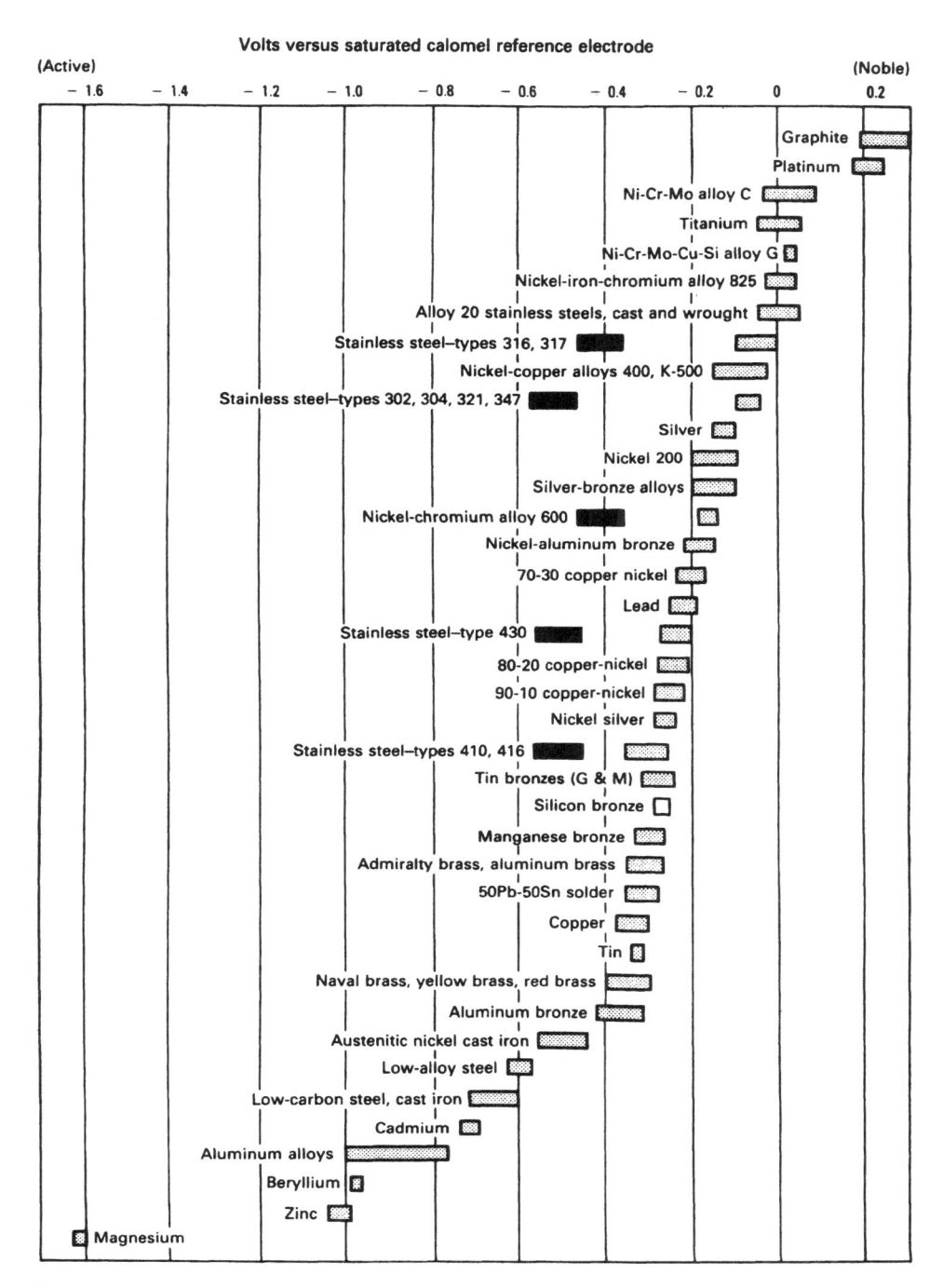

Fig. 13.2 Graphical version of the galvanic series for seawater. The material with the most negative, or anodic, corrosion potential has a tendency to suffer accelerated corrosion when electrically connected to a material with a more positive, or noble, potential. Dark boxes indicate active behavior of active-passive alloys

Table 13.5 Tabular version (no specific potential values given) of the galvanic series in seawater at 25 °C (77 °F)

With certain exceptions, this series is broadly applicable in natural waters and in uncontaminated atmospheres.

Corroded end (anodic, or least noble)
Magnesium
Magnesium alloys
Zinc
Galvanized steel or galvanized wrought iron
Aluminum alloys
  5052, 3004, 3003, 1100, 6053, in this order
Cadmium
Aluminum alloys
  2117, 2017, 2024, in this order
Low-carbon steel
Wrought iron
Cast iron
Ni-Resist (high-nickel cast iron)
Type 410 stainless steel (active)
50-50 lead-tin solder
Type 304 stainless steel (active)
Type 316 stainless steel (active)
Lead
Tin
Copper alloy C28000 (Muntz metal, 60% Cu)
Copper alloy C67500 (manganese bronze A)
Copper alloys C46400, C46500, C46600, C46700 (naval brass)
Nickel 200 (active)
Inconel alloy 600 (active)
Hastelloy alloy B
Chlorimet 2

Copper alloy C27000 (yellow brass, 65% Cu)
Copper alloys C44300, C44400, C44500 (admiralty brass)
Copper alloys C60800, C61400 (aluminum bronze)
Copper alloy C23000 (red brass, 85% Cu)
Copper C11000 (ETP copper)
Copper alloys C65100, C65500 (silicon bronze)
Copper alloy C71500 (copper nickel, 30% Ni)
Copper alloy C92300, cast (leaded tin bronze G)
Copper alloy C92200, cast (leaded tin bronze M)
Nickel 200 (passive)
Inconel alloy 600 (passive)
Monel alloy 400
Type 410 stainless steel (passive)
Type 304 stainless steel (passive)
Type 316 stainless steel (passive)
Incoloy alloy 825
Inconel alloy 625
Hastelloy alloy C
Chlorimet 3
Silver
Titanium
Graphite
Gold
Platinum
Protected end (cathodic, or most noble)

Table 13.6 Chemical resistance of cast iron to various environments

This table was compiled from data supplied by material manufacturers. Their nomenclature was condensed into key symbols: A, acceptable—excellent resistance, fully resistant, suitable, recommended, excellent compatibility, fully compatible; Q, questionable—good resistance, minor effect, moderate effect, slight effect, slight attack, fair resistance; N, not recommended—severe effect, unsatisfactory, not acceptable, do not use. Temperature conditions are outlined in the footnotes.

Material	Resistance rating	Material	Resistance rating
Acetaldehyde(a)	Q	Acid, boric(c)	N
Acetamide	N	Acid, butyric(a)	N
Acetate solvents	N	Acid, carbolic (phenol)(a)	N
Acetone(a)(c)	A	Acid, carbonic	N
Acetylene(a)	A	Acid, chloroacetic(a)	N
Acetylene tetrabromide	N	Acid, chlorosulfonic	N
Acid, acetic (50% unaerated)(a)	N	Acid, chromic (5%)(a)	N
Acid, acetic (50% unaerated)(c)	N	Acid, chromic (10%)(c)	N
Acid, acetic (100% unaerated)(a)	N	Acid, chromic (50%)(c)	N
Acid, acetic (100% unaerated)(c)	N	Acid, citric (15%)(a)	N
Acid, acetic anhydride(a)	A	Acid, citric (15%)(c)	N
Acid, acetic anhydride(c)	A	Acid, citric (conc.)(c)	N
Acid, acetic, vapor	Q	Acid, cresylic	Q
Acid, arsenic(c)	N	Acids, fatty	N
Acid, benzene sulfonic	N	Acid, fluoroboric	N
Acid, benzoic(a)	N	Acid, fluorosilicic	N

(continued)

Table 13.6 (continued)

Material	Resistance rating	Material	Resistance rating
Acid, formic(a)(b)	N	Ammonium chloride (50%)(a)(c)	N
Acid, glacial acetic	N	Ammonium hydroxide (10%)	A
Acid, hydrogromic(c)	N	Ammonium hydroxide (46.5%)	A
Acid, hydrochloric (20%)(a)	N	Ammonium nitrate(a)	A
Acid, hydrochloric (37%)(a)	N	Ammonium oxalate(a)	N
Acid, hydrochloric (all)(b)	N	Ammonium persulfate(a)	N
Acid, hydrocyanic	N	Ammonium phosphate	Q
Acid, hydrofluoric (up to 50%)	N	Ammonium sulfate (sat.)(a)(c)	Q
Acid, hydrofluoric (50–100%)	N	Ammonium thiosulfate	N
Acid, lactic (5%)(a)	N	Amyl acetate	Q
Acid, lactic (5%)(b)	N	Amyl alcohol	A
Acid, lactic (10%)(b)(c)	N	Aniline (sat.)(a)	Q
Acid, nitric (conc.)(a)	N	Aniline dyes	Q
Acid, nitric (fuming)(a)	N	Aniline oil	A
Acid, oleic (5%)(a)	N	Anise oil	N
Acid, oxalic (10%)(a)	Q	Antifreeze, Dowgard	A
Acid, phosphoric (crude)	N	Antifreeze, Hubbard-Hall	A
Acid, phosphoric (1%)(a)	N	Antifreeze, Permaguard	A
Acid, phosphoric (10%)(a)	N	Antifreeze, Prestone	A
Acid, phosphoric (50%)	N	Antifreeze, Pyro-Permanent	A
Acid, phosphoric (pure)	N	Antifreeze, Pyro-Super	A
Acid, picric water solution	A	Antifreeze, Shell Zone	A
Acid, stearic (conc.)(c)	Q	Antifreeze, Texaco P.T.	A
Acid, sulfuric (5%)(a)	N	Antifreeze, Telar	A
Acid, sulfuric (5%)(c)	N	Antifreeze, Valvolene	A
Acid, sulfuric (10%)	N	Antifreeze, Zerex	A
Acid, sulfuric (30%)(a)	N	Aromatic hydrocarbons	A
Acid, sulfuric (30%)(c)	N	Asphalt	A
Acid, sulfuric (75%)	N	ASTM oils No. 1, No. 2, No. 3	A
Acid, sulfuric (conc.)(a)	N	Automotive gasoline	A
Acid, sulfuric (conc.)(c)	N	Aviation gasoline	A
Acid, sulfuric (fuming)(a)	N	Barbecue sauce	N
Acid, sulfurous(b)	N	Barium chloride (sat.)(a)	Q
Acid, tannic (10%)(a)(b)	Q	Barium hydroxide(a)	A
Acid, tartaric(a)(b)	Q	Barium nitrate(b)	A
Acid, trichloroacetic (50%)	N	Beef extract	N
Acrylonitrile	Q	Beer	N
Alcohol, amyl(a)	A	Beet sugar syrups	A
Alcohol, benzyl	A	Benzaldehyde	A
Alcohol, butyl(a)	N	Benzene(a)	A
Alcohol, diacetone	A	Benzine (gasoline)	A
Alcohol, ethyl	A	Benzol (benzene)(b)	A
Alcohol, hexyl	A	Benzyl chloride	N
Alcohol, isobutyl	Q	Borax	A
Alcohol, isopropyl	Q	Boron fuels	N
Alcohol, methyl	A	Brake fluid	A
Alcohol, octyl	A	Brewery slop	A
Alcohol, propyl	Q	Brine	N
Aluminum chloride (5%)(a)	N	Butane	A
Aluminum hydroxide (sat.)(a)	A	Butanol (butyl alcohol)	A
Aluminum oxide	N	Butter	N
Aluminum sulfate (sat.)(a)	N	Buttermilk	N
Amines	N	Butyl acetate(a)	A
Ammonia, anhydrous (liquid)(a)	A	Butyraldehyde	A
Ammonia liquors	A	Calcium bisulfate	N
Ammonia nitrate	A	Calcium chloride(a)	Q
Ammonium bicarbonate(b)	Q	Calcium hydroxide (10%)(c)	A
Ammonium bifluoride	N	Calcium hypochlorite (2%)(a)	N
Ammonium carbonate(a)(c)	A	Calcium hypochlorite (20% on plastics)	N

(continued)

Table 13.6 (continued)

Material	Resistance rating	Material	Resistance rating
Calcium sulfate(a)	A	Fish oil	N
Calgon	N	Formaldehyde (Formaline)	Q
Cane sugar liquors	A	Freon 11	Q
Carbon dioxide	N	Freon 12	A
Carbon disulfide	A	Freon 22	Q
Carbon monoxide	N	Fruit juices	N
Carbon tetrachloride	Q	Furfural	A
Carbonated beverages	N	Gasoline	A
Castor oil	A	Gelatin	N
Catsup	N	Ginger oil	N
Cellulube	A	Glucose	A
Chlorinated lime	N	Glue	A
Chlorine (anhydrous liquid)	Q	Glycerin (glycerol)	A
Chloroacetone	N	Gold monocyanide	N
Chlorobenzene(a)	N	Grapefruit oil	N
Chlorobromomethane	N	Grape juice	N
Chlorobutadiene	N	Grease	A
Chloroform(a)	N	Heptane	A
Chlorox (bleach)	N	Hexane	A
Chocolate syrup	N	Honey	A
Cider	N	Hydraulic fluids	A
Cinnamon oil	N	Hydrazine (water-base)	Q
Citric oils	N	Hydrazine (alcohol-base)	Q
Clove oil	N	Hydrogen peroxide(a)	N
Coconut oil	A	Hydrogen peroxide (10%)	N
Cod liver oil	N	Hydrogen sulfide(a)	N
Copper chloride	N	Ink	N
Copper cyanide (sat.)(c)	N	Iodine	N
Copper sulfate (5%)(a)	N	Isopropyl alcohol	A
Copper sulfate (sat.)(c)	N	Jet fuel (JPI-JP6)	A
Corn oil	A	Kerosene	A
Cottonseed oil	A	Kerosene & naphtha	A
Cream	N	Lacquers	Q
Creosols	Q	Lard(a)	A
Creosote oil (coal tar)(b)	A	Larvacide	A
Cutting oil (water-soluble)	A	Lime	A
Cutting oil (sulfur-base)	A	Linseed oil	A
Cyclohexane	A	Lithium bromide	A
Developing solutions (hypos)	N	Lube oil SAE 10, 20, 30, etc.	A
Dibenzyl ether	N	Lubricating oils	A
Dibromochloropropane	N	Magnesium chloride (5%)(a)	N
Dibutyl ether	N	Magnesium hydroxide(a)	A
Diesel fuel	A	Magnesium nitrate	N
De-ester synthetic lubricants	A	Magnesium oxide	A
Diphenyl oxides	A	Magnesium sulfate (5%)(b)	A
Distillery wort	Q	Mayonnaise	N
Ether compounds	Q	Melamine resins	N
Ethyl acetate	A	Mercuric chloride	N
Ethyl chloride	Q	Mercury	A
Ethyl ether	Q	Methanol	A
Ethylene chloride	Q	Methyl alcohol	A
Ethylene dichloride	A	Methyl chloride	A
Ethylene glycol	A	Methylene bromide	A
Ethylene oxide	N	Methylene chloride	A
Fatty acids	Q	Milk	N
Ferric chloride	N	Mineral oil	A
Ferric sulfate	N	Molasses	A
Ferrous chloride	N	Multicircuit etch	N
Ferrous sulfate	N	Mustard	N

(continued)

Table 13.6 (continued)

Material	Resistance rating	Material	Resistance rating
Naphtha	A	Propane	A
Naphthalene	A	Propylene glycol	A
Nickel chloride(a)	N	Pyridine	A
Nickel sulfate(a)	N	Rapeseed oil	A
Nitro benzene	Q	Salad dressing	N
Oil, aniline	A	Sesame seed oil	A
Oil, ASTM No. 1, No. 3	A	Shellac	A
Oil, bone	A	Silica gel	A
Oil, castor	A	Silicone X527	A
Oil, Chevron	A	Silver nitrate	N
Oil, citric	N	Soap solutions(a)	A
Oil, coconut	A	Soda ash (sodium carbonate)(a)	A
Oil, corn	A	Sodium arsenite	N
Oil, cottonseed	A	Sodium bicarbonate(b)	Q
Oil, creosote	A	Sodium bisulfate	N
Oil, diester synthetic lubricating	A	Sodium bisulfite	N
Oil, Dromus	A	Sodium carbonate	Q
Oil, hydraulic	A	Sodium chloride	A
Oil, linseed	A	Sodium chromate	A
Oil, mineral(a)(b)	A	Sodium cyanide	A
Oil, olive	A	Sodium hydroxide (15%)(a)	A
Oil, pale	A	Sodium hydroxide (20%)	A
Oil, palm	A	Sodium hydroxide (50%)(a)	Q
Oil, peanut	A	Sodium hypochlorite	N
Oil, Pella	A	Sodium metaphosphate	N
Oil, pine	Q	Sodium nitrite	A
Oil, rapeseed	A	Sodium perborate	Q
Oil, red	Q	Sodium peroxide(c)	N
Oil, Royal Triton	A	Sodium phosphate mono	N
Oil, sesame seed	A	Sodium phosphate DI	A
Oil, Shell Dieselene	A	Sodium phosphate TRI	A
Oil, silicone	A	Sodium polyphosphate	A
Oil, soybean	A	Sodium silicate	A
Oil, sperm	A	Sodium sulfate (conc.)(a)	A
Oil, sulfur-base cutting	A	Sodium sulfide (sat.)(a)	A
Oil, turbine	A	Sodium sulfite (5%)(a)	A
Oil, vegetable	A	Sodium thiosulfate	Q
Oil, water-soluble cutting	A	Sodium tripolyphosphate	N
Paint (with xylene)	A	Sorghum	A
Palm oil	A	Soybean oil	A
Peanut oil	A	Soy sauce	N
Perchloroethylene	A	Sperm oil	A
Phenol (carbolic acid)	N	Stannic chloride(a)	N
Photographic developer	N	Stannic fluoborate	N
Pine oil	Q	Starch	Q
Pipeline cleaner	A	Stoddard solvent	A
Potassium bicarbonate	A	Sulfate liquors	N
Potassium carbonate(a)	A	Sulfur-base cutting oil	A
Potassium chloride(a)	A	Sulfur chloride	N
Potassium chromate	A	Sulfur dioxide (dry)	A
Potassium cyanide	A	Tetrachloroethane	Q
Potassium dichromate	A	Tetraethyl lead	A
Potassium hydroxide (5%)(b)	Q	Thionyl chloride	N
Potassium hydroxide (50%)(b)	Q	Toluene	A
Potassium hydroxide (50%)(c)	Q	Toothpaste	N
Potassium permanganate	A	Transformer oil	A
Potassium phosphate	N	Transmission fluid	A
Potassium sulfate (5%)(a)	A	Trichloroethane	A
Potassium sulfate (5%)(b)	A	Trichloroethylene	Q

(continued)

Table 13.6 (continued)

Material	Resistance rating	Material	Resistance rating
Trichloropane	A	Water, distilled	N
Trichloropropane	A	Water, mine	Q
Turpentine	A	Water, salt(a)	N
Urine	A	Whiskey & wines	N
Varnish	Q	White liquor	Q
Vegetable juices	N	Xylene	A
Vegetable oil	Q	Zinc chloride(a)	N
Vinegar(a)	Q	Zinc hydrosulfite	N
Water, tap (to 180 °F)	Q	Zinc sulfate (sat.)	Q
Water, boiling	Q		

Temperature conditions: (a) room ambient, to 100 °F; (b) medium temperature, 100–200 °F; (c) high temperature, >200 °F; if no temperature is listed, use room ambient. Source: Oberdorfer Pump Division, Syracuse, NY

Table 13.7 Corrosion resistance of carbon steel to various environments

Environment	Resistance rating	Environment	Resistance rating
Acetate solvents	S	Carbon bisulfide	P
Acetic acid, all strengths	P	Carbonic acid	P
Acetic anhydride	P	Carbon tetrachloride	S
Alum	P	Cellulose acetate	S
Aluminum chloride	P	Chloroacetic acid	P
Aluminum sulfate + H_2SO_4	P	Chlorinated water	P
Ammonium chloride	P	Chlorine dioxide	P
Ammonium fluoride	P	Chlorine gas, wet	P
Ammonium hydroxide	S	Chromic acid	P
Ammonium nitrate	S	Citric acid	P
Ammonium phosphate	P	Copper nitrate	P
Ammonium sulfate	S	Copper silver nitrate	P
Ammonium sulfate + H_2SO_4	P	Copper sulfate	P
Aniline dyes	S	Copper sulfate + 10% H_2SO_4	P
Aniline hydrochloride	P	Cupric chloride	P
Anodizing solutions	P	Cuprous chloride	P
Antimony trichloride	P	Ethylene dichloride	P
Arsenic acid	P	Fatty acids	S
Barium chloride	S	Ferric chloride	P
Barium nitrate	S	Ferric ferrocyanide	P
Barium sulfate	S	Ferric nitrate	P
Benzoic acid	P	Ferric sulfate	P
Black liquor	P	Ferric sulfate + 10% H_2SO_4	P
Boric acid	P	Ferrous sulfate	P
Brine, acid	P	Ferrous sulfate + 10% H_2SO_4	P
Brine, alkaline	G	Formaldehyde	P
Bromine, dry	P	Formic acid	P
Bromine, wet	P	Glycerin, crude	P
Cadmium sulfate	S	HCl waste pickle liquor	P
Calcium bisulfate	P	Hydrochloric acid (<150 °F)	P
Calcium bisulfite + H_2SO_4	P	Hydrochloric acid (>150 °F)	P
Calcium chloride	S	Hydrofluoric acid	P
Calcium hydroxide (lime)	S	Hydrofluosilicic acid	P
Calcium hypochlorite	P	Hydrogen peroxide	S
Calcium phosphate	S	Hypochlorite bleach	P

(continued)

Table 13.7 (continued)

Environment	Resistance rating	Environment	Resistance rating
Iodine, dry	S	Sodium bicarbonate	S
Lactic acid	P	Sodium bichromate	G
Lead acetate	P	Sodium bisulfate	P
Lead nitrate	P	Sodium bisulfite	S
Lead sulfide	P	Sodium chlorate	P
Lithophone	S	Sodium chloride	S
Magnesium chloride	S	Sodium ferricyanide	S
Magnesium sulfate	S	Sodium hydroxide	S
Maleic acid	S	Sodium hydroxide, fused	P
Malic acid	P	Sodium hypochlorite	P
Manganese chloride	S	Sodium nitrate	S
Mercuric chloride	P	Sodium perchlorate	P
Mercuric nitrate	P	Sodium phosphate	G
Mercuric sulfate	P	Sodium sulfate	S
Mercurous sulfate	P	Sodium sulfide	P
Metal plating solutions	P	Sodium sulfite	P
Mine water	P	Sodium thiosulfate	P
Mixed acid	P	Stannic chloride	P
Nickel chloride	P	Stannous chloride	P
Nickel ammonium sulfate	P	Stearic acid	S
Nitric acid, all strengths	P	Sulfite liquors	S
Nitric acid + 3–5% HF	P	Sulfite liquors + H_2SO_4	P
Nitrobenzene	S	Sulfur	S
Oleic acid	P	Sulfur chloride	P
Oleum	S	Sulfur dioxide	S
Oxalic acid	P	Sulfuric acid, sat. With SO_2	P
Phenol	S	Sulfuric acid, up to 100 °F	P
Phosphoric acid + 2% H_2SO_4 1% HF	P	Sulfuric acid, 5% to boiling	P
Phosphoric acid, all strengths	P	Sulfuric acid, 60–100% 176 °F	P
Picric acid	P	Sulfurous acid	P
Phthalic acid	S	Sugar solutions	S
Potassium bisulfate	P	Tannic acid	P
Potassium chloride	S	Tar and ammonia	S
Potassium hydroxide	S	Tartaric acid	S
Potassium iodide	S	Titanic sulfate	P
Potassium nitrate	S	Toluene	S
Potassium sulfate	S	Zinc chloride	P
Pyridine sulfate	P	Zinc sulfate	P
Seawater	S		

E, excellent—virtually unattacked under all conditions; G, good—generally acceptable with a few limitations; S, satisfactory—suitable under many conditions, not recommended for the remainder (consult a steel manufacturer for details); P, poor—unsuitable under all conditions. Source: The Duriron Company, Inc., Dayton, OH

Table 13.8 Corrosion of structural steels in various atmospheric environments

Type of atmosphere	Time, yr	Structural carbon steel μm	Structural carbon steel mils	Structural copper steel μm	Structural copper steel mils	UNS K11510(a) μm	UNS K11510(a) mils	UNS K11430(b) μm	UNS K11430(b) mils	UNS K11630(c) μm	UNS K11630(c) mils	UNS K11576(d) μm	UNS K11576(d) mils
Industrial (Newark, NJ)	3.5	84	3.3	66	2.6	33	1.3	46	1.8	36	1.4	56	2.2
	7.5	104	4.1	81	3.2	38	1.5	53	2.1	43	1.7	…	…
	15.5	135	5.3	102	4.0	46	1.8	…	…	53	2.1	…	…
Semi-industrial (Monroeville, PA)	1.5	56	2.2	43	1.7	28	1.1	36	1.4	30	1.2	41	1.6
	3.5	94	3.7	64	2.5	30	1.2	53	2.1	36	1.4	61	2.4
	7.5	130	5.1	81	3.2	36	1.4	61	2.4	43	1.7	…	…
	15.5	185	7.3	119	4.7	46	1.8	…	…	46	1.8	…	…
Semi-industrial (South Bend, PA)	1.5	46	1.8	36	1.4	25	1.0	33	1.3	25	1.0	38	1.5
	3.5	74	2.9	56	2.2	33	1.3	48	1.9	38	1.5	61	2.4
	7.5	117	4.6	81	3.2	46	1.8	69	2.7	48	1.9	…	…
	15.5	178	7.0	122	4.8	56	2.2	…	…	64	2.5	…	…
Rural (Potter County, PA)	2.5	…	…	33	1.3	20	0.8	30	1.2	…	…	…	…
	3.5	51	2.0	43	1.7	28	1.1	36	1.4	30	1.2	46	1.8
	7.5	76	3.0	64	2.5	33	1.3	38	1.5	38	1.5	…	…
	15.5	119	4.7	97	3.8	36	1.4	…	…	51	2.0	…	…
Moderate marine (Kure Beach, NC, 250 m, or 800 ft, from ocean)	0.5	23	0.9	20	0.8	15	0.6	20	0.8	18	0.7	25	1.0
	1.5	58	2.3	48	1.9	28	1.1	43	1.7	30	1.2	43	1.7
	3.5	124	4.9	84	3.3	46	1.8	64	2.5	48	1.9	56	2.2
	7.5	142	5.6	114	4.5	64	2.5	94	3.7	74	2.9	…	…
Severe marine (Kure Beach, NC, 25 m, or 80 ft, from ocean)	0.5	183	7.2	109	4.3	56	2.2	97	3.8	28	1.1	18	0.7
	2.0	914	36.0	483	19.0	84	3.3	310	12.2	…	…	53	2.1
	3.5	1448	57.0	965	38.0	…	…	729	28.7	99	3.9	99	3.9
	5.0	…	(e)	…	(e)	493	19.4	986	38.8	127	5.0	…	…

(a) ASTM A242 (type 1). (b) ASTM A588 (grade A). (c) ASTM A514 (type B) and A517 (grade B). (d) ASTM A514 (type F) and A517 (grade F). (e) Specimen corroded completely away.
Source: *Proc. 1st International Congress on Metallic Corrosion*, Butterworths, 1962, p 276–285

Table 13.9 Relative corrosion resistance of standard (AISI) stainless steels

The "X" notations indicate that a specific stainless steel type may be considered as resistant to the corrosive environment categories.

Type No.	UNS No.	Mild atmospheric and fresh water	Atmospheric Industrial	Atmospheric Marine	Salt water	Chemical Mild	Chemical Oxidizing	Reducing
201	S20100	X	X	X		X	X	
202	S20200	X	X	X		X	X	
205	S20500	X	X	X		X	X	
301	S30100	X	X	X		X	X	
302	S30200	X	X	X		X	X	
302B	S30215	X	X	X		X	X	
303	S30300	X	X			X		
303 Se	S30323	X	X			X		
304	S30400	X	X	X		X	X	
304L	S30403	X	X	X		X	X	
	S30430	X	X	X		X	X	
304N	S30451	X	X	X		X	X	
305	S30500	X	X	X		X	X	
308	S30800	X	X	X		X	X	
309	S30900	X	X	X		X	X	
309S	S30908	X	X	X		X	X	
310	S31000	X	X	X		X	X	
310S	S31008	X	X	X		X	X	
314	S31400	X	X	X		X	X	
316	S31600	X	X	X	X	X	X	X
316F	S31620	X	X	X	X	X	X	X
316L	S31603	X	X	X	X	X	X	X
316N	S31651	X	X	X	X	X	X	X
317	S31700	X	X	X		X	X	X
317L	S31703	X	X	X	X	X	X	
321	S32100	X	X	X		X	X	
329	S32900	X	X	X	X	X	X	X
330	N08330	X	X	X	X	X	X	X
347	S34700	X	X	X		X	X	
348	S34800	X	X	X		X	X	
384	S38400	X	X	X		X	X	
403	S40300	X				X		
405	S40500	X				X		
409	S40900	X				X		
410	S41000	X				X		
414	S41400	X				X		
416	S41600	X						
416 Se	S41623	X						
420	S42000	X						
420F	S42020	X						
422	S42200	X						
429	S42900	X	X			X	X	
430	S43000	X	X			X	X	
430F	S43020	X	X			X		
430F Se	S43023	X	X			X		
431	S43100	X	X	X		X		
434	S43400	X	X	X		X	X	
436	S43600	X	X	X		X	X	
440A	S44002	X				X		
440B	S44003	X						
440C	S44004	X						
442	S44200	X	X			X	X	
446	S44600	X	X	X		X	X	
	S13800	X	X			X	X	
	S15500	X	X	X		X	X	
	S17400	X	X	X		X	X	
	S17700	X	X	X		X	X	

Source: *Steel Products Manual: Stainless and Heat Resisting Steels,* American Iron and Steel Institute, Washington, DC, Dec 1974

Table 13.10 Relative corrosion resistance of standard stainless steel grades for different environments

Environment	Grades(a)
Acids	
Hydrochloric acid	Stainless is not generally recommended except when solutions are very dilute and at room temperature (pitting may occur).
Mixed acids	There is usually no appreciable attack on type 304 or 316 as long as sufficient nitric acid is present.
Nitric acid	Type 304L and 430 and some higher-alloy stainless grades have been used.
Phosphoric acid	Type 304 is satisfactory for storing cold phosphoric acid up to 85% and for handling concentrations up to 5% in some unit processes of manufacture. Type 316 is more resistant and is generally used for storing and manufacture if the fluorine content is not too high. Type 317 is somewhat more resistant than type 316. At concentrations ≤85%, the metal temperature should not exceed 100 °C (212 °F) with type 316 and slightly higher with type 317. Oxidizing ions inhibit attack.
Sulfuric acid	Type 304 can be used at room temperature for concentrations >80 to 90%. Type 316 can be used in contact with sulfuric acid ≤10% at temperatures ≤50 °C (120 °F) if the solutions are aerated; the attack is greater in air-free solutions. Type 317 may be used at temperatures as high as 65 °C (150 °F) with ≤5% concentration. The presence of other materials may markedly change the corrosion rate. As little as 500 to 2000 ppm of cupric ions make it possible to use type 304 in hot solutions of moderate concentration. Other additives may have the opposite effect.
Sulfurous acid	Type 304 may be subject to pitting, particularly if some sulfuric acid is present. Type 316 is usable at moderate concentrations and temperatures.
Bases	
Ammonium hydroxide, sodium hydroxide, caustic solutions	Steels in the 300 series generally have good corrosion resistance at virtually all concentrations and temperatures in weak bases, such as ammonium hydroxide. In stronger bases, such as sodium hydroxide, there may be some attack, cracking, or etching in more concentrated solutions and/or at higher temperatures. Commercial-purity caustic solutions may contain chlorides, which will accentuate any attack and may cause pitting of type 316, as well as type 304.
Organics	
Acetic acid	Acetic acid is seldom pure in chemical plants but generally includes numerous and varied minor constituents. Type 304 is used for a wide variety of equipment including stills, base heaters, holding tanks, heat exchangers, pipelines, valves, and pumps for concentrations ≤99% at temperatures ≤~50 °C (120 °F). Type 304 is also satisfactory—if small amounts of turbidity or color pickup can be tolerated—for room-temperature storage of glacial acetic acid. Types 316 and 317 have the broadest range of usefulness, especially if formic acid is also present or if solutions are unaerated. Type 316 is used for fractionating equipment, for 30–99% concentrations where type 304 cannot be used, for storage vessels, pumps, and process equipment handling glacial acetic acid, which would be discolored by type 304. Type 316 is likewise applicable for parts having temperatures >50 °C (120 °F), for dilute vapors, and for high pressures. Type 317 has somewhat greater corrosion resistance than type 316 under severely corrosive conditions. None of the stainless steels has adequate corrosion resistance to glacial acetic acid at the boiling temperature or at superheated vapor temperatures.
Aldehydes	Type 304 is generally satisfactory.
Amines	Type 316 is usually preferred to type 304.
Cellulose acetate	Type 304 is satisfactory for low temperatures, but type 316 or type 317 is needed for high temperatures.
Formic acids	Type 304 is generally acceptable at moderate temperatures, but type 316 is resistant to all concentrations at temperatures up to boiling.
Esters	With regard to corrosion, esters are comparable to organic acids.
Fatty acids	Type 304 is resistant to fats and fatty acids ≤~150 °C (300 °F), but type 316 is needed at 150–260 °C (300–500 °F), and type 317, at higher temperatures.
Paint vehicles	Type 316 may be needed if exact color and lack of contamination are important.

(continued)

Table 13.10 (continued)

Organics (continued)

Phthalic anhydride	Type 316 is usually used for reactors, fractionating columns, traps, baffles, caps, and piping.
Soaps	Type 304 is used for parts such as spray towers, but type 316 may be preferred for spray nozzles and flake-drying belts to minimize off-color product.
Synthetic detergents	Type 316 is used for preheat, piping, pumps, and reactors in catalytic hydrogenation of fatty acids to give salts of sulfonated high-molecular alcohols.
Tall oil (pulp and paper industry)	Type 304 has only limited use in tall-oil distillation service. High rosin acid streams can be handled by type 316L with a minimum molybdenum content of 2.75%. Type 316 can also be used in the more corrosive high fatty acid streams at temperatures ≤245 °C (475 °F), but type 317 will probably be required at higher temperatures.
Tar	Tar distillation equipment is almost all type 316 because coal tar has a high chloride content; type 304 does not have adequate resistance to pitting.
Urea	Type 316L is generally required.

Pharmaceuticals

Type 316 is usually selected for all parts in contact with the product because of its inherent corrosion resistance and greater assurance of product purity.

(a) The stainless steels mentioned may be considered for use in the indicated environments. Additional information or corrosion expertise may be necessary prior to use in some environments; for example, some impurities may cause localized corrosion (such as chlorides causing pitting or stress-corrosion cracking of some grades).

Table 13.11 Relative ratings of resistance to general corrosion and to SCC of wrought aluminum alloys

No.	Alloy Temper	Resistance to corrosion General(a)	SCC(b)	No.	Alloy Temper	Resistance to corrosion General(a)	SCC(b)
1060	All	A	A	3004	All	A	A
1100	All	A	A	3105	All	A	A
1350	All	A	A	4032	T6	C	B
2011	T3, T4, T451	D(c)	D	5005	All	A	A
	T8	D	B	5050	All	A	A
2014	O	...	...	5052	All	A	A
	T3, T4, T451	D(c)	C	5056	O, H11, H12, H32, H14, H34	A(d)	B(d)
	T6, T651, T6510, T6511	D	C		H18, H38	A(d)	C(d)
2017	T4, T451	D(c)	C		H192, H392	B(d)	D(d)
2018	T61	...	...	5083	All	A(d)	B(d)
2024	O	...	...	5086	O, H32, H116	A(d)	A(d)
	T4, T3, T351, T3510, T3511, T361	D(c)	C		H34, H36, H38, H111	A(d)	A(d)
	T6, T861, T81, T851, T8510, T8511	D	B	5154	All	A(d)	A(d)
	T72	...	...	5252	All	A	A
2025	T6	D	C	5254	All	A(d)	A(d)
2036	T4	C	...	5454	All	A	A
2117	T4	C	A	5456	All	A(d)	B(d)
2218	T61, T72	D	C	5457	O	A	A
2219	O	...	...	5652	All	A	A
	T31, T351, T3510, T3511, T37	D(c)	C	5657	All	A	A
	T81, T851, T8510, T8511, T87	D	B	6053	O	...	...
2618	T61	D	C		T6, T61	A	A
3003	All	A	A	6061	O	B	A
					T4, T451, T4510, T4511	B	B
					T6, T651, T652, T6510, T6511	B	A
				6063	All	A	A

(continued)

Table 13.11 (continued)

No.	Alloy Temper	Resistance to corrosion General(a)	SCC(b)	No.	Alloy Temper	Resistance to corrosion General(a)	SCC(b)
6066	O	C	A	6262	T6, T651, T6510, T6511, T9	B	A
	T4, T4510, T4511, T6,		B	6463	All	A	A
	T6510, T6511	C		7001	O	C(c)	C
6070	T4, T4511, T6	B	B	7075	T6, T651, T652, T6510,		C
6101	T6, T63, T61, T64	A	A		T6511	C(c)	
6151	T6, T652	...	...		T73, T7351	C	B
6201	T81	A	A	7178	T6, T651, T6510, T6511	C(c)	B

(a) Ratings are relative and in decreasing order of merit, based on exposure to NaCl solution by intermittent spraying or immersion. Alloys with A and B ratings can be used in industrial and seacoast atmospheres without protection. Alloys with C, D, and E ratings generally should be protected, at least on faying surfaces. (b) SCC ratings are based on service experience and on laboratory tests of specimens exposed to alternate immersion in 3.5% NaCl solution. A, no known instance of failure in service or in laboratory tests; B, no known instance of failure in service; limited failure in laboratory tests of short transverse specimens; C, service failures when sustained tension stress acts in short-transverse direction relative to grain structure; limited failures in laboratory tests of long transverse specimens; D, limited service failures when sustained stress acts in longitudinal or long-transverse direction relative to grain structure. (c) In relatively thick sections, the rating would be E. (d) This rating may be different for material held at elevated temperatures for long periods.

Table 13.12 Relative ratings of resistance to general corrosion and to SCC of cast aluminum alloys

No.	Alloy Temper	Resistance to corrosion General(a)	SCC(b)	No.	Alloy Temper	Resistance to corrosion General(a)	SCC(b)
Sand castings				**Permanent mold castings**			
208.0	F	B	B	242.0	T571, T61	D	C
224.0	T7	C	B	308.0	F	C	B
240.0	F	D	C	319.0	F	C	B
242.0	All	D	C		T6	C	C
A242.0	T75	D	C	332.0	T5	C	B
249.0	T7	C	B	336.0	T551, T65	C	B
295.0	All	C	C	354.0	T61, T62	C	A
319.0	F, T5	C	B	355.0	All	C	A
	T6	C	C	C355.0	T61	C	A
355.0	All	C	A	356.0	All	B	A
C355.0	T6	C	A	A356.0	T61	B	A
356.0	T6, T7, T71, T51	B	A	F356.0	All	B	A
A356.0	T6	B	A	A357.0	T61	B	A
443.0	F	B	A	358.0	T6	B	A
512.0	F	A	A	359.0	All	B	A
513.0	F	A	A	B443.0	F	B	A
514.0	F	A	A	A444.0	T4	B	A
520.0	T4	A	C	513.0	F	A	A
535.0	F	A	A	705.0	T5	B	B
B535.0	F	A	A	707.0	T5	B	C
705.0	T5	B	B	711.0	T5	B	A
707.0	T5	B	C	713.0	T5	B	B
710.0	T5	B	B	850.0	T5	C	B
712.0	T5	B	C	851.0	T5	C	B
713.0	T5	B	B	852.0	T5	C	B
771.0	T6	C	C	**Die castings**			
850.0	T5	C	B	360.0	F	C	A
851.0	T5	C	B	A360.0	F	C	A
852.0	T5	C	B				

(continued)

Table 13.12 (continued)

No.	Alloy Temper	Resistance to corrosion General(a)	SCC(b)	No.	Alloy Temper	Resistance to corrosion General(a)	SCC(b)
Die castings (continued)				**Die castings (continued)**			
364.0	F	C	A	A413.0	F	C	A
380.0	F	E	A	C443.0	F	B	A
A380.0	F	E	A	518.0	F	A	A
383.0	F	E	A	**Rotor metal(c)**			
384.0	F	E	A	100.1	...	A	A
390.0	F	E	A	150.1	...	A	A
392.0	F	E	A	170.1	...	A	A
413.0	F	C	A				

(a) Relative ratings of general corrosion resistance are in decreasing order of merit, based on exposures to NaCl solution by intermittent spray or immersion. (b) Relative ratings of resistance to SCC are based on service experience and on laboratory tests of specimens exposed to alternate immersion in 3.5% NaCl solution. A, no known instance of failure in service when properly manufactured; B, failure not anticipated in service from residual stresses or from design and assembly stresses below about 45% of the minimum guaranteed yield strength given in applicable specifications; C, failures have occurred in service with either this specific alloy/temper combination or with alloy/temper combinations of this type; designers should be aware of the potential SCC problem that exists when these alloys and tempers are used under adverse conditions. (c) For electric motor rotors

Table 13.13 Relative SCC ratings for wrought products of high-stength aluminum alloys

The letters (ratings) A, B, C, and D have the following significance: A, very high: no record of service problems; SCC not anticipated in general applications. B, high: no record of service problems; SCC not anticipated at stresses of the magnitude caused by solution heat treatment. Precautions must be taken to avoid high sustained tensile stresses (exceeding 50% of the minimum specified yield strength) produced by any combination of sources including heat treatment, straightening, forming, fit-up, and sustained service loading. C, intermediate: stress-corrosion cracking not anticipated if total sustained tensile stress is maintained below 25% of minimum specified yield strength. This rating is designated for the short-transverse direction in products used primarily for high resistance to exfoliation corrosion in relatively thin structures, where appreciable stresses in the short-transverse direction are unlikely. D, low: failure due to SCC is anticipated in any application involving sustained tensile stress in the designated test direction. This rating is currently designated only for the short-transverse direction in certain products.

No.	Alloy Temper	Test direction(b)	Rolled plate	Rod and bar(c)	Extruded shapes	Forgings
2011	T3, T4	L	(d)	B	(d)	(d)
		LT	(d)	D	(d)	(d)
		ST	(d)	D	(d)	(d)
2011	T8	L	(d)	A	(d)	(d)
		LT	(d)	A	(d)	(d)
		ST	(d)	A	(d)	(d)
2014	T6	L	A	A	A	B
		LT	B(e)	D	B(e)	B(e)
		ST	D	D	D	D
2024	T3, T4	L	A	A	A	(d)
		LT	B(e)	D	B(e)	(d)
		ST	D	D	D	(d)

(continued)

Table 13.13 (continued)

No.	Alloy Temper	Test direction(b)	Rolled plate	Rod and bar(c)	Extruded shapes	Forgings
2024	T6	L	(d)	A	(d)	A
		LT	(d)	B	(d)	A(e)
		ST	(d)	B	(d)	D
2024	T8	L	A	A	A	A
		LT	A	A	A	A
		ST	B	A	B	C
2048	T851	L	A	(d)	(d)	(d)
		LT	A	(d)	(d)	(d)
		ST	B	(d)	(d)	(d)
2124	T85	L	A	(d)	(d)	(d)
		LT	A	(d)	(d)	(d)
		ST	B	(d)	(d)	(d)
2219	T3, T37	L	A	(d)	A	(d)
		LT	B	(d)	B	(d)
		ST	D	(d)	D	(d)
2219	T6, T8	L	A	A	A	A
		LT	A	A	A	A
		ST	A	A	A	A
6061	T6	L	A	A	A	A
		LT	A	A	A	A
		ST	A	A	A	A
7005	T53, T63	L	(d)	(d)	A	A
		LT	(d)	(d)	A(e)	A(e)
		ST	(d)	(d)	D	D
7039	T63, T64	L	A	(d)	A	(d)
		LT	A(e)	(d)	A(e)	(d)
		ST	D	(d)	D	(d)
7049	T73	L	A	(d)	A	A
		LT	A	(d)	A	A
		ST	A	(d)	B	A
7049	T76	L	(d)	(d)	A	(d)
		LT	(d)	(d)	A	(d)
		ST	(d)	(d)	C	(d)
7149	T73	L	(d)	(d)	A	A
		LT	(d)	(d)	A	A
		ST	(d)	(d)	B	A
7050	T736	L	A	(d)	A	A
		LT	A	(d)	A	A
		ST	B	(d)	B	B
7050	T76	L	A	A	A	(d)
		LT	A	B	A	(d)
		ST	C	B	C	(d)
7075	T6	L	A	A	A	A
		LT	B(e)	D	B(e)	B(e)
		ST	D	D	D	D
7075	T73	L	A	A	A	A
		LT	A	A	A	A
		ST	A	A	A	A

(continued)

Table 13.13 (continued)

No.	Alloy Temper	Test direction(b)	Rolled plate	Rod and bar(c)	Extruded shapes	Forgings
7075	T736	L	(d)	(d)	(d)	A
		LT	(d)	(d)	(d)	A
		ST	(d)	(d)	(d)	B
7075	T76	L	A	(d)	A	(d)
		LT	A	(d)	A	(d)
		ST	C	(d)	C	(d)
7175	T736	L	(d)	(d)	(d)	A
		LT	(d)	(d)	(d)	A
		ST	(d)	(d)	(d)	B
7475	T6	L	A	(d)	(d)	(d)
		LT	B(e)	(d)	(d)	(d)
		ST	D	(d)	(d)	(d)
7475	T73	L	A	(d)	(d)	(d)
		LT	A	(d)	(d)	(d)
		ST	A	(d)	(d)	(d)
7475	T76	L	A	(d)	(d)	(d)
		LT	A	(d)	(d)	(d)
		ST	C	(d)	(d)	(d)
7178	T6	L	A	(d)	A	(d)
		LT	B(e)	(d)	B(e)	(d)
		ST	D	(d)	D	(d)
7178	T76	L	A	(d)	A	(d)
		LT	A	(d)	A	(d)
		ST	C	(d)	C	(d)
7079	T6	L	A	(d)	A	A
		LT	B(e)	(d)	B(e)	B(e)
		ST	D	(d)	D	D

(a) Ratings apply to standard mill products in the types of tempers indicated and also in Tx5x and Tx5xx (stress-relieved) tempers and may be invalidated in some cases by use of nonstandard thermal treatments, or mechanical deformation at room temperature, by the user. (b) Test direction refers to orientation of direction in which stress is applied relative to the directional grain structure typical of wrought alloys, which for extrusions and forgings may not be predictable on the basis of the cross-sectional shape of the product: L, longitudinal; LT, long transverse; ST, short transverse. (c) Sections with width-to-thickness ratios equal to or less than two, for which there is no distinction between LT and ST properties. (d) Rating not established because product not offered commercially. (e) Rating is one class lower for thicker sections: extrusions, 25 mm (1 in.) and thicker; plate and forgings, 38 mm (1.5 in.) and thicker

Table 13.14 Weathering data for 0.89 mm (0.035 in.) thick aluminum alloy sheet after 20-year exposure

No.	Alloy Temper	Corrosion rate mm/yr	μ in./yr	Average depth of attack μm	mils	Maximum depth of attack m	mils	Loss in tensile strength, %
Phoenix, AZ (desert)								
1100	H14	76	3.0	8	0.3	18	0.7	0
2017	T3	76	3.0	23	0.9	51	2.0	0
2017	T3, alclad	13	0.5	10	0.4	23	0.9	0

(continued)

Table 13.14 (continued)

No.	Alloy Temper	Corrosion rate mm/yr	μin./yr	Average depth of attack μm	mils	Maximum depth of attack m	mils	Loss in tensile strength, %
Phoenix, AZ (desert) (continued)								
3003	H14	13	0.5	5	0.2	10	0.4	0
6051	T4	13	0.5	28	1.1	74	2.9	0
State College, PA (rural)								
1100	H14	76	3.0	36	1.4	89	3.5	3
2017	T3	102	4.0	25	1.0	81	3.2	2
2017	T3, alclad	76	3.0	10	0.4	25	1.0	0
3003	H14	89	3.5	23	0.9	56	2.2	3
6051	T4	76	3.0	23	0.9	96	3.8	0
Sand Hook, NJ (seacoast)								
1100	H14	279	11.0	96	3.8	231	9.1	3
2017	T3	...	...	43	1.7	132	5.2	10
2017	T3, alclad	...	...	23	0.9	33	1.3	...
3003	H14	356	14.0	36	1.4	84	3.3	...
6051	T4	343	13.5	58	2.3	137	5.4	9
La Jolla, CA (seacoast)								
1100	H14	584	23.0	102	4.0	356	14.0	8
2017	T3	2260	89.0	147	5.8	515	20.3	20
2017	T3, alclad	584	23.0	33	1.3	74	2.9	0
3003	H14	610	24.0	107	4.2	259	10.2	7
6051	T4	775	30.5	84	3.3	307	12.1	20
New York, NY (industrial)								
1100	H14	749	29.5	89	3.5	213	8.4	7
2017	T3	1260	49.6	51	2.0	180	7.1	7
2017	T3, alclad	762	30.0	28	1.1	36	1.4	0
3003	H14	965	38.0	51	2.0	163	6.4	8
6051	T4	914	36.0	74	2.9	170	6.7	12

Source: STP 174, ASTM, 1956, p 21

Table 13.15 Corrosion ratings of wrought copper alloys in various corrosive media

This table is intended to serve only as a general guide to the behavior of copper and copper alloys in corrosive environments. It is impossible to cover in a simple tabulation the performance of a material for all possible variations of temperature, concentration, velocity, impurity content, degree of aeration, and stress. The ratings are based on general performance; they should be used with caution, and then only for the purpose of screening candidate alloys. The letters E, G, F, and P have the following significance: E, excellent: resists corrosion under almost all conditions of service. G, good: some corrosion will take place, but satisfactory service can be expected under all but the most severe conditions. F, fair: corrosion rates are higher than for the G classification, but the metal can be used if needed for a property other than corrosion resistance and if either the amount of corrosion does not cause excessive maintenance expense or the effects of corrosion can be lessened, such as by use of coatings or inhibitors. P, poor: corrosion rates are high, and service is generally unsatisfactory.

Corrosive medium	Coppers	Low-zinc brasses	High-zinc brasses	Special brasses	Phosphor bronzes	Aluminum bronzes	Silicon bronzes	Copper nickels	Nickel silvers
Acetate solvents	E	E	G	E	E	E	E	E	E
Acetic acid(a)	E	E	P	P	E	E	E	E	G
Acetone	E	E	E	E	E	E	E	E	E
Acetylene(b)	P	P	(b)	P	P	P	P	P	P
Alcohols(a)	E	E	E	E	E	E	E	E	E
Aldehydes	E	E	F	F	E	E	E	E	E
Alkylamines	G	G	G	G	G	G	G	G	G
Alumina	E	E	E	E	E	E	E	E	E
Aluminum chloride	G	G	P	P	G	G	G	G	G
Aluminum hydroxide	E	E	E	E	E	E	E	E	E
Aluminum sulfate and alum	G	G	P	G	G	G	G	G	G
Ammonia, dry	E	E	E	E	E	E	E	E	E
Ammonia, moist(c)	P	P	P	P	P	P	P	F	P
Ammonium chloride(c)	P	P	P	P	P	P	P	F	P
Ammonium hydroxide(c)	P	P	P	P	P	P	P	F	P
Ammonium nitrate(c)	P	P	P	P	P	P	P	F	P
Ammonium sulfate(c)	F	F	F	F	F	F	F	G	F
Aniline and aniline dyes	F	F	F	F	F	F	F	F	F
Asphalt	E	E	E	E	E	E	E	E	E
Atmosphere									
Industrial(c)	E	E	E	E	E	E	E	E	E
Marine	E	E	E	E	E	E	E	E	E
Rural	E	E	E	E	E	E	E	E	E

(continued)

Table 13.15 (continued)

Corrosive medium	Coppers	Low-zinc brasses	High-zinc brasses	Special brasses	Phosphor bronzes	Aluminum bronzes	Silicon bronzes	Copper nickels	Nickel silvers
Barium carbonate	E	E	E	E	E	E	E	E	E
Barium chloride	G	G	F	F	G	G	G	G	G
Barium hydroxide	E	E	G	E	E	E	E	E	E
Barium sulfate	E	E	E	E	E	E	E	E	E
Beer(a)	E	E	G	E	E	E	E	E	E
Beet-sugar syrup(a)	E	E	G	E	E	E	E	E	E
Benzene, benzine, benzol	E	E	E	E	E	E	E	E	E
Benzoic acid	E	E	E	E	E	E	E	E	E
Black liquor, sulfate process	P	P	P	P	P	P	P	G	P
Bleaching powder (wet)	G	G	P	G	G	G	P	G	G
Borax	E	E	E	E	E	E	E	E	E
Bordeaux mixture	E	E	G	E	E	E	E	E	E
Boric acid	E	E	G	E	E	E	E	E	E
Brines	G	G	P	G	G	G	G	E	E
Bromine									
Dry	E	E	E	E	E	E	E	E	E
Moist	G	G	P	F	G	G	G	G	G
Butane(d)	E	E	E	E	E	E	E	E	E
Calcium bisulfate	G	G	P	G	G	G	G	G	G
Calcium chloride	G	G	F	G	G	G	G	G	G
Calcium hydroxide	E	E	G	E	E	E	E	E	E
Calcium hypochlorite	G	G	P	G	G	G	G	G	G
Cane-sugar syrup(a)	E	E	E	E	E	E	E	E	E
Carbolic acid (phenol)	F	G	P	G	G	G	G	G	G
Carbonated beverages(a)(e)	E	E	E	E	E	E	E	E	E
Carbon dioxide									
Dry	E	E	E	E	E	E	E	E	E
Moist(a)(e)	E	E	E	E	E	E	E	E	E
Carbon tetrachloride									
Dry	E	E	E	E	E	E	E	E	E

(continued)

Table 13.15 (continued)

Corrosive medium	Coppers	Low-zinc brasses	High-zinc brasses	Special brasses	Phosphor bronzes	Aluminum bronzes	Silicon bronzes	Copper nickels	Nickel silvers
Carbon tetrachloride (continued)									
Moist	G	G	F	G	E	E	E	E	E
Castor oil	E	E	E	E	E	E	E	E	E
Chlorine									
Dry	E	E	E	E	E	E	E	E	E
Moist	F	F	P	F	F	F	F	G	F
Chloroacetic acid	G	E	P	F	G	G	G	G	G
Chloroform, dry	E	E	E	E	E	E	E	E	E
Chromic acid	P	P	P	P	P	P	P	P	P
Citric acid(a)	E	E	F	E	E	E	E	E	E
Copper, chloride	F	F	P	F	F	F	F	F	F
Copper nitrate	F	F	P	F	F	F	F	F	F
Copper sulfate	G	G	P	G	G	G	G	E	G
Corn oil(a)	E	E	G	E	E	E	E	E	E
Cottonseed oil(a)	E	E	G	E	E	E	E	E	E
Creosote	E	E	G	E	E	E	E	E	E
Dowtherm "A"	E	E	E	E	E	E	E	E	E
Ethanol amine	G	G	G	G	G	G	G	G	G
Ethers	E	E	E	E	E	E	E	E	E
Ethyl acetate (esters)	E	E	G	E	E	E	E	E	E
Ethylene glycol	E	E	G	E	E	E	E	E	E
Ferric chloride	P	P	P	P	P	P	P	P	P
Ferric sulfate	P	P	P	P	P	P	P	P	P
Ferrous chloride	G	G	P	G	G	G	G	G	G
Ferrous sulfate	G	G	P	G	G	G	G	G	G
Formaldehyde (aldehydes)	E	E	G	E	E	E	E	E	E
Formic acid	G	G	P	F	G	G	G	G	G
Freon									
Dry	E	E	E	E	E	E	E	E	E

(continued)

Table 13.15 (continued)

Corrosive medium	Coppers	Low-zinc brasses	High-zinc brasses	Special brasses	Phosphor bronzes	Aluminum bronzes	Silicon bronzes	Copper nickels	Nickel silvers
Freon (continued)									
Moist	E	E	E	E	E	E	E	E	E
Fuel oil									
Light	E	E	E	E	E	E	E	E	E
Heavy	E	E	G	E	E	E	E	E	E
Furfural	E	E	F	E	E	E	E	E	E
Gasoline	E	E	E	E	E	E	E	E	E
Gelatin(a)	E	E	E	E	E	E	E	E	E
Glucose(a)	E	E	E	E	E	E	E	E	E
Glue	E	E	G	E	E	E	E	E	E
Glycerin	E	E	G	E	E	E	E	E	E
Hydrobromic acid	F	F	P	F	F	F	F	F	F
Hydrocarbons	E	E	E	E	E	E	E	E	E
Hydrochloric acid (muriatic)	F	F	P	F	F	F	F	F	F
Hydrocyanic acid									
Dry	E	E	E	E	E	E	E	E	E
Moist	P	P	P	P	P	P	P	P	P
Hydrofluoric acid									
Anhydrous	G	G	P	G	G	G	G	G	G
Hydrated	F	F	P	F	F	F	F	F	F
Hydrofluosilicic acid	G	G	P	G	G	G	G	G	G
Hydrogen(d)	E	E	E	E	E	E	E	E	E
Hydrogen peroxide									
Up to 10%	G	G	F	G	G	G	G	G	G
Over 10%	P	P	P	P	P	P	P	P	P
Hydrogen sulfide									
Dry	E	E	E	E	E	E	E	E	E
Moist	P	P	F	F	P	P	P	P	F
Kerosine	E	E	E	E	E	E	E	E	E

(continued)

Table 13.15 (continued)

Corrosive medium	Coppers	Low-zinc brasses	High-zinc brasses	Special brasses	Phosphor bronzes	Aluminum bronzes	Silicon bronzes	Copper nickels	Nickel silvers
Ketones	E	E	E	E	E	E	E	E	E
Lacquers	E	E	E	E	E	E	E	E	E
Lacquer thinners (solvents)	E	E	E	E	E	E	E	E	E
Lactic acid(a)	E	E	F	E	E	E	E	E	E
Lime	E	E	E	E	E	E	E	E	E
Lime sulfur	P	P	F	F	P	P	P	F	F
Linseed oil	G	G	G	G	G	G	G	G	G
Lithium compounds	G	G	P	F	G	G	G	E	E
Magnesium chloride	G	G	F	F	G	G	G	G	G
Magnesium hydroxide	E	E	G	E	E	E	E	E	E
Magnesium sulfate	E	E	G	E	E	E	E	E	E
Mercury or mercury salts	P	P	P	P	P	P	P	P	P
Milk(a)	E	E	G	E	E	E	E	E	E
Molasses	E	E	G	E	E	E	E	E	E
Natural gas(d)	E	E	E	E	E	E	E	E	E
Nickel chloride	F	F	P	F	F	F	F	F	F
Nickel sulfate	F	F	P	F	F	F	F	F	F
Nitric acid	P	P	P	P	P	P	P	P	P
Oleic acid	G	G	F	G	G	G	G	G	G
Oxalic acid(g)	E	E	P	E	E	E	E	E	E
Oxygen(h)	E	E	E	E	E	E	E	E	E
Palmitic acid	G	G	F	G	G	G	G	G	G
Paraffin	E	E	E	E	E	E	E	E	E
Phosphoric acid	G	G	P	F	G	G	G	G	G
Picric acid	P	P	P	P	P	P	P	P	P
Potassium carbonate	E	E	E	E	E	E	E	E	E
Potassium chloride	G	G	P	F	G	G	G	E	E
Potassium cyanide	P	P	P	P	P	P	P	P	P
Potassium dichromate (acid)	P	P	P	P	P	P	P	P	P

(continued)

Table 13.15 (continued)

Corrosive medium	Coppers	Low-zinc brasses	High-zinc brasses	Special brasses	Phosphor bronzes	Aluminum bronzes	Silicon bronzes	Copper nickels	Nickel silvers
Potassium hydroxide	G	G	F	G	G	G	G	E	E
Potassium sulfate	E	E	G	E	E	E	E	E	E
Propane(d)	E	E	E	E	E	E	E	E	E
Rosin	E	E	E	E	E	E	E	E	E
Seawater	G	G	F	E	G	E	G	E	E
Sewage	E	E	F	E	E	E	E	E	E
Silver salts	P	P	P	P	P	P	P	P	P
Soap solution	E	E	E	E	E	E	E	E	E
Sodium bicarbonate	E	E	G	E	E	E	E	E	E
Sodium bisulfate	G	G	F	G	G	G	G	E	E
Sodium carbonate	E	E	G	E	E	E	E	E	E
Sodium chloride	G	G	P	F	G	G	G	E	E
Sodium chromate	E	E	E	E	E	E	E	E	E
Sodium cyanide	P	P	P	P	P	P	P	P	P
Sodium dichromate (acid)	P	P	P	P	P	P	P	P	P
Sodium hydroxide	G	G	F	G	G	G	G	E	E
Sodium hypochlorite	G	G	P	G	G	G	G	G	G
Sodium nitrate	G	G	P	F	G	G	G	E	G
Sodium peroxide	F	F	P	F	F	F	F	G	G
Sodium phosphate	E	E	G	E	E	E	E	E	E
Sodium silicate	E	E	G	E	E	E	E	E	E
Sodium sulfate	E	E	F	E	E	E	E	E	E
Sodium sulfide	P	P	F	F	P	P	P	F	F
Sodium thiosulfate	P	P	F	F	P	P	P	F	F
Steam	E	E	F	E	E	E	E	E	E
Stearic acid	E	E	F	E	E	E	E	E	E
Sugar solutions	E	E	G	E	E	E	E	E	E
Sulfur									
Solid	G	G	E	G	G	G	G	E	G

(continued)

Table 13.15 (continued)

Corrosive medium	Coppers	Low-zinc brasses	High-zinc brasses	Special brasses	Phosphor bronzes	Aluminum bronzes	Silicon bronzes	Copper nickels	Nickel silvers
Sulfur (continued)									
Molten	P	P	P	P	P	P	P	P	P
Sulfur chloride									
Dry	E	E	E	E	E	E	E	E	E
Moist	P	P	P	P	P	P	P	P	P
Sulfur dioxide									
Dry	E	E	E	E	E	E	E	E	E
Moist	G	G	P	G	G	G	G	F	F
Sulfur trioxide (dry)	E	E	E	E	E	E	E	E	E
Sulfuric acid									
80–95%(i)	G	G	P	F	G	G	G	G	G
40–80%(i)	F	F	F	P	F	F	F	F	F
40%(i)	G	G	P	F	G	G	G	G	G
Sulfurous acid	G	G	P	G	G	G	G	F	F
Tannic acid	E	E	E	E	E	E	E	E	E
Tartaric acid(a)	E	E	G	E	E	E	E	E	E
Toluene	E	E	E	E	E	E	E	E	E
Trichloracetic acid	G	G	P	F	G	G	G	G	G
Trichlorethylene									
Dry	E	E	E	E	E	E	E	E	E
Moist	G	G	F	G	E	E	E	E	E
Turpentine	E	E	E	E	E	E	E	E	E
Varnish	E	E	E	E	E	E	E	E	E
Vinegar(a)	E	E	P	F	E	E	E	E	G
Water									
Acidic mine	F	F	P	F	G	F	F	P	F
Potable	E	E	G	E	E	E	E	E	E
Condensate(c)	E	E	E	E	E	E	E	E	E
Wetting agents(j)	E	E	E	E	E	E	E	E	E

(continued)

Table 13.15 (continued)

Corrosive medium	Coppers	Low-zinc brasses	High-zinc brasses	Special brasses	Phosphor bronzes	Aluminum bronzes	Silicon bronzes	Copper nickels	Nickel silvers
Whiskey(a)	E	E	E	E	E	E	E	E	E
White water	G	G	G	E	E	E	E	E	E
Zinc chloride	G	G	P	G	G	G	G	G	G
Zinc sulfate	E	E	P	E	E	E	E	E	E

(a) Copper and copper alloys are resistant to corrosion by most food products. Traces of copper may be dissolved and affect taste or color of the products. In such cases, copper alloys are often tin coated. (b) Acetylene forms an explosive compound with copper when moisture or certain impurities are present and the gas is under pressure. Alloys containing less than 65% Cu are satisfactory; when the gas is not under pressure, other copper alloys are satisfactory. (c) Precautions should be taken to avoid SCC. (d) At elevated temperatures, hydrogen will react with tough pitch copper, causing failure by embrittlement. (e) Where air is present, corrosion rate may be increased. (f) Below 150 °C (300 °F), corrosion rate is very low; above this temperature, corrosion is appreciable and increases rapidly with temperature. (g) Aeration and elevated temperature may increase corrosion rate substantially. (h) Excessive oxidation may begin above 120 °C (250 °F). If moisture is present, oxidation may begin at lower temperatures. (i) Use of high-zinc brasses should be avoided in acids because of the likelihood of rapid corrosion by dezincification. Copper, low-zinc brasses, phosphor bronzes, silicon bronzes, aluminum bronzes, and copper nickels offer good resistance to corrosion by hot and cold dilute H_2SO_4 and to corrosion by cold concentrated H_2SO_4. Intermediate concentrations of H_2SO_4 are sometimes more corrosive to copper alloys than either concentrated or dilute acid. Concentrated H_2SO_4 may be corrosive at elevated temperatures due to breakdown of acid and formation of metallic sulfides and sulfur dioxide, which cause localized pitting. Tests indicate that copper alloys may undergo pitting in 90 to 95% H_2SO_4 at about 50 °C (122 °F), in 80% acid at about 70 °C (160 °F), and in 60% acid at about 100 °C (212 °F). (j) Wetting agents may increase corrosion rates of copper and copper alloys slightly to substantially when carbon dioxide or oxygen is present by preventing formation of a film on the metal surface and by combining (in some instances) with the dissolved copper to produce a green, insoluble compound.

Table 13.16 Corrosion ratings of cast copper alloys in various media

The letters A, B, and C have the following significance: A, recommended; B, acceptable; C, not recommended

Corrosive medium	Copper	Tin bronze	Leaded tin bronze	High-leaded tin bronze	Leaded red brass	Leaded semi-red brass	Leaded yellow brass	Leaded high-strength yellow brass	High-strength yellow brass	Aluminum bronze	Leaded nickel brass	Leaded nickel bronze	Silicon bronze	Silicon brass
Acetate solvents	B	A	A	A	A	A	B	A	A	A	A	A	A	B
Acetic acid														
20%	A	C	B	C	B	C	C	C	C	A	C	A	A	B
50%	A	C	B	C	B	C	C	C	C	A	C	B	A	B
Glacial	A	A	A	C	A	C	C	C	C	A	B	B	A	A
Acetone	A	A	A	A	A	A	A	A	A	A	A	A	A	A
Acetylene(a)	C	C	C	C	C	C	C	C	C	C	C	C	C	C
Alcohols(b)	A	A	A	A	A	A	A	A	A	A	A	A	A	A
Aluminum chloride	C	C	C	C	C	C	C	C	C	C	C	C	C	C
Aluminum sulfate	B	B	B	B	B	C	C	C	C	B	C	C	A	A
Ammonia, moist gas	C	C	C	C	C	C	C	C	C	C	C	C	C	C
Ammonia, moisture-free	A	A	A	A	A	A	A	A	A	A	A	A	A	A
Ammonium chloride	C	C	C	C	C	C	C	C	C	C	C	C	C	C
Ammonium hydroxide	C	C	C	C	C	C	C	C	C	C	C	C	C	C
Ammonium nitrate	C	C	C	C	C	C	C	C	C	C	C	C	C	C
Ammonium sulfate	B	B	B	B	B	C	C	C	C	B	C	C	A	A
Aniline and aniline dyes	C	C	C	C	C	C	C	C	C	C	C	C	C	C
Asphalt	A	A	A	A	A	A	A	A	A	A	A	A	A	A
Barium chloride	A	A	A	A	A	C	C	C	C	A	A	A	A	A
Barium sulfide	C	C	C	C	C	C	C	C	B	C	C	C	C	C
Beer(b)	A	A	B	B	B	B	C	A	A	A	C	A	A	B
Beet-sugar syrup	A	A	B	B	B	A	A	B	B	A	A	A	B	B
Benzine	A	A	A	A	A	A	A	A	A	A	A	A	A	A
Benzol	A	A	A	A	A	A	A	A	A	A	A	A	A	A
Boric acid	A	A	A	A	A	A	A	B	A	A	A	A	A	A
Butane	A	A	A	A	A	A	A	A	A	A	A	A	A	A
Calcium bisulfite	A	A	B	B	B	C	C	C	C	A	B	A	A	B

(continued)

Table 13.16 (continued)

Corrosive medium	Copper	Tin bronze	Leaded tin bronze	High-leaded tin bronze	Leaded red brass	Leaded semi-red brass	Leaded yellow brass	Leaded high-strength yellow brass	High-strength yellow brass	Aluminum bronze	Leaded nickel brass	Leaded nickel bronze	Silicon bronze	Silicon brass
Calcium chloride (acid)	B	B	B	B	B	B	C	C	C	A	C	C	A	C
Calcium chloride (alkaline)	C	C	C	C	C	C	C	C	C	A	C	A	C	B
Calcium hydroxide	C	C	C	C	C	C	C	C	C	B	C	C	C	C
Calcium hypochlorite	C	C	B	B	B	C	C	C	C	B	C	C	C	C
Cane-sugar syrups	A	A	B	A	B	A	A	A	A	A	A	A	A	B
Carbonated beverages(b)	A	C	C	C	C	C	C	C	C	A	C	C	A	C
Carbon dioxide, dry	A	A	A	A	A	A	A	A	A	A	A	A	A	A
Carbon dioxide, moist(b)	B	B	B	C	B	B	C	C	C	A	C	A	A	B
Carbon tetrachloride, dry	A	A	A	A	A	A	A	A	A	A	A	A	A	A
Carbon tetrachloride, moist	B	B	B	B	B	B	B	B	B	B	B	B	A	A
Chlorine, dry	A	A	A	A	A	A	A	A	A	A	A	A	A	A
Chlorine, moist	C	C	B	B	B	C	C	C	C	C	C	C	C	C
Chromic acid	C	C	C	C	C	C	C	C	C	C	C	C	C	C
Citric acid	A	A	A	A	A	A	A	A	A	A	A	A	A	A
Copper sulfate	B	A	B	B	B	C	C	C	C	B	B	B	A	B
Cottonseed oil(b)	A	A	A	A	A	A	A	A	A	A	A	B	A	A
Creosote	B	B	B	B	B	A	A	A	A	A	B	B	B	B
Ethers	A	A	A	A	A	A	A	A	A	A	A	A	A	A
Ethylene glycol	A	A	A	A	A	A	A	A	A	A	A	A	A	A
Ferric chloride, sulfate	C	C	C	C	C	C	C	C	C	C	C	C	C	C
Ferrous chloride, sulfate	C	C	C	C	C	C	C	C	C	C	C	C	C	C
Formaldehyde	A	A	A	A	A	A	A	A	A	A	A	A	A	A
Formic acid	A	A	A	A	A	B	B	B	B	A	B	B	B	C
Freon	A	A	A	A	A	A	A	A	A	A	A	A	A	B
Fuel oil	A	A	A	A	A	A	A	A	A	A	A	A	A	A
Furfural	A	A	A	A	A	A	A	A	A	A	A	A	A	A
Gasoline	A	A	A	A	A	A	A	A	A	A	A	A	A	A
Gelatin(b)	A	A	A	A	A	A	A	A	A	A	A	A	A	A

(continued)

Table 13.16 (continued)

Corrosive medium	Copper	Tin bronze	Leaded tin bronze	High-leaded tin bronze	Leaded red brass	Leaded semi-red brass	Leaded yellow brass	Leaded high-strength yellow brass	High-strength yellow brass	Aluminum bronze	Leaded nickel brass	Leaded nickel bronze	Silicon bronze	Silicon brass
Glucose	A	A	A	A	A	A	A	A	A	A	A	A	A	A
Glue	A	A	A	A	A	A	A	A	A	A	A	A	A	A
Glycerin	A	A	A	A	A	A	A	A	A	A	A	A	A	A
Hydrochloric or muriatic acid	C	C	C	C	C	C	C	C	C	B	C	C	C	C
Hydrofluoric acid	B	B	B	B	B	B	B	B	B	A	B	B	B	B
Hydrofluosilicic acid	B	B	B	B	B	C	C	C	C	B	C	C	B	C
Hydrogen	A	A	A	A	A	A	A	A	A	A	A	A	A	A
Hydrogen peroxide	C	C	C	C	C	C	C	C	C	C	C	C	C	C
Hydrogen sulfide, dry	C	C	C	C	C	C	C	C	C	B	C	C	B	C
Hydrogen sulfide, moist	C	C	C	C	C	C	C	C	C	B	C	C	C	C
Lacquers	A	A	A	A	A	A	A	A	A	A	A	A	A	A
Lacquer thinners	A	A	A	A	A	A	A	A	A	A	A	A	A	A
Lactic acid	A	A	A	A	A	C	C	C	C	A	C	C	A	C
Linseed oil	A	A	A	A	A	A	A	A	A	A	A	A	A	A
Liquors														
Black liquor	B	B	B	B	B	C	C	C	C	B	C	C	B	B
Green liquor	C	C	C	C	C	C	C	C	C	B	C	C	C	B
White liquor	C	C	C	C	C	C	C	C	C	A	C	C	C	B
Magnesium chloride	A	A	A	A	A	A	A	A	C	A	C	C	A	B
Magnesium hydroxide	B	B	B	B	B	B	B	B	B	A	B	B	B	B
Magnesium sulfate	A	A	A	A	A	C	C	C	C	A	C	B	B	B
Mercury, mercury salts	C	C	C	C	C	C	C	C	C	C	C	C	C	C
Milk(b)	A	A	A	A	A	A	A	A	A	A	A	A	A	A
Molasses(b)	A	A	A	A	A	A	A	A	A	A	A	A	A	A
Natural gas	A	A	A	A	A	A	A	A	A	A	A	A	A	A
Nickel chloride	A	A	A	A	A	C	C	C	C	B	C	C	A	C
Nickel sulfate	A	A	A	A	A	C	C	C	C	A	C	C	A	C
Nitric acid	C	C	C	C	C	C	C	C	C	C	C	C	C	C

(continued)

Table 13.16 (continued)

Corrosive medium	Copper	Tin bronze	Leaded tin bronze	High-leaded tin bronze	Leaded red brass	Leaded semi-red brass	Leaded yellow brass	Leaded high-strength yellow brass	High-strength yellow brass	Aluminum bronze	Leaded nickel brass	Leaded nickel bronze	Silicon bronze	Silicon brass
Oleic acid	A	A	B	B	B	C	C	C	C	A	C	A	A	B
Oxalic acid	A	A	B	B	B	C	C	C	C	A	C	A	A	B
Phosphoric acid	A	A	A	A	A	C	C	C	C	A	C	A	A	A
Picric acid	C	C	C	C	C	C	C	C	C	C	C	C	C	C
Potassium chloride	A	A	A	A	A	C	C	C	C	A	C	C	A	C
Potassium cyanide	C	C	C	C	C	C	C	C	C	C	C	C	C	C
Potassium hydroxide	C	C	C	C	C	C	C	C	C	C	C	C	C	C
Potassium sulfate	A	A	A	A	A	C	C	C	C	A	C	C	A	C
Propane gas	A	A	A	A	A	A	A	A	A	A	A	A	A	A
Seawater	A	A	A	A	A	C	C	C	C	A	C	C	B	B
Soap solutions	A	A	A	A	B	C	C	C	C	A	C	C	A	C
Sodium bicarbonate	A	A	A	A	A	A	A	A	A	A	A	A	A	B
Sodium bisulfate	C	C	C	C	C	C	C	C	C	A	C	C	C	C
Sodium carbonate	C	A	A	A	A	C	C	C	C	A	C	C	A	A
Sodium chloride	A	A	A	A	A	B	C	C	C	A	C	C	A	C
Sodium cyanide	C	C	C	C	C	C	C	C	C	B	C	C	C	C
Sodium hydroxide	C	C	C	C	C	C	C	C	C	A	C	C	C	C
Sodium hypochlorite	C	C	C	C	C	C	C	C	C	C	C	C	C	C
Sodium nitrate	B	B	B	B	B	B	B	B	B	A	B	B	A	A
Sodium peroxide	B	B	B	B	B	B	B	B	B	B	B	B	B	B
Sodium phosphate	A	A	A	A	A	A	A	A	A	A	A	A	A	A
Sodium sulfate, silicate	A	A	B	B	B	B	C	C	C	A	C	C	A	B
Sodium sulfide, thiosulfate	C	C	C	C	C	C	C	C	C	B	C	C	C	C
Stearic acid	C	A	A	A	A	A	C	A	A	A	A	A	A	A
Sulfur, solid	C	C	C	C	C	C	C	C	C	A	C	C	C	C
Sulfur chloride	C	C	C	C	C	C	C	C	C	C	C	C	C	C
Sulfur dioxide, dry	A	A	A	A	A	A	A	A	A	A	A	A	A	A
Sulfur dioxide, moist	A	A	A	A	B	C	C	C	C	A	C	C	A	B
Sulfur trioxide, dry	A	A	A	A	A	A	A	A	A	A	A	A	A	A

(continued)

Corrosion Data 187

Table 13.16 (continued)

Corrosive medium	Copper	Tin bronze	Leaded tin bronze	High-leaded tin bronze	Leaded red brass	Leaded semi-red brass	Leaded yellow brass	Leaded high-strength yellow brass	High-strength yellow brass	Aluminum bronze	Leaded nickel brass	Leaded nickel bronze	Silicon bronze	Silicon brass
Sulfuric acid														
78% or less	B	B	B	B	B	C	C	C	C	A	C	C	B	B
78% to 90%	C	C	C	C	C	C	C	C	C	B	C	C	C	C
90% to 95%	C	C	C	C	C	C	C	C	C	B	C	C	C	C
Fuming	C	C	C	C	C	C	C	C	C	A	C	C	C	C
Tannic acid	A	A	A	A	A	A	A	A	A	A	A	A	A	A
Tartaric acid	B	A	A	A	A	A	A	A	A	A	A	A	A	A
Toluene	B	B	A	A	A	B	B	B	A	B	B	B	B	A
Trichlorethylene, dry	A	A	A	A	A	A	A	A	A	A	A	A	A	A
Trichlorethylene, moist	A	A	A	A	A	A	A	A	A	A	A	A	A	A
Turpentine	A	A	A	A	A	A	A	A	A	A	A	A	A	A
Varnish	A	A	A	A	A	A	A	A	A	A	A	A	A	A
Vinegar	A	A	B	B	B	C	C	C	C	B	C	C	C	B
Water, acid mine	C	C	C	C	C	C	C	C	C	C	C	C	C	C
Water, condensate	A	A	A	A	A	A	A	A	A	A	A	A	A	A
Water, potable	A	A	A	A	A	A	B	B	B	A	A	A	A	A
Whiskey(b)	A	A	C	C	C	C	C	C	C	A	C	C	A	C
Zinc chloride	C	C	C	C	C	C	C	C	C	B	C	C	B	C
Zinc sulfate	A	A	A	A	A	C	C	C	C	B	C	A	A	C

(a) Acetylene forms an explosive compound with copper when moist or when certain impurities are present and the gas is under pressure. Alloys containing less than 65% Cu are satisfactory for this use. When gas is not under pressure, other copper alloys are satisfactory. (b) Copper and copper alloys resist corrosion by most food products. Traces of copper may be dissolved and affect taste or color. In such cases, copper metals are often tin coated.

Table 13.17 Atmospheric corrosion of selected copper alloys

Alloy	Altoona, PA µm/yr	Altoona, PA mils/yr	New York, NY µm/yr	New York, NY mils/yr	Key West, FL µm/yr	Key West, FL mils/yr	La Jolla, CA µm/yr	La Jolla, CA mils/yr	State College, PA µm/yr	State College, PA mils/yr	Phoenix, AZ µm/yr	Phoenix, AZ mils/yr
C1100	1.40	0.055	1.38	0.054	0.56	0.022	1.27	0.050	0.43	0.017	0.13	0.005
C12000	1.32	0.052	1.22	0.048	0.51	0.020	1.42	0.056	0.36	0.014	0.08	0.003
C23000	1.88	0.074	1.88	0.074	0.56	0.022	0.33	0.013	0.46	0.018	0.10	0.004
C26000	3.05	0.120	2.41	0.095	0.20	0.008	0.15	0.006	0.46	0.018	0.10	0.004
C52100	2.24	0.088	2.54	0.100	0.71	0.028	2.31	0.091	0.33	0.013	0.13	0.005
C61000	1.63	0.064	1.60	0.063	0.10	0.004	0.15	0.006	0.25	0.010	0.51	0.002
C65500	1.65	0.065	1.73	0.068	...	...	1.38	0.054	0.51	0.020	0.15	0.006
C44200	2.13	0.084	2.51	0.099	...	...	0.33	0.013	0.53	0.021	0.10	0.004
70Cu-29Ni-1Sn(b)	2.64	0.104	2.13	0.084	0.28	0.011	0.36	0.014	0.48	0.019	0.10	0.004

(a) Derived from 20-year exposure tests. Types of atmospheres: Altoona, industrial; New York City, industrial marine; Key West, tropical rural marine; La Jolla, humid marine; State College, northern rural; Phoenix, dry rural. (b) Although obsolete, this alloy indicates the corrosion resistance expected of C71500.

14 Coefficients of Friction

In tribology (the science and technology of friction, lubrication, and wear), the *coefficient of friction* is the dimensionless ratio of the friction force (F) between two bodies to the normal force (N) pressing these bodies together:

$$\mu \text{ (or } f) = (F/N)$$

The *static coefficient of friction* is the coefficient of friction corresponding to the maximum friction force that must be overcome to initiate macroscopic motion between two bodies. The *kinetic coefficient of friction* is the coefficient of friction under conditions of macroscopic relative motion between two bodies. In most cases, a greater force is needed to set a resting body in motion than to sustain the motion; in other words, the static coefficient of friction is usually somewhat greater than the kinetic coefficient of friction. For metals, the friction coefficient typically ranges from 0.03 for a very well lubricated bearing to 0.5 to 0.9 for dry sliding.

The five tables of friction coefficient values in this chapter contain both static and kinetic friction coefficients. They are arranged by material type as follows:

- Table 14.1 Metals on metals
- Table 14.2 Ceramics on various materials
- Table 14.3 Polymers on various materials
- Table 14.4 Coatings on various materials
- Table 14.5 Miscellaneous materials

It should be emphasized that the data in the tables are for unlubricated solids at room temperature and in ambient air. The reference list provided with each table lists both the sources of the data for the table and a brief description of the testing conditions used to generate these data, if such information was available in the reference. If accurate friction information is required for a specific application, the use of carefully simulated conditions or instrumentation of the actual machine should be conducted in lieu of using tabulated values because even a small change in contact conditions (for example, sliding speed or relative humidity for some materials) can result in a marked change in the measured or apparent friction coefficient.

Following the tables, a metal-on-metal compatibility chart is presented. This generalized "map" shows which metals can safely slide against one another and which metal couples should be avoided.

Table 14.1 Friction coefficient data for metals sliding on metals

Metals tested in air at room temperature

Material (Fixed specimen)	Material (Moving specimen)	Test geometry(a)	Friction coefficient Static	Friction coefficient Kinetic	Ref
Ag	Ag	IS	0.50	...	1
	Au	IS	0.53	...	1
	Cu	IS	0.48	...	1
	Fe	IS	0.49	...	1
Al	Al	IS	0.57	...	1
	Ti	IS	0.54	...	1
Al, alloy 6061-T6	Al, alloy 6061-T6	FOF	0.42	0.34	2
	Cu	FOF	0.28	0.23	2
	Steel, 1032	FOF	0.35	0.25	2
	Ti-6Al-4V	FOF	0.34	0.29	2
Au	Ag	IS	0.53	...	1
	Au	IS	0.49	...	1
Brass, 60Cu-40Zn	Steel, tool	POR	...	0.24	3
Cd	Cd	IS	0.79	...	1
	Fe	IS	0.52	...	1
Co	Co	IS	0.56	...	1
	Cr	IS	0.41	...	1
Cr	Co	IS	0.41	...	1
	Cr	IS	0.46	...	1
Cu	Co	IS	0.44	...	1
	Cr	IS	0.46	...	1
	Cu	IS	0.55	...	1
	Fe	IS	0.50	...	1
	Ni	IS	0.49	...	1
	Zn	IS	0.56	...	1
Cu, OFHC	Steel, 4619	BOR	...	0.82	4
Fe	Co	IS	0.41	...	1
	Cr	IS	0.48	...	1
	Fe	IS	0.51	...	1
	Mg	IS	0.51	...	1
	Mo	IS	0.46	...	1
	Ti	IS	0.49	...	1
	W	IS	0.47	...	1
	Zn	IS	0.55	...	1
In	In	IS	1.46	...	1
Mg	Mg	IS	0.69	...	1
Mo	Fe	IS	0.46	...	1
	Mo	IS	0.44	...	1
Nb	Nb	IS	0.46	...	1
Ni	Cr	IS	0.59	...	1
	Ni	IS	0.50	...	1
	Pt	IS	0.64	...	1
Pb	Ag	IS	0.73	...	1
	Au	IS	0.61	...	1
	Co	IS	0.55	...	1
	Cr	IS	0.53	...	1
	Fe	IS	0.54	...	1
	Pb	IS	0.90	...	1
	Steel	SPOF	...	0.80	5
Pt	Ni	IS	0.64	...	1
	Pt	IS	0.55	...	1
Sn	Fe	IS	0.55	...	1
	Sn	IS	0.74	...	1
Steel	Cu	SPOF	...	0.80	5
	Pb	SPOF	...	1.40	5

(continued)

Table 14.1 (continued)

Fixed specimen	Moving specimen	Test geometry(a)	Static	Kinetic	Ref
Steel, 1020	Steel, 4619	BOR	...	0.54	4
Steel, 1032	Al, alloy 6061-T6	FOF	0.47	0.38	2
	Cu	FOF	0.32	0.25	2
	Steel, 1032	FOF	0.31	0.23	2
	Ti-6Al-4V	FOF	0.36	0.32	2
Steel, 52100	Ni$_3$Al, alloy IC-396M	RSOF	...	1.08	6
	Ni$_3$Al, alloy IC-50	RSOF	...	0.70	6
	Steel, 1015 annealed	BOR	...	0.74	7
	Steel, dual-phase DP-80	BOR	...	0.55	7
	Steel, O2 tool	BOR	...	0.49	7
Steel, mild	Steel, mild	BOR	...	0.62	3
Steel, M50 tool	Ni$_3$Al, alloy IC-50	RSOF	...	0.68	6
Steel, stainless	Steel, tool	POR	...	0.53	3
Steel, stainless 304	Cu	FOF	0.23	0.21	2
Stellite	Steel, tool	POR	...	0.60	3
Ti	Al	IS	0.54	...	1
	Steel, 17-4 stainless	POF	0.48	0.48	8
	Ti	POF	0.47	0.40	8
	Ti	FOF	0.55	...	1
	Ti-6Al-4V	POF	0.43	0.36	8
Ti-6Al-4V	Al, alloy 6061-T6	FOF	0.41	0.38	2
	Cu-Al (bronze)	POF	0.36	0.27	8
	Nitronic 60	POF	0.38	0.31	8
	Steel, 17-4 stainless	POF	0.36	0.31	8
	Steel, type 440C stainless	POF	0.44	0.37	8
	Stellite 12	POF	0.35	0.29	8
	Stellite 6	POF	0.45	0.36	8
	Ta	POF	0.53	0.53	8
	Ti-6Al-4V	FOF	0.36	0.30	2
	T-6Al-4V	POF	0.36	0.31	8
W	Cu	IS	0.41	...	1
	Fe	IS	0.47	...	1
	W	IS	0.51	...	1
Zn	Cu	IS	0.56	...	1
	Fe	IS	0.55	...	1
	Zn	IS	0.75	...	1
Zr	Zr	IS	0.63	...	1

(a) Test geometry codes: BOR, flat block pressed against the cylindrical surface of a rotating ring; FOF, flat surface sliding on another flat surface; IS, sliding down an inclined surface; POR, pin sliding against the cylindrical surface of a rotating ring; RSOF, reciprocating, spherically ended pin on a flat surface; SPOF, spherically ended pin on a flat coupon

REFERENCES/TEST CONDITIONS FOR TABLE 14.1

1. E. Rabinowicz, *ASLE Trans.*, Vol 14, 1971, p 198; plate sliding on plate at 50% relative humidity
2. "Friction Data Guide," General Magnaplate Corporation, 1988; TMI Model 98-5 slip and friction tester, 1.96 N (0.200 kgf) load, ground specimens, 54% relative humidity, average of five tests
3. J.F. Archard, *ASME Wear Control Handbook,* M.B. Peterson and W.O. Winer, Ed., American Society of Mechanical Engineers, 1980, p 38; pin-on-rotating ring, 3.9 N (0.40 kgf) load, 1.8 m/s (350 ft/min) velocity

4. A.W. Ruff, L.K. Ives, and W.A. Glaeser, *Fundamentals of Friction and Wear of Materials*, ASM International, 1981, p 235; flat block-on-rotating 35 mm ($1^3/_8$ in.) diameter ring, 10 N (1.02 kgf) load, 0.2 m/s (40 ft/min) velocity
5. F.P. Bowden and D. Tabor, *The Friction and Lubrication of Solids*, Oxford Press, 1986, p 127; sphere-on-flat, unspecified load and velocity
6. P.J. Blau and C.E. DeVore, *Tribol. Int.*, Vol 23 (No. 4), 1990, p 226; reciprocating ball-on-flat, 10 Hz, 25 N (2.6 kgf) load, 10 mm stroke
7. P.J. Blau, *J. Tribology*, Vol 107, 1985, p 483; flat block-on-rotating 35 mm ($1^3/_8$ in.) diameter ring, 133 N (13.6 kgf) load, 5.0 cm/s (2.0 in./s) velocity
8. K.G. Budinski, *Proc. Wear of Materials*, American Society of Mechanical Engineers, 1991, p 289; modified ASTM G 98 galling test procedure

Table 14.2 Friction coefficient data for ceramics sliding on various materials
Specimens tested in air at room temperature

Fixed specimen	Moving specimen	Test geometry(a)	Static	Kinetic	Ref
Ag	Alumina	RPOF	...	0.37	1
	Zirconia	RPOF	...	0.39	1
Al	Alumina	RPOF	...	0.75	1
	Zirconia	RPOF	...	0.63	1
Alumina	Alumina	SPOD	...	0.50	2
	Alumina	SPOD	...	0.52	3
	Alumina	SPOD	...	0.33	4
	WRA(b)	SPOD	...	0.53	5
	WRZTA(c)	SPOD	...	0.50	5
	ZTA(d)	SPOD	...	0.56	5
Boron carbide	Boron carbide	POD	...	0.53	6
Cr	Alumina	RPOF	...	0.50	1
	Zirconia	RPOF	...	0.61	1
Cu	Alumina	RPOF	...	0.43	1
	Zirconia	RPOF	...	0.40	1
Fe	Alumina	RPOF	...	0.45	1
	Zirconia	RPOF	...	0.35	1
Glass, tempered	Al, alloy 6061-T6	FOF	0.17	0.14	7
	Steel, 1032	FOF	0.13	0.12	7
	Teflon(e)	FOF	0.10	0.10	7
Silicon carbide	Silicon carbide	SPOD	...	0.52	6
	Silicon nitride	SPOD	...	0.53	4
	Silicon nitride	SPOD	...	0.71	2
	Silicon nitride	SPOD	...	0.63	3
Silicon nitride	Silicon carbide	SPOD	...	0.54	4
	Silicon carbide	SPOD	...	0.67	2
	Silicon carbide	SPOD	...	0.84	3
	Silicon nitride	SPOD	...	0.17	6
Steel, M50 tool	Boron carbide	POD	...	0.29	6
	Silicon carbide	POD	...	0.29	6
	Silicon nitride	POD	...	0.15	6
	Tungsten carbide	POD	...	0.19	6
Ti	Alumina	RPOF	...	0.42	1
	Zirconia	RPOF	...	0.27	1
Tungsten carbide	Tungsten carbide	POD	...	0.34	6

(a) Test geometry codes: FOF, flat surface sliding on another flat surface; POD, pin on disk (pin tip geometry not given); RPOF, reciprocating pin on flat; SPOD, spherically ended pin on flat disk; SPOF, spherically ended pin on a flat coupon. (b) WRA, silicon carbide whisker-reinforced alumina. (c) WRZTA, silicon carbide whisker-reinforced, zirconia-toughened alumina. (d) ZTA, zirconia-toughened alumina. (e) Teflon, polytetrafluoroethylene

REFERENCES/TEST CONDITIONS FOR TABLE 14.2

1. K. Demizu, R. Wadabayashim, and H. Ishigaki, *Tribol. Trans.*, Vol 33 (No. 4), 1990, p 505; 1.5 mm (0.060 in.) radius pin reciprocating on a flat, 4 N (0.4 kgf) load, 0.17 mm/s (0.0067 in./s) velocity, 50% relative humidity
2. P.J. Blau, Oak Ridge National Laboratory
3. P.J. Blau, Oak Ridge National Laboratory, 1.0 N (0.10 kgf) load and 0.1 m/s (20 ft/min) velocity
4. P.J. Blau, Oak Ridge National Laboratory, 10 N (1.0 kgf) load and 0.1 m/s (20 ft/min) velocity
5. C.S. Yust, *Tribology of Composite Materials*, P.K. Rohatgi, P.J. Blau, and C.S. Yust, Ed., ASM International, 1990, p 27; 9.5 mm ($^3/_8$ in.) diameter sphere-on-disk, 2 to 9 N (0.2 to 0.9 kgf) load, 0.3 m/s (60 ft/min) velocity
6. B. Bhushan and B.K. Gupta, table in *Handbook of Tribology*, McGraw-Hill, 1991; 20 N (2.0 kgf), 3 mm/s (0.12 in./s) velocity
7. "Friction Data Guide," General Magnaplate Corporation, 1988; TMI Model 98-5 slip and friction tester, 1.96 N (0.200 kgf) load, ground specimens, 54% relative humidity, average of five tests

Table 14.3 Friction coefficient data for polymers sliding on various materials
Specimens tested in air at room temperature

Fixed specimen	Moving specimen	Test geometry(b)	Static	Kinetic	Ref
Polymers sliding on polymers					
Acetal	Acetal	TW	0.06	0.07	1
Nylon 6/6	Nylon 6/6	TW	0.06	0.07	1
PMMA	PMMA	NSp	0.80	...	2
Polyester PBT	Polyester PBT	TW	0.17	0.24	1
Polystyrene	Polystyrene	NSp	0.50	...	2
Polyethylene	Polyethylene	NSp	0.20	...	2
Teflon	Teflon	NSp	0.04	...	2
	Teflon	FOF	0.08	0.07	3
Dissimilar pairs with the polymer as the fixed specimen					
Nylon 6 (cast)	Steel, mild	TPOD	...	0.35	4
(extruded)	Steel, mild	TPOD	...	0.37	4
Nylon 6/6	Polycarbonate	TW	0.25	0.04	1
Nylon 6/6 (+ PTFE)	Steel, mild	TPOD	...	0.35	4
PA 66	Steel, 52100	BOR	...	0.57	5
PA 66 (+ 15% PTFE)	Steel, 52100	BOR	...	0.13	5
PA 66 (PTFE/glass)	Steel, 52100	BOR	...	0.31	5
PEEK	Steel, 52100	BOR	...	0.49	5
PEEK (+ 15% PTFE)	Steel, 52100	BOR	...	0.18	5
PEEK (PTFE/glass)	Steel, 52100	BOR	...	0.20	5
PEI	Steel, 52100	BOR	...	0.43	5
PEI (+ 15% PTFE)	Steel, 52100	BOR	...	0.21	5
PEI (PTFE/glass)	Steel, 52100	BOR	...	0.21	5
PETP	Steel, 52100	BOR	...	0.68	5
PETP (+ 15% PTFE)	Steel, 52100	BOR	...	0.14	5
PETP (PTFE/glass)	Steel, 52100	BOR	...	0.18	5
Polyurethane(c)	Steel, mild	TPOD	...	0.51	4
Polyurethane(d)	Steel, mild	TPOD	...	0.35	4
POM	Steel, 52100	BOR	...	0.45	5
POM (+ 15% PTFE)	Steel, 52100	BOR	...	0.21	5
POM (PTFE/glass)	Steel, 52100	BOR	...	0.23	5
PPS	Steel, 52100	BOR	...	0.70	5
PPS (+ 15% PTFE)	Steel, 52100	BOR	...	0.30	5

(continued)

Table 14.3 (continued)

Fixed specimen	Moving specimen	Test geometry(b)	Static	Kinetic	Ref
Dissimilar pairs with the polymer as the fixed specimen (continued)					
PPS (PTFE/glass)	Steel, 52100	BOR	...	0.39	5
Teflon	Al, alloy 6061-T6	FOF	0.24	0.19	3
	Cr plate	FOF	0.09	0.08	3
	Cu	FOF	0.13	0.11	3
	Ni (0.001 P)	FOF	0.15	0.12	3
	Steel, 1032	FOF	0.27	0.27	3
	Ti-6Al-4V	FOF	0.17	0.14	3
	TiN (Magnagold)	FOF	0.15	0.12	3
UHMWPE	Steel, mild	TOPD	...	0.14	4
Dissimilar pairs with the polymer as the moving specimen					
Steel, carbon	ABS resin	POF	0.40	0.27	6
Steel, mild	ABS	TW	0.30	0.35	1
	ABS + 15% PTFE	TW	0.13	0.16	1
	Acetal	TW	0.14	0.21	1
Steel, 52100	Acetal	POD	...	0.31	7
	HDPE	POD	...	0.25	7
Steel, carbon	HDPE	POF	0.36	0.23	6
	LDPE	POF	0.48	0.28	6
Steel, 52100	Lexan 101	POD	...	0.60	7
Steel, mild	Nylon (amorphous)	TW	0.23	0.32	1
Steel, carbon	Nylon 6	POF	0.54	0.37	6
Steel, mild	Nylon 6	TW	0.22	0.26	1
Steel, carbon	Nylon 6/6	POF	0.53	0.38	6
Steel, mild	Nylon 6/6	TW	0.20	0.28	1
Steel, carbon	Nylon 6/10	POF	0.53	0.38	6
Steel, mild	Nylon 6/10	TW	0.23	0.31	1
	Nylon 6/12	TW	0.24	0.31	1
	PEEK (Victrex)	TW	0.20	0.25	1
Steel, carbon	Phenol formaldehyde	POF	0.51	0.44	6
Steel, 52100	PMMA	POD	...	0.68	7
Steel, carbon	PMMA	POF	0.64	0.50	6
Steel, mild	Polycarbonate	TW	0.31	0.38	1
	Polyether PBT	TW	0.19	0.25	1
	Polyethylene	TW	0.09	0.13	1
Steel, carbon	Polyimide	POF	0.46	0.34	6
	Polyoxylmethylene	POF	0.30	0.17	6
	Polypropylene	POF	0.36	0.26	6
Steel, mild	Polypropylene	TW	0.08	0.11	1
Steel, carbon	Polystyrene	POF	0.43	0.37	6
Steel, mild	Polystyrene	TW	0.28	0.32	1
	Polysulfone	TW	0.29	0.37	1
Steel, carbon	PVC	POF	0.53	0.38	6
	PTFE	POF	0.37	0.09	6
Al, alloy 6061-T6	Teflon	FOF	0.19	0.18	3
Cr plate	Teflon	FOF	0.21	0.19	3
Glass, tempered	Teflon	FOF	0.10	0.10	3
Ni (0.001 P)	Teflon	FOF	0.22	0.19	3
Steel, 1032	Teflon	FOF	0.18	0.16	3
Ti-6Al-4V	Teflon	FOF	0.23	0.21	3
TiN (Magnagold)	Teflon	FOF	0.16	0.11	3

(a) ABS, acrylonitrile butadiene styrene; HDPE, high-density polyethylene; LDPE, low-density polyethylene; Lexan, trademark of the General Electric Co. (polycarbonate); nylon, one of a group of polyamide resins (see also PA); PA, polyamide; PBT, polybutylene terephthalate; PEEK, polyetheretherketone; PEI, polyetherimide; PETP, polyethylene terephthalate; PMMA, polymethylmethacrylate; POM, polyoxymethylene; PPS, polyphenylene sulfide; PTFE, polytetrafluoroethylene; PVC, polyvinyl chloride; UHMWPE, ultra high molecular weight polyethylene; Magnagold, product of General Magnaplate, Inc.; Teflon, trademark of E.I. Du Pont de Nemours & Co., Inc. (PTFE). (b) Test geometry codes: BOR, flat block-on-rotating ring; FOF, flat surface sliding on another flat surface; NSp, not specified; POD, pin on disk; POF, pin on flat; TPOD, triple pin-on-disk; TW, thrust washer test. (c) Green polyurethane. (d) Cream-colored polyurethane

REFERENCES/TEST CONDITIONS FOR TABLE 14.3

1. "Lubricomp® Internally-Lubricated Reinforced Thermoplastics and Fluoropolymer Composites," Bulletin 254-688, ICI Advanced Materials; thrust washer apparatus, 0.28 MPa (40 psi), 0.25 m/s (50 ft/min), after running-in for one full rotation
2. F.P. Bowden and D. Tabor, Appendix IV, *The Friction and Lubrication of Solids*, Oxford Press, 1986; unspecified testing conditions
3. "Friction Data Guide," General Magnaplate Corporation, 1988; TMI Model 98-5 slip and friction tester, 1.96 N (0.200 kgf) load, ground specimens, 54% relative humidity, average of five tests
4. J.M. Thorp, *Tribol. Int.*, Vol 15 (No. 2), 1982, p 69; three-pin-on-rotating disk apparatus, 0.1 m/s (20 ft/min)
5. J.W.M. Mens and A.W.J. de Gee, *Wear*, Vol 149, 1991, p 255; flat block-on-rotating ring, 1.5 MPa (0.22 ksi) pressure, 150 N (15 kgf) load, 0.1 m/s (20 ft/min) velocity
6. R.P. Steijn, *Metall. Eng. Quart.*, Vol 7, 1967, p 9; 12.7 mm (0.500 in.) diameter ball-on-flat, 9.8 N (1.0 kgf) load, 0.01 mm/s (4 \ 10–4 in./s) velocity
7. N.P. Suh, *Tribophysics*, Prentice-Hall, 1986, p 226; pin-on-disk, 4.4 N (0.45 kgf) load, 3.3 cm/s (1.3 in./s) velocity, 65% relative humidity

Table 14.4 Friction coefficient data for coatings sliding on various materials
Specimens tested in air at room temperature

Fixed specimen	Moving specimen	Test geometry(a)	Friction coefficient Static	Kinetic	Ref
Al, alloy 6061-T6	Cr plate	FOF	0.27	0.22	1
	Ni (0.001 P) plate	FOF	0.33	0.25	1
	TiN (Magnagold)(c)	FOF	0.25	0.22	1
Au, electroplate	60Pd-40Ag, plate	POF	...	2.40	2
	60Pd-40Au, plate	POF	...	0.30	2
	70Au-30Ag, plate	POF	...	3.00	2
	80Pd-20Au, plate	POF	...	1.80	2
	99Au-1Co, plate	POF	...	2.40	2
	Au plate	POF	...	2.80	2
	Au-0.6Co, plate	POF	...	0.40	2
	Pd plate	POF	...	0.60	2
Cr plate	Al, alloy 6061-T6	FOF	0.20	0.19	1
	Ni (0.001 P) plate	FOF	0.19	0.17	1
	Steel, 1032	FOF	0.20	0.17	1
	Teflon(b)	FOF	0.21	0.19	1
	Ti-6Al-4V	FOF	0.38	0.33	1
Niobium carbide, coating	Niobium carbide, coating	FOF	0.19	0.13	3
Ni (0.001 P) plate	Al, alloy 6061-T6	FOF	0.26	0.23	1
	Cr plate	FOF	0.41	0.36	1
	Ni (0.001 P) plate	FOF	0.32	0.28	1
	Steel, 1032	FOF	0.35	0.31	1
	Steel, D2 tool	FOF	0.43	0.33	1
	Teflon(b)	FOF	0.22	0.19	1
	TiN (Magnagold)(c)	FOF	0.33	0.26	1
Steel	Cu film on steel	SPOD	0.30	...	4
	In film on Ag	SPOD	0.10	...	4
	In film on steel	SPOD	0.08	...	4
	Pb film on Cu	SPOD	0.18	...	4
Steel, 1032	Cr plate	FOF	0.25	0.21	1
	Ni (0.001 P) plate	FOF	0.37	0.30	1

(continued)

Table 14.4 (continued)

Fixed specimen	Moving specimen	Test geometry(a)	Friction coefficient Static	Friction coefficient Kinetic	Ref
Steel, 1032 (continued)					
	TiN (Magnagold)(c)	FOF	0.31	0.28	1
Steel, type 440C stainless	TiC on type 304 stainless	POD	0.12	0.17	5
	TiN on type 304 stainless	POD	0.50	0.75	5
Steel, bearing	Chrome carbide	POD	...	0.79	6
	SiC (CVD)(d)	POD	...	0.23	6
	TiC (CVD)(d)	POD	...	0.25	6
	TiN (CVD)(d)	POD	...	0.49	6
Steel, stainless	Al_2O_3, plasma-sprayed	Ams	...	0.13–0.20	7
	Cr plate	Ams	...	0.30–0.38	7
	Cr_2O_3, plasma-sprayed	Ams	...	0.14–0.15	7
	TiO_2, plasma sprayed	Ams	...	0.10–0.15	7
	WC-12 Co, plasma-sprayed	Ams	...	0.11–0.13	7
Teflon(b)	Cr plate	FOF	0.09	0.08	1
	Ni (0.001 P) plate	FOF	0.15	0.12	1
	TiN (Magnagold)(c)	FOF	0.15	0.12	1
TiC on type 440C stainless steel	Al	POD	0.50	0.85	5
	Ti	POD	0.65	0.80	5
	TiC on type 440C stainless steel	POD	0.22	0.20	5
	TiN on type 440C stainless steel	POD	0.25	0.20	5
TiN on type 440C stainless steel	Al	POD	0.27	0.40	5
	Steel, type 304 stainless	POD	0.29	0.41	5
	Ti	POD	0.50	0.76	5
	TiC on type 440C stainless steel	POD	0.05	0.06	5
	TiN on type 440C stainless steel	POD	0.65	0.45	5
TiN (Magnagold)(c)	Al, alloy 6061-T6	FOF	0.30	0.26	1
	Steel, 1032	FOF	0.38	0.31	1
	Teflon(b)	FOF	0.16	0.11	1
	Ti-6Al-4V	FOF	0.26	0.23	1
	TiN (Magnagold)(c)	FOF	0.25	0.21	1

(a) Ams, Amsler circumferential, rotating disk-on-disk machine; FOF, flat surface sliding on another flat surface; POD, pin on disk; POF, pin on flat; SPOD, spherically ended pin-on-flat disk. (b) Teflon is a registered trademark of E.I. Du Pont de Nemours & Co., Inc. (polytetrafluoroethylene). (c) Magnagold is a product of General Magnaplate, Inc. (d) CVD, chemical vapor deposition

REFERENCES/TEST CONDITIONS FOR TABLE 14.4

1. "Friction Data Guide," General Magnaplate Corporation, 1988; TMI Model 98-5 slip and friction tester, 1.96 N (0.200 kgf) load, ground specimens, 54% relative humidity, average of five tests
2. M. Antler and E.T. Ratcliff, *Proc. Holm Conference on Electrical Contacts*, 1982, p 19; sphere-on-reciprocating flat, 0.49 N (0.050 kgf) load, 1.0 mm/s (0.039 in./s) velocity
3. M.J. Manjoine, *Bearing and Seal Design in Nuclear Power Machinery*, American Society of Mechanical Engineers, 1967; flat plate-on-flat plate, 28 MPa (4.1 ksi) contact pressure, 0.25 mm/s (0.010 in./s) velocity
4. F.P. Bowden and D. Tabor, *The Friction and Lubrication of Solids*, Oxford Press, 1986, p 127; sphere-on-flat, low-speed sliding, 39.2 N (4 kgf) load
5. B. Bhushan and B.K. Gupta, *Handbook of Tribology*, McGraw-Hill, 1991, Table 14.16a; pin-on-disk, 12 N (1.2 kgf) load, 14 to 16 cm/s (0.55 to 0.63 in./s) velocity
6. B. Bhushan and B.K. Gupta, *Handbook of Tribology*, McGraw-Hill, 1991, Table 14.65; pin-on-disk, 5 N (0.5 kgf) load, 1.0 cm/s (0.39 in./s) velocity, 50% relative humidity
7. B. Bhushan and B.K. Gupta, *Handbook of Tribology*, McGraw-Hill, 1991, Table 14.12; Amsler disk machine, 400 rev/min, 250 N (26 kgf) load

Table 14.5 Friction coefficient data for miscellaneous materials

Specimens tested in air at room temperature

Material (Fixed specimen)	Material (Moving specimen)	Test geometry(a)	Friction coefficient Static	Friction coefficient Kinetic	Ref
Brick	Wood	UnSp	0.6	...	1
Cotton thread	Cotton thread	UnSp	0.3	...	1
Diamond	Diamond	UnSp	0.1	...	1
Explosives(b)					
HMX(c)	Glass	RPOF	...	0.55	2
PETN(d)	Glass	RPOF	...	0.40	2
RDX(e)	Glass	RPOF	...	0.35	2
Lead azide [Pb(N3)2]	Glass	RPOF	...	0.28	2
Silver azide (AgN3)	Glass	RPOF	...	0.40	2
Glass, tempered	Al, alloy 6061-T6	FOF	0.17	0.14	3
	Steel, 1032	FOF	0.13	0.12	3
	Teflon(f)	FOF	0.10	0.10	3
Glass, thin fiber	Brass	StOD	...	0.16–0.26	4
	Graphite	StOD	...	0.15	4
	Porcelain	StOD	...	0.36	4
	Steel, stainless	StOD	...	0.31	4
	Teflon(f)	StOD	...	0.10	4
Glass, clean	Glass (clean)	UnSp	0.9–1.0	...	1
Graphite, molded	Al, alloy 2024	FOF	0.16	...	5
	Al, alloy 2219	FOF	0.22	...	5
	Graphite, extruded	FOF	0.20	0.17	5
	Graphite, molded	FOF	0.18	0.14	5
	Inconel X-750(g)	FOF	0.16	...	5
	Steel, type 304 stainless	FOF	0.18	...	5
	Steel, type 347 stainless	FOF	0.19	...	5
Graphite (clean)	Graphite (clean)	UnSp	0.10	...	1
Graphite (outgassed)	Graphite (outgassed)	UnSp	0.5–0.8	...	1
Hickory wood, waxed	Snow	UnSp	...	0.14	6
Ice	Bronze	UnSp	...	0.02	6
	Ebonite	UnSp	...	0.02	6
	Ice	UnSp	0.05–0.15	...	6
	Ice	UnSp	...	0.02	6
	Ice	FOF	>0.01	>0.01	3
Leather	Metal (clean)	UnSp	0.6	...	1
Metal	Glass (clean)	UnSp	0.5–0.7	...	1
Mica (cleaved)	Mica (cleaved)	UnSp	1.0	...	1
Mica (contaminated)	Mica (contaminated)	UnSp	0.2–0.4	...	1
Nylon fibers	Nylon fibers	UnSp	0.15–0.25	...	1
Paper, copier	Paper, copier	FOF	0.28	0.26	3
Sapphire	Sapphire	UnSp	0.2	...	1
Silk fibers	Silk fibers	UnSp	0.2–0.3	...	1
Steel (clean)	Graphite	UnSp	0.1	...	1
Wood (clean)	Metals	UnSp	0.2–0.6	...	1
	Wood (clean)	UnSp	0.25–0.5	...	1

(a) FOF, flat surface sliding on another flat surface; RPOF, reciprocating pin-on-flat; StOD, strand wrapped over a drum; UnSp, unspecified method. (b) Explosives reported here were tested as reciprocating, single-crystal, flat-ended pin-on-moving flat. (c) HMX, cyclotetramethylene tetranitramine; (d) PETN, pentaerithritol tetranitrate. (e) RDX, cyclotrimethylene trinitramine. (f) Teflon is a registered trademark of E.I. Du Pont de Nemours & Co., Inc. (g) Inconel is a product of INCO, Inc.

REFERENCES/TEST CONDITIONS FOR TABLE 14.5

1. F.P. Bowden and D. Tabor, Appendix IV, *The Friction and Lubrication of Solids*, Oxford Press, 1986; method unspecified
2. J.K.A. Amuzu, B.J. Briscoe, and M.M. Chaudhri, *J. Phys. D, Appl. Phys.*, Vol 9, 1976, p 133; reciprocating, single-crystal flat sliding on smooth fired glass surfaces, range 5 to 20 gf (0.049 to 0.1962 N load), 0.20 mm/s (0.08 in./s) velocity
3. "Friction Data Guide," General Magnaplate Corporation, 1988; TMI Model 98-5 slip and friction tester, 1.96 N (0.200 kgf) load, ground specimens, 54% relative humidity, average of five tests
4. P.K. Gupta, *J. Am. Ceram. Soc.*, Vol 74 (No. 7), 1991, p 1692; strand lying on a rotating drum, 1.96 N (0.200 kgf) load, 8.5 mm/s (0.33 in./s) velocity
5. M.J. Manjoine, *Bearing and Seal Design in Nuclear Power Machinery*, American Society of Mechanical Engineers, 1967; flat plate-on-flat plate, 28 MPa (4.1 ksi) contact pressure, 0.25 mm/s (0.010 in./s) velocity
6. F.P. Bowden and D. Tabor, *The Friction and Lubrication of Solids*, Oxford Press, 1986; unspecified method, 4.0 m/s (790 ft/min) at 0 °C

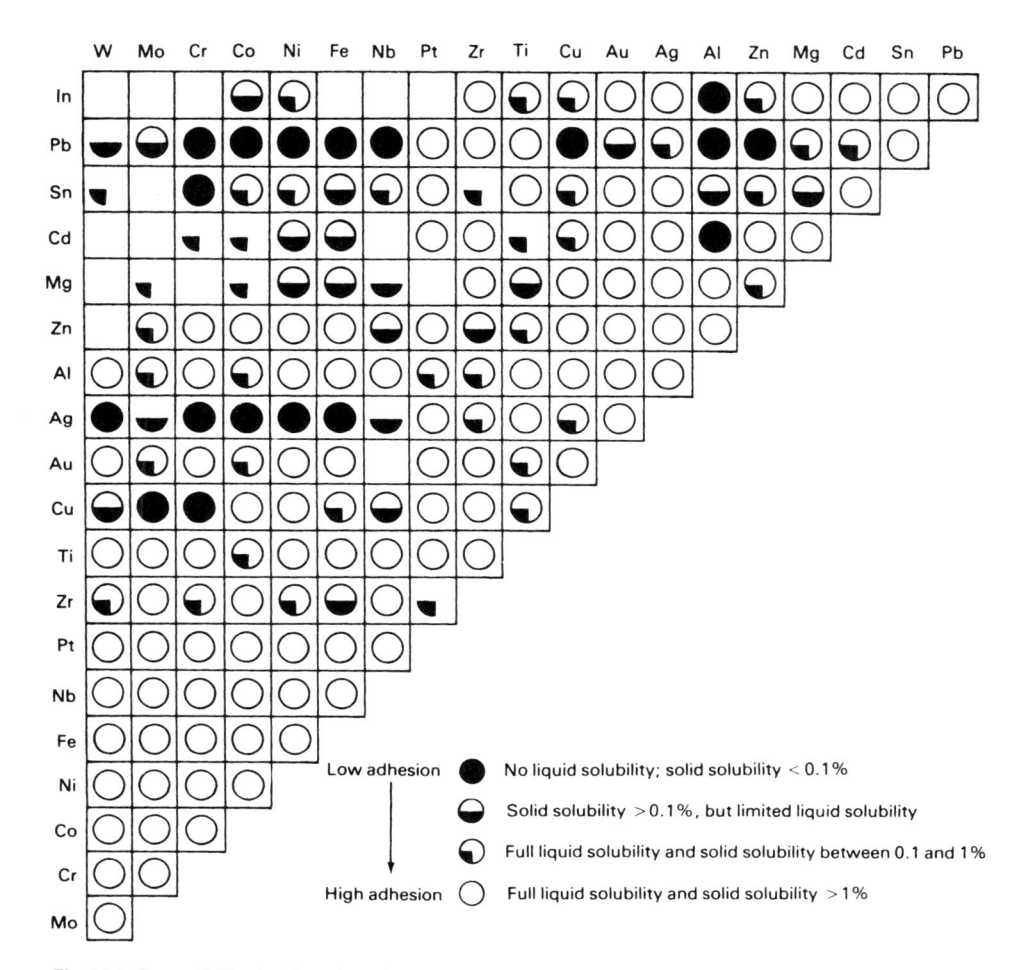

Fig. 14.1 Compatibility chart for selected metal combinations derived from binary equilibrium diagrams. Chart indicates the degree of expected adhesion (and thus friction) between the various metal combinations. The friction is greater between two metals that form alloy solutions or alloy compounds with each other than if the two are mutually insoluble. Source: E. Rabinowicz, Determination of Compatibility of Metals through Static Friction Tests, *ASLE Trans.*, Vol 14, 1971, p 198-205

15 Engineering/Scientific Constants

Table 15.1 Fundamental physical constants

Quantity	Symbol	Numerical value(a)	Units
Speed of light (in vacuum)	c	299,792,458 (exact)	$m \cdot s^{-1}$
Electronic charge	e	1.6027733 (49)	10^{-19} C
Planck's constant	h	6.6260755 (40)	10^{-34} J $\cdot$ s
Avogadro constant (number)	N_A	6.0221367 (36)	10^{23} mol^{-1}
Atomic mass unit	amu or u	1.6605402 (10)	10^{-27} kg
Electron rest mass	m_e	9.1093897 (54)	10^{-31} kg
		5.4859903 (13)	10^{-4} u
Proton rest mass	m_p	1.6726231 (10)	10^{-27} kg
		1.007276470 (12)	u
Neutron rest mass	m_n	1.6749286 (10)	10^{-27} kg
		1.008664904 (14)	u
Faraday constant	F	9.6485309 (29)	10^4 C $\cdot$ mol^{-1}
Electron magnetic moment	μ_e	9.2847701 (31)	10^{-24} J $\cdot$ T^{-1}
Molar gas constant	R	8.314510 (70)	J $\cdot$ mol^{-1} $\cdot$ K^{-1}
		8.205784 (69)	10^{-5} m^3 $\cdot$ atm $\cdot$ mol^{-1} $\cdot$ K^{-1}
Molar volume of ideal gas at STP(b)	V_m	22.41410 (19)	10^{-3} m^3 $\cdot$ mol^{-1}
Boltzmann constant	k	1.380658 (12)	10^{-23} J $\cdot$ K^{-1}
Standard gravity (gravitational acceleration)	g	9.80665	m $\cdot$ s^{-2}
Absolute temperature	$T_{0°C}$	273.150 ± 0.010	K

(a) The numbers in parentheses are the one-standard-deviation uncertainties in the last digits of the quoted value computed on the basis of internal consistency. (b) STP = standard temperature and pressure (0 °C at 1 atm, or 760 torr). Source: *J. Res. National Bureau of Standards*, Vol 92 (No. 2), 1987, p 85–95

16 Metric Practice Guide

Table 16.1 SI prefixes—names and symbols

Exponential expression	Multiplication factor	Prefix	Symbol
10^{18}	1 000 000 000 000 000 000	exa	E
10^{15}	1 000 000 000 000 000	peta	P
10^{12}	1 000 000 000 000	tera	T
10^{9}	1 000 000 000	giga	G
10^{6}	1 000 000	mega	M
10^{3}	1 000	kilo	k
10^{2}	100	hecto(a)	h
10^{1}	10	deka(a)	da
10^{0}	1	BASE UNIT	
10^{-1}	0.1	deci(a)	d
10^{-2}	0.01	centi(a)	c
10^{-3}	0.001	milli	m
10^{-6}	0.000 001	micro	μ
10^{-9}	0.000 000 001	nano	n
10^{-12}	0.000 000 000 001	pico	p
10^{-15}	0.000 000 000 000 001	femto	f
10^{-18}	0.000 000 000 000 000 001	atto	a

(a) Nonpreferred. Prefixes should be selected in steps of 10^3 so that the resultant number before the prefix is between 0.1 and 1000. These prefixes should not be used for units of linear measurement, but may be used for higher order units. For example, the linear measurement, decimeter, is nonperferred, but square decimeter is acceptable.

Table 16.2 Base, supplementary, and derived SI units

Measure	Unit	Symbol	Formula
Base units			
Amount of substance	mole	mol	
Electric current	ampere	A	
Length	meter	m	
Luminous intensity	candela	cd	
Mass	kilogram	kg	
Thermodynamic temperature	kelvin	K	
Time	second	s	
Supplementary units			
Plane angle	radian	rad	
Solid angle	steradian	sr	
Derived units			
Absorbed dose	gray	Gy	J/kg
Acceleration	meter per second squared	m/s^2	
Activity (of radionuclides)	becquerel	Bq	
Angular acceleration	radian per second squared	rad/s^2	
Angular velocity	radian per second	rad/s	
Area	square meter	m^2	
Concentration (of amount of substance)	mole per cubic meter	mol/m^3	
Current density	ampere per square meter	A/m^2	
Density, mass	kilogram per cubic meter	kg/m^3	
Dose equivalent	sievert	Sv	J/kg
Electric capacitance	farad	F	C/V
Electric charge density	coulomb per cubic meter	C/m^3	
Electric conductance	siemens	S	A/V
Electric field strength	volt per meter	V/m	
Electric flux density	coulomb per square meter	C/m^2	
Electric potential, potential difference, electromotive force	volt	V	W/A
Electric resistance	ohm	Ω	V/A
Energy, work, quantity of heat	joule	J	N · m
Energy density	joule per cubic meter	J/m^3	
Entropy	joule per kelvin	J/K	

Measure	Unit	Symbol	Formula
Derived units (continued)			
Force	newton	N	kg · m/s^2
Frequency	hertz	Hz	1/s
Heat capacity	joule per kelvin	J/K	
Heat flux density	watt per square meter	W/m^2	
Illuminance	lux	lx	lm/m^2
Inductance	henry	H	Wb/A
Irradiance	watt per square meter	W/m^2	
Luminance	candela per square meter	cd/m^2	
Luminous flux	lumen	lm	
Magnetic field strength	ampere per meter	A/m	
Magnetic flux	weber	Wb	V · s
Magnetic flux density	tesla	T	Wb/m^2
Molar energy	joule per mole	J/mol	
Molar entropy	joule per mole kelvin	J/mol · K	
Molar heat capacity	joule per mole kelvin	J/mol · K	
Moment of force	newton meter	N · m	
Permeability	henry per meter	H/m	
Permittivity	farad per meter	F/m	
Power, radiant flux	watt	W	J/s
Pressure, stress	pascal	Pa	N/m^2
Quantity of electricity, electric charge	coulomb	C	A · s
Radiance	watt per square meter steradian	W/m^2 · sr	
Radiant intensity	watt per steradian	W/sr	
Specific heat capacity	joule per kilogram kelvin	J/kg · K	
Specific energy	joule per kilogram	J/kg	
Specific entropy	joule per kilogram kelvin	J/kg · K	
Specific volume	cubic meter per kilogram	M^3/kg	
Surface tension	newton per meter	N/m	
Thermal conductivity	watt per meter kelvin	W/m · K	
Velocity	meter per second	m/s	
Viscosity, dynamic	pascal second	Pa · s	
Viscosity, kinematic	square meter per second	m^2/s	
Volume	cubic meter	m^3	
Wavenumber	1 per meter	1/m	

Metric Practice Guide 203

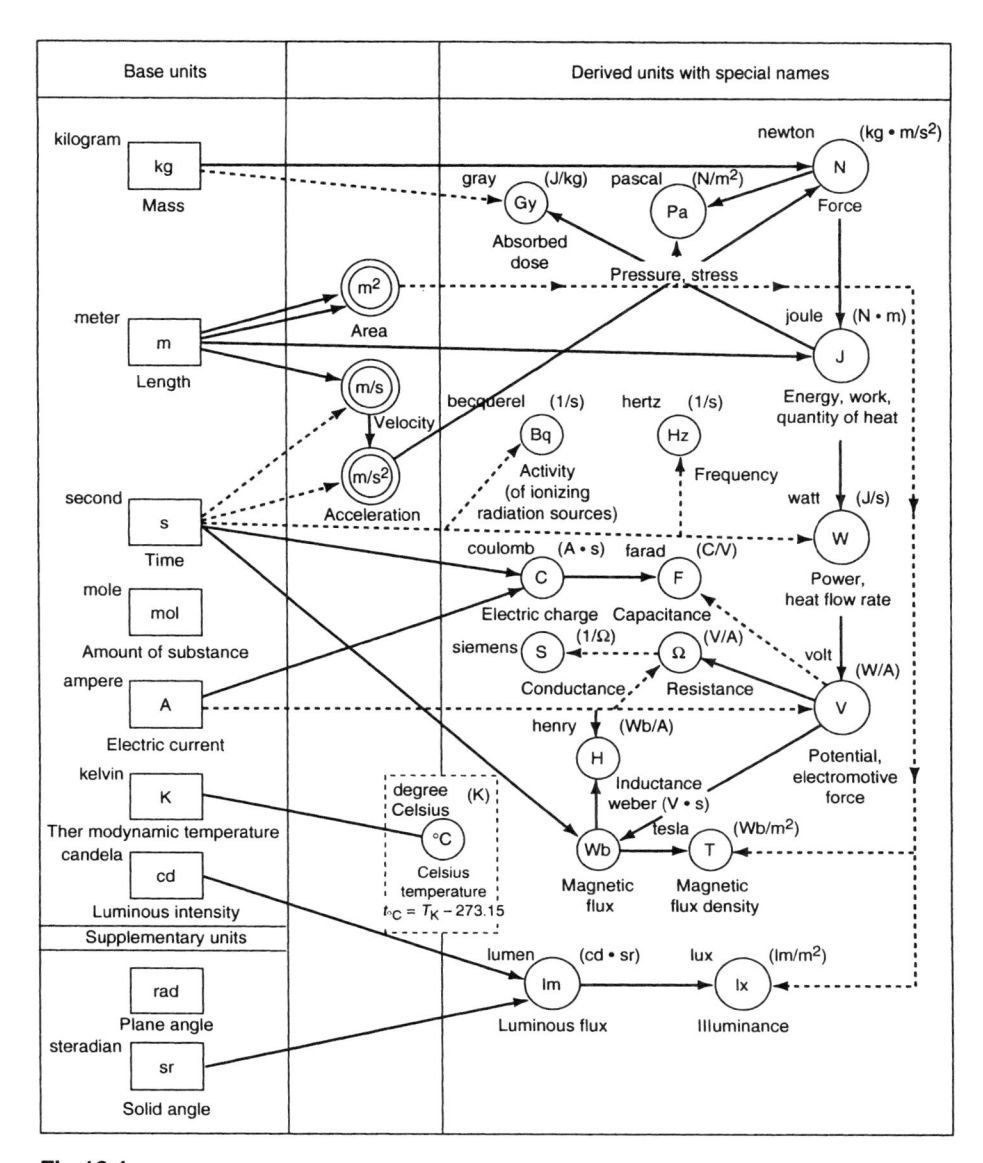

Fig 16.1 Graphic relationships of SI units with names. Solid lines indicate multiplication; broken (dashed) lines indicate division. Source: *Perry's Chemical Engineers' Handbook,* 6th ed., McGraw-Hill Book Company, 1984

Table 16.3 Conversion factors classified according to the quantity/property of interest

To convert from	to	multiply by	To convert from	to	multiply by
Angle			**Length**		
degree	rad	1.745 329 E − 02	Å	nm	1.000 000 E − 01
Area			μin.	μm	2.540 000 E − 02
in.^{2}	mm^2	6.451 600 E + 02	mil	μm	2.540 000 E + 01
in.^{2}	cm^2	6.451 600 E + 00	in.	mm	2.540 000 E + 01
in.^{2}	m^2	6.451 600 E − 04	in.	cm	2.540 000 E + 00
ft^2	m^2	9.290 304 E − 02	ft	m	3.048 000 E − 01
Bending moment or torque			yd	m	9.144 000 E − 01
lbf · in.	N · m	1.129 848 E − 01	mile	km	1.609 300 E + 00
lbf · ft	N · m	1.355 818 E + 00	**Mass**		
Kgf · m	N · m	9.806 650 E + 00	oz	kg	2.834 952 E − 02
ozf · in.	N · m	7.061 552 E − 03	lb	kg	4.535 924 E − 01
Bending moment or torque per unit length			ton (short, 2000 lb)	kg	9.071 847 E + 02
lbf · in./in.	N · m/m	4.448 222 E + 00	ton (short, 2000 lb)	Kg × 10^3(a)	9.071 847 E − 01
Lbf · ft/in.	N · m/m	5.337 866 E + 01	ton (long, 2240 lb)	kg	1.016 047 E + 03
Current density			**Mass per unit area**		
A/in.^{2}	A/cm^2	1.550 003 E − 01	oz/in.^{2}	kg/m^2	4.395 000 E + 01
A/in.^{2}	A/mm^2	1.550 003 E − 03	oz/ft^2	kg/m^2	3.051 517 E − 01
A/ft^2	A/m^2	1.076 400 E + 01	oz/yd^2	kg/m^2	3.390 575 E − 02
Electricity and magnetism			lb/ft^2	kg/m^2	4.882 428 E + 00
gauss	T	1.000 000 E − 04	**Mass per unit length**		
maxwell	μWb	1.000 000 E − 02	lb/ft	kg/m	1.488 164 E + 00
mho	S	1.000 000 E + 00	lb/in.	kg/m	1.785 797 E + 01
Oersted	A/m	7.957 700 E + 01	**Mass per unit time**		
Ω · cm	Ω · m	1.000 000 E − 02	lb/h	kg/s	1.259 979 E − 04
Ω circular-mil/ft	μΩ · m	1.662 426 E − 03	lb/min	kg/s	7.559 873 E − 03
Energy (impact, other)			lb/s	kg/s	4.535 924 E − 01
ft · lbf	J	1.355 818 E + 00	**Mass per unit volume (includes density)**		
Btu (thermochemical)	J	1.054 350 E + 03	g/cm^3	kg/m^3	1.000 000 E + 03
cal (thermochemical)	J	4.184 000 E + 00	lb/ft^3	g/cm^3	1.601 846 E − 02
kW · h	J	3.600 000 E + 06	lb/ft^3	kg/m^3	1.601 846 E + 01
W · h	J	3.600 000 E + 03	lb/in.^{3}	g/cm^3	2.767 990 E + 01
Flow rate			lb/in.^{3}	kg/m^3	2.767 990 E + 04
ft^3/h	L/min	4.719 475 E − 01	**Power**		
ft^3/min	L/min	2.831 000 E + 01	Btu/s	kW	1.055 056 E + 00
gal/h	L/min	6.309 020 E − 02	Btu/min	kW	1.758 426 E − 02
gal/min	L/min	3.785 412 E + 00	Btu/h	W	2.928 751 E − 01
Force			erg/s	W	1.000 000 E − 07
lbf	N	4.448 222 E + 00	ft · lbf/s	W	1.355 818 E + 00
kip (1000 lbf)	N	4.448 222 E + 03	ft · lbf/min	W	2.259 697 E − 02
tonf	kN	8.896 443 E + 00	ft · lbf/h	W	3.766 161 E − 04
kgf	N	9.806 650 E + 00	hp (550 ft · lbf/s)	kW	7.456 999 E − 01
Force per unit length			hp (electric)	kW	7.460 000 E − 01
lbf/ft	N/m	1.459 390 E + 01	**Power density**		
lbf/in.	N/m	1.751 268 E + 02	W/in.^{2}	W/m^2	1.550 003 E + 03
Fracture toughness			**Pressure (fluid)**		
ksi√in.	MPa√m	1.098 800 E + 00	atm (standard)	Pa	1.013 250 E + 05
Heat content			bar	Pa	1.000 000 E + 05
			in. Hg (32 °F)	Pa	3.386 380 E + 03
			in. Hg (60 °F)	Pa	3.376 850 E + 03
Btu/lb	kJ/kg	2.326 000 E + 03	lbf/in.^{2} (psi)	Pa	6.894 757 E + 03
cal/g	kJ/kg	4.186 800 E + 00	torr (mm Hg, 0 °C)	Pa	1.333 220 E + 02
Heat input			**Specific heat**		
J/in.	J/m	3.937 008 E + 01	Btu/lb · °F	J/kg · K	4.186 800 E + 03
kJ/in.	kJ/m	3.937 008 E + 01	cal/g · °C	J/kg · K	4.186 800 E + 03

(continued)

Metric Practice Guide 205

Table 16.3 (continued)

To convert from	to	multiply by	To convert from	to	multiply by
Stress (force per unit area)			**Velocity (continued)**		
tonf/in.^{2} (tsi)	MPa	1.378 951 E + 01	in./s	m/s	2.540 000 E − 02
kgf/mm^2	MPa	9.806 650 E + 00	km/h	m/s	2.777 778 E − 01
ksi	MPa	6.894 757 E + 00	mph	km/h	1.609 344 E + 00
lbf/in.^{2} (psi)	MPa	6.894 757 E − 03	**Velocity of rotation**		
MN/m^2	MPa	1.000 000 E + 00	rev/min (rpm)	rad/s	1.047 164 E − 01
Temperature			rev/s	rad/s	6.283 185 E + 00
°F	°C	5/9 · (°F − 32)	**Viscosity**		
°R	°K	5/9	poise	Pa · s	1.000 000 E + 01
Temperature interval			strokes	m^2/s	1.000 000 E − 04
°F	°C	5/9	ft^2/s	m^2/s	9.290 304 E − 02
Thermal conductivity			in.^{2}/s	mm^2/s	6.451 600 E + 02
Btu · in./s · ft^2 · °F	W/m · K	5.192 204 E + 02	**Volume**		
Btu/ft · h · °F	W/m · K	1.730 735 E + 00	in.^{3}	m^3	1.638 706 E − 05
Btu · in./h · ft^2 · °F	W/m · K	1.442 279 E − 01	ft^3	m^3	2.831 685 E − 02
cal/cm · s · °C	W/m · K	4.184 000 E + 02	fluid oz	m^3	2.957 353 E − 05
Thermal expansion			gal (U.S. liquid)	m^3	3.785 412 E − 03
in./in. · °C	m/m · K	1.000 000 E + 00	**Volume per unit time**		
in./in. · °F	m/m · K	1.800 000 E + 00	ft^3/min	m^3/s	4.719 474 E − 04
Velocity			ft^3/s	m^3/s	2.831 685 E − 02
ft/h	m/s	8.466 667 E − 05	in.^{3}/min	m^3/s	2.731 177 E − 07
ft/min	m/s	5.080 000 E − 03	**Wavelength**		
ft/s	m/s	3.048 000 E − 01	Å	nm	1.000 000 E − 01

(A) kg × 10^3 = 1 metric tonne

Table 16.4 Alphabetical listing of common conversion factors

Conversion factors are written as a number greater than one and less than ten with six or fewer decimal places. This number is followed by the letter E (for exponent), a plus or minus symbol, and two digits which indicate the power of 10 by which the number must be multiplied to obtain the correct value. For example:
  3.523 907 E − 02 is 3.523 907 × 10^{-2} or 0.035 239 07
An asterisk (*) after the sixth decimal place indicates that the conversion factor is exact and that all subsequent digits are zero. All other conversion factors have been rounded off.

To convert from	to	Multiply by
abampere	ampere (A)	1.000 000* E + 01
abcoulomb	coulomb (C)	1.000 000* E + 01
abfarad	farad (F)	1.000 000* E + 09
abhenry	henry (H)	1.000 000* E − 09
abmho	siemens (S)	1.000 000* E + 09
abohm	ohm (Ω)	1.000 000* E − 09
abvolt	volt (V)	1.000 000* E − 08
ampere hour	coulomb (C)	3.600 000* E + 03
angstrom	meter (m)	1.000 000* E − 10
atmosphere, standard	pascal (Pa)	1.013 250* E + 05
atmosphere, technical (= 1 kgf/cm^2)	pascal (Pa)	9.806 650* E + 04
bar	pascal (Pa)	1.000 000* E + 05
barn	square meter (m^2)	1.000 000* E − 28
barrel (for petroleum, 42 gal)	cubic meter (m^3)	1.589 873 E − 01
British thermal unit (International Table)	joule (J)	1.055 056 E + 03
British thermal unit (mean)	joule (J)	1.055 87 E + 03
British thermal unit (thermochemical)	joule (J)	1.054 350 E + 03
British thermal unit (39 °F)	joule (J)	1.059 67 E + 03
British thermal unit (59 °F)	joule (J)	1.054 80 E + 03
British thermal unit (60 °F)	joule (J)	1.054 68 E + 03
	(continued)	

Table 16.4 (continued)

To convert from	to	Multiply by
Btu (International Table) · ft/(h · ft² · °F) (thermal conductivity)	watt per meter kelvin [W/(m · K)]	1.730 735 E + 00
Btu (thermochemical) · ft/(h · ft² · °F)) (thermal conductivity)	watt per meter kelvin [W/(m · K)]	1.729 577 E + 00
Btu (International Table) · in./(h · ft² · °F) (thermal conductivity)	watt per meter kelvin [W/(m · K)]	1.442 279 E − 01
Btu (thermochemical) · in./(h · ft² · °F) (thermal conductivity)	watt per meter kelvin [W/(m · K)]	1.441 314 E − 01
Btu (International Table) · in./s · ft² · °F) (thermal conductivity)	watt per meter kelvin [W/(m · K)]	5.192 204 E + 02
Btu (thermochemical) · in./(s · ft² · °F) (thermal conductivity)	watt per meter kelvin [W/(m · K)]	5.188 732 E + 02
Btu (International Table)/h	watt (W)	2.930 711 E − 01
Btu (International Table)/s	watt (W)	1.055 056 E + 03
Btu (thermochemical)/h	watt (W)	2.928 751 E − 01
Btu (thermochemical)/min	watt (W)	1.757 250 E + 01
Btu (thermochemical)/s	watt (W)	1.054 350 E + 03
Btu (International Table)/ft²	joule per square meter (J/m²)	1.135 653 E + 04
Btu (thermochemical)/ft²	joule per square meter (J/m²)	1.134 893 E + 04
Btu (International Table)/(ft² · s)	watt per square meter (W/m²)	1.135 653 E + 04
Btu (International Table)/(ft² · h)	watt per square meter (W/m²)	3.154 591 E + 00
Btu (thermochemical)/(ft² · h)	watt per square meter (W/m²)	3.152 481 E + 00
Btu (thermochemical)/(ft² · min)	watt per square meter (W/m²)	1.891 489 E + 02
Btu (thermochemical)/(ft² · s)	watt per square meter (W/m²)	1.134 893 E + 04
Btu (thermochemical)/(in.² · s)	watt per square meter (W/m²)	1.634 246 E + 06
Btu (International Table)/(h · ft² · °F) (thermal conductance)	watt per square meter kelvin [W/(m² · K)]	5.678 263 E + 00
Btu (thermochemical)/(h · ft² · °F) (thermal conductance)	watt per square meter kelvin [W/(m² · K)]	5.674 466 E + 00
Btu (International Table)/(s · ft² · °F)	watt per square meter kelvin [W/(m² · K)]	2.044 175 E + 04
Btu (thermochemical)/(s · ft² · °F)	watt per square meter kelvin [W/m² · K)]	2.042 808 E + 04
Btu (International Table)/lb	joule per kilogram (J/kg)	2.326 000* E + 03
Btu (thermochemical)/lb	joule per kilogram (J/kg)	2.324 444 E + 03
Btu (International Table)/(lb · °F) (heat capacity)	joule per kilogram kelvin [J/(kg · K)]	4.186 800* E + 03
Btu (thermochemical)/(lb · °F) (heat capacity)	joule per kilogram kelvin [J/(kg · K)]	4.184 000* E + 03
Btu (International Table)/ft³	joule per cubic meter (J/m³)	3.725 895 E + 04
Btu (thermochemical)/ft³	joule per cubic meter (J/m³)	3.723 402 E + 04
bushel (U.S.)	cubic meter (m³)	3.523 907 E − 02
calorie (International Table)	joule (J)	4.186 800* E + 00
calorie (mean)	joule (J)	4.190 02 E + 00
calorie (thermochemical)	joule (J)	4.184 000* E + 00
calorie (15 °C)	joule (J)	4.185 80 E + 00
calorie (20 °C)	joule (J)	4.181 90 E + 00
calorie (kilogram, International Table)	joule (J)	4.186 800* E + 03
calorie (kilogram, mean)	joule (J)	4.190 02 E + 03
calorie (kilogram, thermochemical)	joule (J)	4.184 000* E + 03
cal (thermochemical)/cm²	joule per square meter (J/m²)	4.184 000* E + 04
cal (International Table)/g	joule per kilogram (J/kg)	4.186 800* E + 03
cal (thermochemical)/g	joule per kilogram (J/kg)	4.184 000* E + 03
cal (International Table)/(g · °C)	joule per kilogram kelvin [J/(kg · K)]	4.186 800* E + 03
cal (thermochemical)/(g · °C)	joule per kilogram kelvin [J/(kg · K)]	4.184 000* E + 03
cal (thermochemical)/min	watt (W)	6.973 333 E − 02
cal (thermochemical)/s	watt (W)	4.184 000* E + 00
cal (thermochemical)/(cm² · s)	watt per square meter (W/m²)	4.184 000* E + 04
cal (thermochemical)/(cm² · min)	watt per square meter (W/m²)	6.973 333 E + 02
cal (thermochemical)/(cm² · s)	watt per square meter (W/m²)	4.184 000* E + 04
cal (thermochemical)/(cm · s · °C)	watt per meter kelvin [W/(m · K)]	4.184 000* E + 02
cd/in.²	candela per square meter (cd/m²)	1.550 003 E + 03
carat (metric)	kilogram (kg)	2.000 000* E − 04
centimeter of mercury (0 °C)	pascal (Pa)	1.333 22 E + 03
centimeter of water (4 °C)	pascal (Pa)	9.806 38 E + 01
centipoise (dynamic viscosity)	pascal second (Pa · s)	1.000 000* E − 03
centistokes (kinematic viscosity)	square meter per second (m²/s)	1.000 000* E − 06
circular mil	square meter (m²)	5.067 075 E − 10
curie	becquerel (Bq)	3.700 000* E + 10

(continued)

Table 16.4 (continued)

To convert from	to	Multiply by
degree (angle)	radian (rad)	1.745 329 E−02
degree Celsius	kelvin (K)	$T_K = t_{°C} + 273.15$
degree Fahrenheit	degree Celsius (°C)	$t_{°C} = (t_{°F} − 32)/1.8$
degree Fahrenheit	kelvin (K)	$T_K = (t_{°F} + 459.67)/1.8$
degree Rankine	kelvin (K)	$T_K = T_{°R}/1.8$
°F · h · ft^2/Btu (International Table) (thermal resistance)	kelvin square meter per watt (K · m^2/W)	1.761 102 E−01
°F · h · ft^2/Btu (thermochemical) (thermal resistance)	kelvin square meter per watt (K · m^2/W)	1.762 280 E−01
°F · h · ft^2/[Btu (International Table) · in.] (thermal resistivity)	kelvin meter per watt (K · m/W)	6.933 471 E+00
°F · h · ft^2/[Btu (thermochemical) · in.] (thermal resistivity)	kelvin meter per watt (K · m/W)	6.938 113 E+00
denier	kilogram per meter (kg/m)	1.111 111 E−07
dyne	newton (N)	1.000 000* E−05
dyne · cm	newton meter (N · m)	1.000 000* E−07
dyne/cm^2	pascal (Pa)	1.000 000* E−01
electronvolt	joule (J)	1.602 19 E−19
EMU (electromagnetic units) of capacitance	farad (F)	1.000 000* E+09
EMU of current	ampere (A)	1.000 000* E+01
EMU of electric potential	volt (V)	1.000 000* E−08
EMU of inductance	henry (H)	1.000 000* E−09
EMU of resistance	ohm (Ω)	1.000 000* E−09
ESU (electrostatic units) of capacitance	farad (F)	1.112 650 E−12
ESU of current	ampere (A)	3.335 6 E−10
ESU of electric potential	volt (V)	2.997 9 E+02
ESU of inductance	henry (H)	8.987 554 E+11
ESU of resistance	ohm (Ω)	8.987 554 E+11
erg	joule (J)	1.000 000* E−07
erg/(cm^2 · s)	watt per square meter (W/m^2)	1.000 000* E−03
erg/s	watt (W)	1.000 000* E−07
faraday (based on carbon-12)	coulomb (C)	9.648 70 E+04
faraday (chemical)	coulomb (C)	9.649 57 E+04
faraday (physical)	coulomb (C)	9.652 19 E+04
fluid ounce (U.S.)	cubic meter (m^3)	2.957 353 E−05
foot	meter (m)	3.048 000* E−01
ft^2	square meter (m^2)	9.290 304* E−02
ft^2/h (thermal diffusivity)	square meter per second (m^2/s)	2.580 640* E−05
ft^2/s	square meter per second (m^2/s)	9.290 304* E−02
ft^3	cubic meter	2.831 685 E−02
ft^3/min	cubic meter per second (m^3/s)	4.719 474 E−04
ft^3/s	cubic meter per second (m^3/s)	2.831 685 E−02
ft/h	meter per second (m/s)	8.466 667 E−05
ft/min	meter per second (m/s)	5.080 000* E−03
ft/s	meter per second (m/s)	3.048 000* E−01
ft/s^2	meter per second squared (m/s^2)	3.048 000* E−01
footcandle	lux (lx)	1.076 391 E+01
footlambert	candela per square meter (cd/m^2)	3.426 259 E+00
ft · lbf	joule (J)	1.355 818 E+00
ft · lbf/h	watt (W)	3.766 161 E−04
ft · lbf/min	watt (W)	2.259 697 E−02
ft · lbf/s	watt (W)	1.355 818 E+00
ft-poundal	joule (J)	4.214 011 E−02
g, standard free fall	meter per second squared (m/s^2)	9.806 650* E+00
gal	meter per second squared (m/s^2)	1.00 000* E−02
gallon (Canadian liquid)	cubic meter (m^3)	4.546 090 E−03
gallon (U.K. liquid)	cubic meter (m^3)	4.546 092 E−03
gallon (U.S. dry)	cubic meter (m^3)	4.404 884 E−03
gallon (U.S. liquid)	cubic meter (m^3)	3.785 412 E−03
gallon (U.S. liquid) per day	cubic meter per second (m^3/s)	4.381 264 E−08
gallon (U.S. liquid) per minute	cubic meter per second (m^3/s)	6.309 020 E−05
gallon (U.S. liquid) per hp · h (SFC, specific fuel consumption)	cubic meter per joule (m^3/J)	1.410 089 E−09
gauss	tesla (T)	1.000 000* E−04
gilbert	ampere (A)	7.957 747 E−01
grain	kilogram (kg)	6.479 891* E−05
grain/gal (U.S. liquid)	kilogram per cubic meter (kg/m^3)	1.711 806 E−02
gram	kilogram (kg)	1.000 000* E−03
g/cm^3	kilogram per cubic meter (kg/m^3)	1.000 000* E+03

(continued)

Table 16.4 (continued)

To convert from	to	Multiply by
gf/cm^2	pascal (Pa)	9.806 650* E+01
hectare	square meter (m^2)	1.000 000* E+04
horsepower (550 ft · lbf/s)	watt (W)	7.456 999 E+02
horsepower (boiler)	watt (W)	9.809 50 E+03
horsepower (electric)	watt (W)	7.460 000* E+02
horsepower (metric)	watt (W)	7.354 99 E+02
horsepower (water)	watt (W)	7.460 43 E+02
horsepower (U.K.)	watt (W)	7.457 0 E+02
inch	meter (m)	2.540 000* E−02
inch of mercury (32 °F)	pascal (Pa)	3.386 38 E+03
inch of mercury (60 °F)	pascal (Pa)	3.376 85 E+03
inch of water (39.2 °F)	pascal (Pa)	2.490 82 E+02
inch of water (60 °F)	pascal (Pa)	2.488 4 E+02
in.^{2}	square meter (m^2)	6.451 600* E−04
in.^{3} (volume)	cubic meter (m^3)	1.638 706 E−05
in.^{3}/min	cubic meter per second (m^3/s)	2.731 177 E−07
in./s	meter per second (m/s)	2.540 000* E−02
in./s^2	meter per second squared (m/s^2)	2.540 000* E−02
kelvin	degree Celsius (°C)	$t_°C = T_K − 273.15$
kilocalorie (International Table)	joule (J)	4.186 800* E+03
kilocalorie (mean)	joule (J)	4.190 02 E+03
kilocalorie (thermochemical)	joule (J)	4.184 000* E+03
kilocalorie (thermochemical)/min	watt (W)	6.973 333 E+01
kilocalorie (thermochemical)/s	watt (W)	4.184 000* E+03
kilogram-force (kgf)	newton (N)	9.806 650* E+00
kgf · m	newton meter (N · m)	9.806 650* E+00
kgf · s^2/m (mass)	kilogram (kg)	9.806 650* E+00
kgf/cm^2	pascal (Pa)	9.806 650* E+04
kgf/m^2	pascal (Pa)	9.806 650* E+00
kgf/mm^2	pascal (Pa)	9.806 650* E+06
km/h	meter per second (m/s)	2.777 778 E−01
kilopond (1 kp = 1 kgf)	newton (N)	9.806 650* E+00
kW · h	joule (J)	3.600 000* E+06
kip (1000 lbf)	newton (N)	4.448 222 E+03
kip/in.^{2} (ksi)	pascal (Pa)	6.894 757 E+06
lambert	candela per square meter (cd/m^2)	$1/\pi$ * E+04
lambert	candela per square meter (cd/m^2)	3.183 099 E+03
liter	cubic meter (m^3)	1.000 000* E−03
lm/ft^2	lumen per square meter (lm/m^2)	1.076 391 E+01
maxwell	weber (Wb)	1.000 000* E−08
mho	siemens (S)	1.000 000* E+00
microinch	meter (m)	2.540 000* E−08
micron (use preferred term micrometer)	meter (m)	1.000 000* E−06
mil	meter (m)	2.540 000* E−05
mile (international)	meter (m)	1.609 344* E+03
mile (U.S. statute)	meter (m)	1.609 347 E+03
mile (international nautical)	meter (m)	1.852 000* E+03
mile (U.S. nautical)	meter (m)	1.852 000* E+03
mi^2 (international)	square meter (m^2)	2.589 988 E+06
mi^2 (U.S. statute)	square meter (m^2)	2.589 998 E+06
mi/h (international)	meter per second (m/s)	4.470 400* E−01
mi/h (international)	kilometer per hour (km/h)	1.609 344* E+00
mi/min (international)	meter per second (m/s)	2.682 240* E+01
mi/s (international)	meter per second (m/s)	1.609 344* E+03
millibar	pascal (Pa)	1.000 000* E+02
millimeter of mercury (0 °C)	pascal (Pa)	1.333 22 E+02
minute (angle)	radian (rad)	2.908 882 E−04
oersted	ampere per meter (A/m)	7.957 747 E+01
ohm centimeter	ohm meter (Ω · m)	1.000 000* E−02
ohm circular-mil per foot	ohm meter (Ω · m)	1.662 426 E−09
ounce (avoirdupois)	kilogram (kg)	2.834 952 E−02
ounce (troy or apothecary)	kilogram (kg)	3.110 348 E−02
ounce (U.K. fluid)	cubic meter (m^3)	2.841 306 E−05
ounce (U.S. fluid)	cubic meter (m^3)	2.957 353 E−05
ounce-force	newton (N)	2.780 139 E−01
ozf · in.	newton meter (N · m)	7.061 552 E−03

(continued)

Table 16.4 (continued)

To convert from	to	Multiply by
oz (avoirdupois)/gal (U.K. liquid)	kilogram per cubic meter (kg/m^3)	6.236 023 E + 00
oz (avoirdupois)/gal (U.S. liquid)	kilogram per cubic meter (kg/m^3)	7.489 152 E + 00
oz (avoirdupois)/in.^{3}	kilogram per cubic meter (kg/m^3)	1.729 994 E + 03
oz (avoirdupois)/ft^2	kilogram per square meter (kg/m^2)	3.051 517 E − 01
oz (avoirdupois)/yd^2	kilogram per square meter (kg/m^2)	3.390 575 E − 02
pint (U.S. dry)	cubic meter (m^3)	5.506 105 E − 04
pint (U.S. liquid)	cubic meter (m^3)	4.731 765 E − 04
poise (absolute viscosity)	pascal second (Pa · s)	1.000 000* E − 01
pound (lb avoirdupois)	kilogram (kg)	4.535 924 E − 01
pound (troy or apothecary)	kilogram (kg)	3.732 417 E − 01
lb · ft^2 (moment of inertia)	kilogram square meter (kg · m^2)	4.214 011 E − 02
lb · in.^{2} (moment of inertia)	kilogram square meter (kg · m^2)	2.926 397 E − 04
lb/ft · h	pascal second (Pa · s)	4.133 789 E − 04
lb/ft · s	pascal second (Pa · s)	1.488 164 E + 00
lb/ft^2	kilogram per square meter (kg/m^2)	4.882 428 E + 00
lb/ft^3	kilogram per cubic meter (kg/m^3)	1.601 846 E + 01
lb/gal (U.K. liquid)	kilogram per cubic meter (kg/m^3)	9.977 637 E + 01
lb/gal (U.S. liquid)	kilogram per cubic meter (kg/m^3)	1.198 264 E + 02
lb/h	kilogram per second (kg/s)	1.259 979 E − 04
lb/hp · h (SFC, specific fuel consumption)	kilogram per joule (kg/J)	1.689 659 E − 07
lb/in.^{3}	kilogram per cubic meter (kg/m^3)	2.767 990 E + 04
lb/min	kilogram per second (kg/s)	7.559 873 E − 03
lb/s	kilogram per second (kg/s)	4.535 924 E − 01
lb/yd^3	kilogram per cubic meter (kg/m^3)	5.932 764 E − 01
poundal	newton (N)	1.382 550 E − 01
poundal/ft^2	pascal (Pa)	1.488 164 E + 00
poundal · s/ft^2	pascal second (Pa · s)	1.488 164 E + 00
pound-force (lbf)	newton (N)	4.448 222 E + 00
lbf · ft	newton meter (N · m)	1.355 818 E + 00
lbf · ft/in.	newton meter per meter (N · m/m)	5.337 866 E + 01
lbf · in.	newton meter (N · m)	1.129 848 E − 01
lbf · in./in.	newton meter per meter (N · m/m)	4.448 222 E + 00
lbf · s/ft^2	pascal second (Pa · s)	4.788 026 E + 01
lbf · s/in.^{2}	pascal second (Pa · s)	6.894 757 E + 03
lbf/ft	newton per meter (N/m)	1.459 390 E + 01
lbf/ft^2	pascal (Pa)	4.788 026 E + 01
lbf/in.	newton per meter (N/m)	1.751 268 E + 02
lbf/in.^{2} (psi)	pascal (Pa)	6.894 757 E + 03
lbf/lb (thrust/weight [mass] ratio)	newton per kilogram (N/kg)	9.806 650 E + 00
quart (U.S. dry)	cubic meter (m^3)	1.101 221 E − 03
quart (U.S. liquid)	cubic meter (m^3)	9.463 529 E − 04
rad (absorbed dose)	gray (Gy)	1.000 000* E − 02
rem (dose equivalent)	sievert (Sv)	1.000 000* E − 02
roentgen	coulomb per kilogram (C/kg)	2.58 000* E − 04
rpm (r/min)	radian per second (rad/s)	1.047 198 E − 01
second (angle)	radian (rad)	4.848 137 E − 06
statampere	ampere (A)	3.335 640 E − 10
statcoulomb	coulomb (C)	3.335 640 E − 10
statfarad	farad (F)	1.112 650 E − 12
stathenry	henry (H)	8.987 554 E + 11
statmho	siemens (S)	1.112 650 E − 12
statohm	ohm (Ω)	8.987 554 E + 11
statvolt	volt (V)	2.997 925 E + 02
stokes (kinematic viscosity)	square meter per second (m^2/s)	1.000 000* E − 04
ton (assay)	kilogram (kg)	2.916 667 E − 02
ton (long, 2240 lb)	kilogram (kg)	1.016 047 E + 03
ton (metric)	kilogram (kg)	1.000 000* E + 03
ton (nuclear equivalent of TNT)	joule (J)	4.184 E + 09
ton (register)	cubic meter (m^3)	2.831 685 E + 00
ton (short, 2000 lb)	kilogram (kg)	9.071 847 E + 02
ton (long)/yd^3	kilogram per cubic meter (kg/m^3)	1.328 939 E + 03
ton (short)/yd^3	kilogram per cubic meter (kg/m^3)	1.186 553 E + 03
ton (short)/h	kilogram per second (kg/s)	2.519 958 E − 01
ton-force (2000 lbf)	newton (N)	8.896 443 E + 03
tonne	kilogram (kg)	1.000 000* E + 03
torr (mmHg, 0 °C)	pascal (Pa)	1.333 22 E + 02
W · h	joule (J)	3.600 000* E + 03
	(continued)	

Table 16.4 (continued)

To convert from	to	Multiply by
W · s	joule (J)	1.000 000* E + 00
W/cm^2	watt per square meter (W/m^2)	1.000 000* E + 04
W/in.^{2}	watt per square meter (W/m^2)	1.550 003 E + 03
yard	meter (m)	9.144 000* E − 01
yd^2	square meter (m^2)	8.361 274 E − 01
yd^3	cubic meter (m^3)	7.645 549 E − 01
yd^3/min	cubic meter per second (m^3/s)	1.274 258 E − 02

Source: Adapted from ASTM E380, "Standard Practice for Use of the International System of Units (SI)—(The Modernized Metric System)"

17 Sheet Metal and Wire Gages

Table 17.1 Sheet metal gage thickness conversions

Gage	Diameter in.	Diameter mm	Gage	Diameter in.	Diameter mm
30	0.0120	0.3048	16	0.0598	1.5189
29	0.0135	0.3429	15	0.0673	1.7094
28	0.0149	0.3785	14	0.0747	1.8974
27	0.0164	0.4166	13	0.0897	2.2784
26	0.0179	0.4547	12	0.1046	2.6568
25	0.0109	0.5309	11	0.1196	3.0378
24	0.0239	0.6071	10	0.1345	3.4163
23	0.0269	0.6833	9	0.1495	3.7973
22	0.0299	0.7595	8	0.1644	4.1758
21	0.0329	0.8357	7	0.1793	4.5542
20	0.0359	0.9119	6	0.1943	4.9352
19	0.0418	1.0617	5	0.2092	5.3137
18	0.0478	1.2141	4	0.2242	5.6947
17	0.0538	1.3665	3	0.2391	6.0731

Table 17.2 Wire gage diameter conversions

U.S. steel wire gage No.	Diameter in.	Diameter mm	U.S. steel wire gage No.	Diameter in.	Diameter mm
7/0's	0.4900	12.447	10	0.1350	3.429
6/0's	0.4615	11.7221	11	0.1205	3.0607
5/0's	0.4305	10.9347	12	0.1055	2.6797
4/0's	0.3938	10.0025	13	0.0915	2.3241
3/0's	0.3625	9.2075	14	0.0800	2.032
2/0's	0.3310	8.4074	15	0.0720	1.8389
0	0.3065	7.7851	16	0.0625	1.5875
1	0.2830	7.1882	17	0.0540	1.3716
2	0.2625	6.6675	18	0.0475	1.2065
3	0.2437	6.1899	19	0.0410	1.0414
4	0.2253	5.7226	20	0.0348	0.8839
5	0.2070	5.2578	21	0.0317	0.8052
6	0.1920	4.8768	22	0.0286	0.7264
7	0.1770	4.4958	23	0.0258	0.6553
8	0.1620	4.1148	24	0.0230	0.5842
9	0.1483	3.7668	25	0.0204	0.5182

(continued)

Table 17.2 (continued)

U.S. steel wire gage No.	Diameter in.	Diameter mm	U.S. steel wire gage No.	Diameter in.	Diameter mm
26	0.0181	0.4597	34	0.0104	0.2642
27	0.0173	0.4394	35	0.0095	0.2413
28	0.0162	0.4115	36	0.0090	0.2286
29	0.0150	0.381	37	0.0085	0.2159
30	0.0140	0.3556	38	0.0080	0.2032
31	0.0132	0.3353	39	0.0075	0.1905
32	0.0128	0.3251	40	0.0070	0.1778
33	0.0118	0.2997	41	0.0066	0.1678

18 Pipe Dimensions

Table 18.1 Dimensions of welded and seamless pipe manufactured in the United States
All units are given in inches. See Table 18.2 for heavier wall thicknesses.

Nominal pipe size, in.	Outside diameter	Schedule 5S	Schedule 10S	Schedule 10	Schedule 20	Schedule 30	Schedule standard	Schedule 40
1/8	0.405	...	0.049	...	...	...	0.068	0.068
1/4	0.540	...	0.065	...	...	...	0.088	0.088
3/8	0.675	...	0.065	...	...	...	0.091	0.091
1/2	0.840	0.065	0.083	...	...	...	0.109	0.109
3/4	1.050	0.065	0.083	...	...	...	0.113	0.113
1	1.315	0.065	0.109	...	...	...	0.133	0.133
1 1/4	1.660	0.065	0.109	...	...	...	0.140	0.140
1 1/2	1.900	0.065	0.109	...	...	...	0.145	0.145
2	2.375	0.065	0.109	...	...	...	0.154	0.154
2 1/2	2.875	0.083	0.120	...	...	...	0.203	0.203
3	3.5	0.083	0.120	...	...	...	0.216	0.216
3 1/2	4.0	0.083	0.120	...	...	...	0.226	0.226
4	4.5	0.083	0.120	...	...	...	0.237	0.237
5	5.563	0.109	0.134	...	...	...	0.258	0.258
6	6.625	0.109	0.134	...	...	...	0.280	0.280
8	8.625	0.109	0.148	...	0.250	0.277	0.322	0.322
10	10.75	0.134	0.165	...	0.250	0.307	0.365	0.365
12	12.75	0.156	0.180	...	0.250	0.330	0.375	0.406
14 O.D.	14.0	0.156	0.188	0.250	0.312	0.375	0.375	0.438
16 O.D.	16.0	0.165	0.188	0.250	0.312	0.375	0.375	0.500
18 O.D.	18.0	0.165	0.188	0.250	0.312	0.438	0.375	0.562
20 O.D.	20.0	0.188	0.218	0.250	0.375	0.500	0.375	0.594
22 O.D.	22.0	0.188	0.218	0.250	0.375	0.500	0.375	...
24 O.D.	24.0	0.218	0.250	0.250	0.375	0.562	0.375	0.688
26 O.D.	26.0	...	...	0.312	0.500	...	0.375	...
28 O.D.	28.0	...	...	0.312	0.500	0.625	0.375	...
30 O.D.	30.0	0.250	0.312	0.312	0.500	0.625	0.375	...
32 O.D.	32.0	...	...	0.312	0.500	0.625	0.375	0.688
34 O.D.	34.0	...	...	0.312	0.500	0.625	0.375	0.688
36 O.D.	36.0	...	...	0.312	0.500	0.625	0.375	0.750
42 O.D.	42.0	...	...	...	...	...	0.375	...

Source: *The Metals Black Book*, Casti Publishing Inc., 1995, p 716–717

Table 18.2 Dimensions of welded and seamless pipe manufactured in the United States

All units are given in inches. See Table 18.1 for thinner wall thicknesses.

Nominal pipe size, in.	Outside diameter	Schedule 60	Extra strong	Schedule 80	Schedule 100	Schedule 120	Schedule 140	Schedule 160	XX strong
1/8	0.405	...	0.095	0.095	...	...	...	...	...
1/4	0.540	...	0.119	0.119	...	...	...	...	...
3/8	0.675	...	0.126	0.126	...	...	...	...	...
1/2	0.840	...	0.147	0.147	...	...	...	0.188	0.294
3/4	1.050	...	0.154	0.154	...	...	...	0.219	0.308
1	1.315	...	0.179	0.179	...	...	...	0.250	0.358
1 1/4	1.660	...	0.191	0.191	...	...	...	0.250	0.382
1 1/2	1.900	...	0.200	0.200	...	...	...	0.281	0.400
2	2.375	...	0.218	0.218	...	...	...	0.344	0.436
2 1/2	2.875	...	0.276	0.276	...	...	...	0.375	0.552
3	3.5	...	0.300	0.300	...	...	...	0.438	0.600
3 1/2	4.0	...	0.318	0.318	...	...	...	...	...
4	4.5	...	0.337	0.337	...	0.438	...	0.531	0.674
5	5.563	...	0.375	0.375	...	0.500	...	0.625	0.750
6	6.625	...	0.432	0.432	...	0.562	...	0.719	0.864
8	8.625	0.406	0.500	0.500	0.594	0.719	0.812	0.906	0.875
10	10.75	0.500	0.500	0.594	0.719	0.844	1.000	1.125	1.000
12	12.75	0.562	0.500	0.688	0.844	1.000	1.125	1.312	1.000
14 O.D.	14.0	0.594	0.500	0.750	0.938	1.094	1.250	1.406	...
16 O.D.	16.0	0.656	0.500	0.844	1.031	1.219	1.438	1.594	...
18 O.D.	18.0	0.750	0.500	0.938	1.156	1.375	1.562	1.781	...
20 O.D.	20.0	0.812	0.500	1.031	1.281	1.500	1.750	1.9691	...
22 O.D.	22.0	0.875	0.500	1.125	1.375	1.625	1.875	2.125	...
24 O.D.	24.0	0.969	0.500	1.218	1.531	1.812	2.062	2.344	...
26 O.D.	26.0	...	0.500	...	...	...	...	...	...
28 O.D.	28.0	...	0.500	...	...	...	...	...	...
30 O.D.	30.0	...	0.500	...	...	...	...	...	...
32 O.D.	32.0	...	0.500	...	...	...	...	...	...
34 O.D.	34.0	...	0.500	...	...	...	...	...	...
36 O.D.	36.0	...	0.500	...	...	...	...	...	...
42 O.D.	42.0	...	0.500	...	...	...	...	...	...

Source: *The Metals Black Book*, Casti Publishing Inc., 1995, p 717–718

19 Glossary of Abbreviations, Acronyms, and Symbols

Table 19.1 Common abbreviations, acronyms, and symbols found in the materials science literature

a	crack length; crystal lattice length along the a axis; wheel depth of cut in grinding	ACD	annealed cold drawn
a_f	final cross-sectional area	ACGIH	American Conference of Governmental Industrial Hygienists
a_o	original cross-sectional area	ACI	Alloy Casting Institute
A	amp; ampere; area	A/D, ADC	analog-to-digital converter
A	amplitude; area; ratio of the alternating stress amplitude to the mean stress; absorbance	ADCI	American Die Casting Institute
		ADI	austempered ductile iron
		ADTT	average daily truck traffic
Å	angstrom	Ae_{cm}, Ae_1, Ae_3	equilibrium transformation temperatures in steel
AA	Aluminum Association; atomic absorption		
AAC	air carbon arc cutting	AE	acoustic emission
AAR	Association of American Railroads	AECL	Atomic Energy of Canada Limited
AAS	atomic absorption spectroscopy/spectrometry	AECMA	Association Européenne des Constructeurs de Matérial Aérospatial
AASHTO	American Association of State Highway Transportation Officials	AEM	analytical electron microscopy
		AES	Auger electron spectroscopy; atomic emission spectrometry
AAW	air acetylene welding		
AB	arc brazing	AFD	automated forging design
ABS	acrylonitrile-butadiene-styrene; American Bureau of Shipping	AFM	abrasive flow machining; atomic force microscope/microscopy
		AFNOR	Association Francaise de Normalisation
ABST	alpha-beta solution treatment		
ac	alternating current	AFS	American Foundrymen's Society; atomic fluorescence spectrometry
Ac_1	temperature at which austenite begins to form on heating		
		AFWAL	Air Force Wright Aeronautical Laboratories
Ac_3	temperature at which transformation of ferrite to austenite is completed on heating	AGA	American Gas Association
		AGMA	American Gear Manufacturers Association
Ac_{cm}	in hypereutectoid steel, temperature at which cementite completes solution in austenite	AGV	automatic guided vehicle
		AHW	atomic hydrogen welding
		AI	artificial intelligence
AC	acetal; air cooled; adaptive control	AIA	Aerospace Industries Association

(continued)

Table 19.1 (continued)

AIME	American Institute of Mining, Metallurgical and Petroleum Engineers	AWG	American wire gage
		AWJ	abrasive waterjet
AIP	American Institute of Physics	AWM	abrasive waterjet machining
AISC	American Institute of Steel Construction	AWS	American Welding Society
AISE	Association of Iron and Steel Engineers	b	barn; Burgers vector
AISI	American Iron and Steel Institute	b	crystal lattice length along the b axis
AJM	abrasive jet machining	B	magnetic induction; specimen thickness
AKDQ	aluminum-killed drawing quality	bal	balance; remainder
AKS	aluminum-potassium-silicon	bcc	body-centered cubic
ALPID	Analysis of Large Plastic Incremental Deformation (bulk deformation modeling software)	BCIRA	British Cast Iron Research Association
		bct	body-centered tetragonal
		BDT	brittle-ductile transition
AM	analytical modeling	Bé	Baumé (specific-gravity scale)
AMMRC	Army Materials and Mechanics Research Center	BE	backscattered electron
		BET	Brunauer-Emmett-Teller
AMS	Aerospace Material Specification	BF	bright-field (illumination)
amu	atomic mass unit	BHP	Broken Hill Proprietary
ANSI	American National Standards Institute	BI	basicity index
AOAC	Association of Official Analytical Chemists	BMA	butadiene-co-maleic anhydride
AOC	oxygen arc cutting	BMAW	bare metal arc welding
AOCS	American Oil Chemists' Society	BMC	bulk molding compound
AOD	argon oxygen decarburization	BMI	bismaleimide (resin)
AP	armor piercing; atom probe	BNI	bisnadimide
APA	3-aminophenylacetylene	BOP	basic oxygen process
APB	1,3-bis(3-aminophenoxy) benzene; antiphase boundary	BOS	basic oxygen steelmaking
		BPO	benzoyl peroxide
APCVD	atmospheric-pressure chemical vapor deposition	Bq	becquerel
		BS	British Standard
API	American Petroleum Institute	BSCCO	Bi-Sr-Ca-Cu-O
APM	atom probe microanalysis	BST	beta solution treatment
APMI	American Powder Metallurgy Institute	BTDE	benzophenone tetracarboxylic acid dimethyl ester
APT	ammonium paratungstate		
APU	auxiliary power unit (space shuttle)	Btu	British thermal unit
AQ	as quenched	BUE	built-up edge
AQL	acceptable quality levels	BUS	broken-up structure
Ar_1	temperature at which transformation to ferrite or to ferrite plus cementite is completed on cooling	BWG	Birmingham wire gage
		BWR	boiling water reactor(s)
		c	crystal lattice length along the c axis; velocity (speed of light)
Ar_3	temperature at which transformation of austenite to ferrite begins on cooling		
		C	coulomb; cementite
Ar_{cm}	temperature at which cementite begins to precipitate from austenite on cooling	C	capacitance; heat capacity
		C^*	steady-state creep characterizing parameter
AREA	American Railway Engineering Association	C_t	creep fracture mechanics parameter
ARIP	activated reactive ion plating	CAB	calcium argon blowing
ASA	acrylonitrile-styrene-acrylate terpolymer	CAC	carbon arc cutting
ASCE	American Society of Civil Engineers	CAD/CAM	computer-aided design/computer-aided manufacturing
ASIP	Aircraft Structural Integrity Program		
ASM	ASM International (formerly American Society for Metals)	CAE	computer-aided engineering
		cal	calorie
ASME	American Society of Mechanical Engineers	CAM	computer-aided-manufacturing
ASP	antisegregation process	CANDU	Canadian deuterium uranium (reactor)
AS/RS	automatic storage and retrieval systems	CAP	consolidation by atmospheric pressure
ASTM	American Society for Testing and Materials	CARES	Ceramic Analysis and Reliability Evaluation of Structures
at.%	atomic percent		
ATEM	analytical transmission electron microscope/microscopy	CARP	Committee for Acoustic Emission in Reinforced Plastics
atm	atmospheres (pressure)		
at.ppm	atomic parts per million	CASS	copper-accelerated acetic acid-salt spray (test)
ATR	attenuated total reflectance		

(continued)

Table 19.1 (continued)

CAT	computer-aided tomography	CM	chemical machining (milling)
CAW	carbon arc welding	cmc	critical micelle concentration
CBED	convergent-beam electron diffraction	CMC	ceramic-matrix composite
CBEDP	convergent-beam electron diffraction pattern	CMM	coordinate measuring machine
		CMOS	complementary metal oxide semiconductor
CBN	cubic boron nitride	CN	cyanogen
CC	combined carbon; center cracked	CNB	Chevron notch bend
C-C, C/C	carbon-carbon	CNC	computer numerical control
CCD	charge-coupled device	COD	crack-opening displacement
CCG	creep crack growth	cos	cosine
CCI	crevice corrosion index	cot	cotangent
CCM	Crucible ceramic mold	cP	centipoise
CCPA	Cemented Carbide Producers Association	CP	commercially pure
CCR	conventional controlled rolling	cpm	cycles per minute
CCT	critical crevice temperature; center-cracked tension (specimen); continuous cooling transformation	CPM	Crucible Particle Metallurgy
		cps	cycles per second; counts per second
		CPVC	chlorinated polyvinyl chloride
cd	candela	CPU	central processing unit
CDA	Copper Development Association	CQ	commercial quality
CE	carbon equivalent	CR	cold rolled
CEBAF	continuous electron beam accelerator facility	CRE	carbon removal efficiency
		CRO	cathode-ray oscilloscope
CEN	Comité Européen de Normalisation (European Committee for Standardization)	CRP	carbon-fiber reinforced plastic
		CRR	carbon removal rate
CERCLA	Comprehensive Environmental Response, Compensation, and Liability Act	CRT	cathode-ray tube
		CS	ceramic shell
CERT	constant extension rate test	CSA	Canadian Standards Association
CET	columnar-equiaxed transition	C-SAM	C-mode scanning acoustic microscopy
CFC	corrosion-fatigue cracking	CSD	controlled spray deposition
CFCG	creep-fatigue crack growth	CSOM	confocal scanning optical microscope/microscopy
CFG	creep-feed grinding		
CFRP	carbon fiber reinforced plastic	CSP	compact strip production
CFTA	Committee of Foundry Technical Associations	cSt	centistokes
		$C(t)$	creep fracture mechanics parameter
CG	compacted graphite	CT	computed tomography; compact type (specimen); compact tension (test specimen); continuous transformation
CGA	Compressed Gas Association		
cgs	centimeter-gram-second (system of units)		
CHIP	cold and hot isostatic pressing	CTBN	carboxyl-terminated butadiene acrylonitrile
CHR	conventional hot rolling	CTE	coefficient of thermal expansion
Ci	curie	CTFE	polychlorotrifluoroethylene
CI	compression ignition	CTOA	crack tip opening angle
CIE	Commission Internationale de l'Eclairage (International Commission on Illumination)	CTOD	crack tip opening displacement
		CVD	chemical vapor deposition
		CVI	chemical vapor infiltration (impregnation)
CIM	computer-integrated manufacturing; Canadian Institute for Mining and Metallurgy	CVN	Charpy V-notch (impact test or specimen)
		cw	continuous-wave
CINDAS	Center for Information and Numerical Data Analysis and Synthesis	d	day
		d	depth; diameter; lattice spacing of crystal planes; an operator used in mathematical expressions involving a derivative (denotes rate of change)
CIP	cold isostatic pressing		
CIRCLE	cylindrical internal reflection cell		
CL	confidence limits; cathodoluminescence		
CLA	centerline average; counter-gravity low-pressure casting of air-melted alloys		
		D	diameter; distance
		DAC	digital-to-analog converter
CLAS	counter-gravity low-pressure air-melted sand casting	da/dN	fatigue crack growth rate
		da/dt	crack growth rate per unit time
CLV	counter-gravity low-pressure casting of vacuum-melted alloys	DADPS	diaminodiphenylsulfone
		DAIP	diallyl isophthalate
cm	centimeter	DAP	diallyl phthalate

(continued)

Table 19.1 (continued)

DARPA	Defense Advanced Research Projects Agency	DSA	dispersion-strengthened alloy
		DSC	differential scanning calorimetry
DAS	dendrite arm spacing	DT	drop tower; dynamic tear (test)
dB	decibel	DTA	differential thermal analysis
DBMS	data base management system	DU	depleted uranium
DBTT	ductile-brittle transition temperature	DWT	drop-weight test
dc	direct current	DWTT	drop-weight tear test
DCB	double-cantilever beam	e	natural log base, 2.71828; charge of an electron; engineering strain (see also ε)
DCDT	direct-current displacement transducer		
DCEN	direct current electrode negative	e_1	major engineering strain
DCEP	direct current electrode positive	e_2	minor engineering strain
DCP	direct-current plasma	E	energy; modulus of elasticity (Young's modulus)
DCRF	Die Casting Research Foundation		
DDS	diaminodiphenyl sulfone	E_{cell}	measured cell potential
deca	decahydronaphthalene	E_{corr}	corrosion potential
DESY	Deutsche Electronen Syncrotron	EAA	poly(ethylene-co-acrylic acid)
DETA	diethylene triamine	EAF	electric arc furnace
d.f.	degrees of freedom	EAW	electric arc wire (thermal spray)
DF	dark-field (illumination)	EB	electron beam
Df	dilution factor	EBCHM	electron beam cold-hearth melting
DFB	diffusion brazing	EBHT	electron beam hardening treatment
DFW	diffusion welding	EBIC	electron beam induced current
dhcp	double hexagonal close-packed	EBM	electron beam machining
Di	didymium (mixture of the rare earth elements praseodymium and neodymium)	EBPVD	electron beam physical vapor deposition
		EBW	electron beam welding
diam	diameter	EBW-HV	electron beam welding—high vacuum
DIB	diiodobutane	EBW-MV	electron beam welding—medium vacuum
DIC	differential interference contrast (illumination)	EBW-NV	electron beam welding—nonvacuum
		EC	eddy current
DIN	Deutsche Industrie-Normen (German Industrial Standards)	ECAP	energy-compensated atom probe
		ECEA	end cutting edge angle
DIP	dual-in-line package (electronic component)	ECG	electrochemical grinding
DIS	Draft International Standard; Ductile Iron Society	ECM	electrochemical machining
		EDM	electrical discharge machining
DLC	diamondlike carbon	EDS	energy-dispersive x-ray spectrometry; energy-dispersive spectroscopy
dm	decimeter		
DMAC	dimethyl acetamide	EDTA	ethylenediamine tetraacetic acid
DMF	dimethyl formamide	EDXA	electron dispersive analysis by x-ray
DMSO	dimethyl sulfoxide	EEC	European Economic Community
DMW	dissimilar-metal weld	EELS	electron energy loss spectroscopy
DNC	direct numerical control	EGW	electrogas welding
DOA	dead on arrival	EHD	elastohydrodynamic
DOC	Department of Commerce; depth of cut	EIA	Electronics Industries Association
DoD	Department of Defense	ELI	extra-low interstitial
DOF	device operating failure	ELP	electropolishing
DOT	Department of Transportation	emf	electromotive force
DP	dual phase	EMF	electromagnetic field
DPH	diamond pyramid hardness (Vickers hardness)	EMI	electromagnetic iron; electromagnetic interference
DQ	drawing quality	EMSA	Electron Microscopy Society of America
DQSK	drawing quality special killed	ENAA	epithermal neutron activation
DR	digital radiography; digital radiograph	ENSIP	Engine Structural Integrity Program
DRAM	dynamic random access memory	EP	epoxy; extreme pressure
DRCR	dynamic recrystallization	EPA	Environmental Protection Agency
DRI	direct reduced iron	EPC	evaporative pattern casting
DRS	diffuse reflectance spectroscopy	EPDM	ethylene-propylene-diene monomer
DRX	dynamic recrystallization	EPMA	electron probe x-ray microanalysis; electron probe microanalysis
DS	directional solidification; dip soldering		

(continued)

Table 19.1 (continued)

EPR	ethylene propylene rubber	FN	ferrite number
EPRI	Electric Power Research Institute	FNAA	fast neutron activation analysis
EPS	expanded polystyrene pattern	FOC	chemical flux cutting
Eq	equation	FOLZ	first-order Laue zone
ESC	environmental stress cracking	FOW	forge welding
ESCA	electron spectroscopy for chemical analysis	FP	polycrystalline alumina fiber
ESD	electrostatic discharge; electrospark deposition	FRC	free radical cure
		FRM	fiber-reinforced metals
ESR	electron spin resonance; electroslag remelting	FRP	fiber-reinforced plastic
		FRS	fiber-reinforced superalloys
ESW	electroslag welding	FRTP	fire-refined tough pitch (copper)
et al.	and others	FRW	friction welding
ETP	electrolytic tough pitch (copper); electrolytic tin-plated (steel strip)	FSS	fatigue striation spacing
		FSZ	fully stabilized zirconia
eV	electron volt	ft	foot
EXAFS	extended x-ray absorption fine structure	ftc	footcandle
exp	exponent, exponential	FTIR	Fourier transform infrared (spectroscopy)
EXW	explosion welding	ft-L	footlambert
f	fiber	FTS	Fourier transform spectrometer
f	frequency; transfer function; precipitate volume fraction; coefficient of thermal expansion	FW	flash welding
		FWHM	full width at half maximum
		FZ	fusion zone
F	farad; ferrite; fluorescence	g	gram
F	force; Faraday constant (96,486 C/mol)	g	acceleration due to gravity
FAA	Federal Aviation Administration	G	gauss; graphite
FAC	forced-air cool	G	shear modulus; modulus of rigidity; thermal gradient; Gibbs free energy
FATT	fracture-appearance transition temperature		
FBC	fluidized bed combustion	GA	gas atomization
FC	furnace cool	gal	gallon
FCAW	flux cored arc welding	GAR	grain aspect ratio
fcc	face-centered cubic	GB	grain boundary
FCC	Federal Communications Commission	GBq	gigabecquerel
FCGR	fatigue crack growth rate	GC	grain-coarsened; gas chromatography
fct	face-centered tetragonal	GCHAZ	grain-coarsened heat-affected zone
FDA	Food and Drug Administration	GC-IR	gas chromatography-infrared (spectroscopy)
FDM	finite-difference method		
FEA	finite-element analysis	GC/MS	gas chromatograph/mass spectrometer/spectrometry
FEG	field emission gun		
FEM	finite element modeling/method	gcp	geometrically close-packed
FEP	fluorinated ethylene propylene	GdIG	gadolinium iron garnet
FET	field-effect transistor	gf	gram-force
FFT	fast Fourier transform	GFAAS	graphite furnace atomic absorption spectrometry
FG	flake graphic		
FGD	flue gas desulfurization	GFN	grain fineness number
FGHAZ	fine grain heat-affected zone	GGG	gallium gadolinium garnet
FHWA	Federal Highway Administration	GHz	gigahertz
FIA	flow injection analysis	GJ	gigajoule
FIFO	first in, first out	GMAW	gas metal arc welding
Fig.	figure	GOR	gas-oil ratio (in petroleum production)
FIM	field ion microscopy	GP	Guinier-Preston (zone)
FIOR	fluid iron ore reduction	GPa	gigapascal
FLD	forming limit diagram	GPC	gel permeation chromatography
fm	femtometer	gpd	grams per denier
FM	frequency modulation; full mold; ferromagnet	gr	grain
		Gr	graphite
FM (process)	*fonte mince* (thin iron)	GS	grain size
FMR	ferromagnetic resonance	GSGG	gallium scandium gadolinium garnet
FMS	flexible manufacturing system	GTAW	gas tungsten arc welding

(continued)

Table 19.1 (continued)

GTO	gate turnoff	HV	Vickers hardness (diamond pyramid hardness)
Gy	gray (unit of absorbed radiation)		
h	hour	HVC	hydrovac process
h	Planck's constant	HVOF	high-velocity oxyfuel (thermal spray)
H	Henry	HWR	heavy water reactor(s)
H	enthalpy; Grossmann number; height; magnetic field; magnetic field strength	Hz	hertz
		i	current (measure of number of electrons)
H_a	applied magnetic field	i	current density
H_c	coercive force; critical magnetic field; thermodynamic critical field	i_0	exchange current
		i_{corr}	corrosion current
HAD	high aluminum defect	i_{crit}	critical current for passivation
HAIM	high-frequency acoustic imaging	i_{pass}	passive current
HAZ	heat-affected zone	I	bias current; current; electrical current; emergent intensity; intensity
HB	Brinell hardness		
HBN	hexagonal boron nitride	IACS	International Annealed Copper Standard
HCF	high-cycle fatigue	IASCC	irradiation-assisted stress-corrosion cracking
HCL	hollow cathode lamp		
hcp	hexagonal close-packed	IBAD	ion-beam assisted deposition
HDI	high-density inclusion	IC	integrated circuit; ion chromatography
HDPE	high-density polyethylene	ICB	ionized cluster beam
HEC	hydrogen embrittlement cracking	ICBM	Intercontinental Ballistic Missile
HEM	hydrogen embrittlement	ICFTA	International Committee of Foundry Technical Associations
HEP	high-energy physics		
HERF	high-energy-rate forging/forming	ICHAZ	intercritical heat-affected zone
hexa	hexamethylene tetramine	ICP	inductively coupled plasma
HF	hardenability factor	ICP-AES	inductively coupled plasma atomic emission spectroscopy
HFP	hexafluoropropylene		
HFRSc	Scleroscope hardness (Model C)	ICPE	inductively coupled plasma emission
HFRSd	Scleroscope hardness (Model D)	ICPMS	inductively coupled plasma mass spectrometry
HFRW	high-frequency resistance welding		
HIC	hydrogen-induced cracking	ICR	intensified controlled rolling
HID	high interstitial defect	ID	inside diameter
HIP	hot isostatic pressing	IEEE	Institute of Electrical and Electronics Engineers
HIPS	high-impact polystyrene		
HK	Knoop hardness	IF	interstitial free; isothermal fatigue
HLW	high-level waste	IFI	Industrial Fasteners Institute
HM	high modulus	IGA	intergranular attack; inert-gas atomization
HMX	cyclotetramethylene tetranitrimine	IGP	intergranular penetration
HOLZ	higher order Laue zone	IGSCC	intergranular stress-corrosion cracking
HOPG	highly oriented pyrolytic graphite	IIR	isoprene-isobutylene rubber
hp	horsepower	IIW	International Institute of Welding
HP	high purity; hot pressed	ILZRO	International Lead Zinc Research Organization
HPLC	high-pressure liquid chromatography		
HPSN	hot-pressed silicon nitride	I/M	ingot metallurgy
HPW	hot pressure welding	IMMA	ion microprobe mass analysis
HR	Rockwell hardness (requires scale designation, such as HRC for Rockwell C hardness)	in.	inch
		INCRA	International Copper Research Association
		INPO	Institute for Nuclear Power Operations
HRA	Rockwell "A" hardness	IP	in-phase; ion plating
HRB	Rockwell "B" hardness	IPC	Institute for Interconnecting and Packaging Electronic Circuits
HRE	Rockwell "E" hardness		
HREM	high-resolution electron microscope/microscopy	ipm	inches per minute
		ips	inches per second
HRF	Rockwell "F" hardness	ipt	inch per tooth
HRH	Rockwell "H" hardness	IPTS	International Practical Temperature Scale
HSLA	high-strength, low-alloy (steel)	IR	infrared (radiation); infrared (spectroscopy); injection refining
HSS	high-speed steel		
HTLA	heat-treatable low-alloy (steel)	IRAS	infrared astronomy satellite

(continued)

Table 19.1 (continued)

IRB	infrared brazing	kHz	kilohertz
IRGCHAZ	intercritically reheated grain-coarsened heat-affected zone	km	kilometer
		KMS	Kloeckner metallurgy scrap
IRRAS	infrared reflection absorption spectroscopy	kN	kilonewton
IRS	infrared soldering	kPa	kilopascal
IS	induction soldering	ksi	kips (1000 lbf) per square inch
ISA	Instrument Society of America	kV	kilovolt
ISCC	intergranular stress-corrosion cracking	kW	kilowatt
ISM	induction skull melting	l	length
ISO	International Organization for Standardization	L	liter; longitudinal
		L	length
ISRI	Institute of Scrap Recycling Industries	L_0	initial length
IT	isothermal transformation	LAMMA	laser microscope mass analysis
ITER	international thermonuclear experimental reactor	LAST	lowest anticipated service temperature
		lb	pound
ITS	International Temperature Scale	LBE	lance bubble equilibrium
IVD	ion vapor deposited	lbf	pound force
IW	induction welding	LBM	laser beam machining
J	joule	LBW	laser beam welding
J_{ec}	Jominy equivalent cooling	LBZ	local brittle zone
J_{eh}	Jominy equivalent hardness	LCD	liquid crystal display
J	crack growth energy release rate (fracture mechanics)	LCF	low-cycle fatigue
		LCP	large coil program; liquid crystal polymer
J_c	critical current density	LCSM	laser confocal scanning microscope/microscopy
JCPDS	Joint Committee on Powder Diffraction Standards		
		LDVD	laser-induced chemical vapor deposition
JIC	Joint Industry Conference	L/D	length-to-diameter ratio
JIS	Japanese Industrial Standard	LDH	limiting dome height
JIT	just in time (manufacturing)	LDR	limiting draw ratio
k	karat	LEC	liquid-encapsulated Czochralski
k	Boltzmann constant; notch sensitivity factor; thermal conductivity	LED	light-emitting diode
		LEED	low-energy electron diffraction
K	Kelvin	LEFM	linear elastic fracture mechanics
K	coefficient of thermal conductivity; bulk modulus of elasticity; stress-intensity factor in linear elastic fracture mechanics	LEISS	low-energy ion-scattering spectroscopy
		LF	ladle furnace
		LF/VD	ladle furnace vacuum degassing
ΔK	stress-intensity factor range	LF/VD-VAD	ladle furnace vacuum degassing and vacuum arc degassing
K_0	crack initiation toughness		
K_a	crack arrest toughness	LIM	liquid injection molding
K_c	plane-stress fracture toughness	LiMCA	liquid metal cleanness analyzer
K_f	fatigue notch factor	LIMS	laser ionization mass spectroscopy
K_I	stress-intensity factor	lm	lumen
K_{Ia}	plane-strain crack arrest toughness	LME	liquid metal embrittlement
K_{Ic}	plane-strain fracture toughness	LMP	Larsen-Miller parameter
K_{Id}	dynamic fracture toughness	LMR	liquid metal refining
K_{IHE}	threshold stress intensity for hydrogen embrittlement	ln	natural logarithm (base e)
		LNG	liquefied natural gas
K_{Iscc}	threshold stress intensity to produce stress-corrosion cracking	log	common logarithm (base 10)
		LOI	limiting oxygen index
K_t	theoretical stress-concentration factor	LOR	loss on reduction
K_{th}	threshold stress intensity factor	LOX	liquid oxygen
KB	kilobyte	LPCVD	low pressure chemical vapor deposition
kbar	kilobar (pressure)	LPE	liquid-phase epitaxy
K-BOP	Kawasaki basic oxygen process	LPG	liquefied petroleum gas
KEK	Japanese atomic energy facility	LRO	long-range order
keV	kiloelectron volt	LSG	low-stress grinding
kg	kilogram	LSI	large-scale integrated (circuit)
kgf	kilogram force	LT	long transverse (direction)

(continued)

Table 19.1 (continued)

LVDT	linear variable differential transformer/transducer	mo	month
		MO, MeO	metal oxide
LWR	light water reactor(s)	MOCVD	metallo-organic chemical vapor deposition
lx	lux	mol%	mole percent
		MOLE	molecular optical laser examiner
m	meter	MOR	modulus of resilience; modulus of rupture
m	strain rate sensitivity factor; mass; molar (solution)	mPa	millipascal
		MPa	megapascal
M	martensite; metal	MPC	Materials Properties Council
M_f	temperature at which martensite formation finishes during cooling	mpg	miles per gallon
		mph	miles per hour
M_s	temperature at which martensite starts to form from austenite on cooling	MPIF	Metal Powder Industries Federation
		MPS	main propulsion system (space shuttle)
M	magnetization; magnification; molar solution; molecular weight	MRS	Materials Research Society
		MSA	Microscopy Society of America
mA	milliampere	MSDS	material safety data sheet
MA	microalloyed	MQG	melt-quench technique
MAA	methacrylic acid	mrem	millirem
MAE	microalloying elements	MRI	magnetic resonance imaging
MAPP, MPS	methyl-acetylene-propadiene (gas)	MRR	material removal rate
MAS	Microbeam Analysis Society	ms	millisecond
max	maximum	MS	megasiemens; mass spectrometry; magnetron sputtering
MB	megabyte		
MBE	molecular beam epitaxy	mSv	millisievert
MC	metal carbide	mT	millitesla
MCR	minimum creep rate	MTF	modulation transfer function
MDA	methylene dianiline	MTI	Materials Technology Institute (of the Chemical Process Industry)
MDI	diphenylmethane-4,4'-diisocyanate		
MDRX	metadynamic recrystallization	MTTF	mean time to failure
ME	Mössbauer effect	mV	millivolt
MEA	monoethanolamine	MV	megavolt
MEK	methyl ethyl ketone	MVEMA	methyl vinyl ether-co-maleic
MEKP	methyl ethyl ketone peroxide	MW	molecular weight
MeV	megaelectronvolt	MWG	music wire gage
MFTF	mirror fusion test facility	n	neutrons
mg	milligram	n	strain-hardening exponent; refractive index
Mg	megagram (metric tonne, or kg × 10^3)	N	Newton
MHD	magnetohydrodynamic (casting)	N	fatigue life (number of cycles); normal solution; speed of rotation (rev/min)
MHz	megahertz		
MIAC	Metals Information Analysis Center	N_f	number of cycles to failure
MIBK	methyl isobutyl ketone	NA	numerical aperture
MIE	metal-induced embrittlement	NAA	neutron activation analysis
MIG	metal inert gas (welding)	NACA	National Advisory Committee for Aeronautics
MIL	military		
MIL-STD	military standard	NACE	National Association of Corrosion Engineers
MIM	metal injection molding		
min	minute; minimum	NADCA	North American Die Casting Association
MINT	metal in-line treatment	NASA	National Aeronautics and Space Administration
mips	million intrusions per second		
MIT	Massachusetts Institute of Technology	NASP	National Aerospace Plane
MJ	megajoule	NBR	acrylonitrile-butadiene rubber
MJR	modified jelly roll	NBS	National Bureau of Standards
mL	milliliter	nC	nanocoulomb
mm	millimeter	NC	numerical control
MM	misch metal	NCCA	National Coil Coaters Association
MMA	methyl methacrylate	ND	normal direction (of a sheet)
MMC	metal-matrix composite	NDE	nondestructive evaluation
MMIC	monolithic microwave integrated	NDI	nondestructive inspection

(continued)

Table 19.1 (continued)

NDT	nil-ductility transition; nondestructive testing	ONERA	Office Nationale d'Etudes et de Recherches Aerospatiales
NDTT	nil ductility transition temperatures	ONIA	Office National Industrial d'Azote
Nd:YAG	neodymium: yttrium-aluminum-garnet (laser)	OP	out-of-phase
		OQ	oil quench
NEMA	National Electrical Manufacturers Association	OQ & T	oil quenched and tempered
		ORNL	Oak Ridge National Laboratory
NFPA	National Fire Prevention Association	OSHA	Occupational Safety and Health Administration
NG	nuclear grade		
NGPA	Natural Gas Producers Association	OSTE	one-step temper embrittlement
NGR	nuclear gamma-ray resonance	OTB	oxygen top blown
NiDI	Nickel Development Institute	OTSG	once-through steam generator
NIR	near infrared	oz	ounce
NIST	National Institute of Standards and Technology	p	page
		p	pressure
nm	nanometer	p_{O_2}	partial pressure of oxygen
NMR	nuclear magnetic resonance	Pa	pascal
NMTP	nonmartensitic transformation product	P	pearlite; poise
No.	number	P	power; pressure; applied load
NPS	American National Standard Straight Pipe Thread	PA	prealloyed; polyamide
		PAC	plasma arc cutting
NPSC	American National Standard Straight Pipe Thread for Couplings	PACVD	plasma-assisted chemical vapor deposition
		PAEK	polyaryletherketone
NPT	American National Standard Taper Pipe Thread	PAI	polyamide-imide
		PAN	polyacrylonitrile
NQR	nuclear quadruple resonance	PAR	polyarylate
NR	natural rubber; neutron radiography	PAS	polyaryl sulfone; photoacoustic spectroscopy
NRC	Nuclear Regulatory Commission		
NRL	Naval Research Laboratories	PAW	plasma arc welding
ns	nanoseconds	PBI	polybenzimidazole
NSR	notch strength ratio	PBT	polybutylene terephthalate
NT	normalized and tempered	p-c	plastic-carbon (replica)
NTHM	net tonne of hot metal	PC	programmable controller; personal computer
NTIS	National Technical Information Service		
NTSB	National Transportation and Safety Board	PCB	printed circuit board
NWTI	National Wood Tank Institute	PCBN	polycrystalline CBN (cubic boron nitride)
O	oil	PCD	polycrystalline diamond
OAW	oxyacetylene welding	PCM	photochemical machining
OBM	oxygen blow method	PD	preferred direction
OD	outside diameter	PDCP	polydicyclopentadiene
ODMR	optical double magnetic resonance	PDF	probability density function
ODS	oxide dispersion strengthened	PDMS	polydimethylsiloxane
Oe	oersted	PE	polyethylene
OECD	Organization for Economic Cooperation and Development	PEA	polyethylene adipate
		PEEK	polyetheretherketone
OES	optical emission spectroscopy	PEI	polyether-imide
OF	oxygen-free	PEK	polyetherketone
OFC	oxyfuel gas cutting	PEL	permissible exposure limits
OFD	oxyfuel detonation (spray)	PES	polyether sulfone
OFHC	oxygen-free high conductivity (copper)	PET	polyethylene terephthalate
OFP	oxyfuel powder (thermal spray)	PETN	pentaerithritol tetranitrate
OFW	oxyfuel gas welding; oxyfuel wire (thermal spray)	PEW	percussion welding
		PFC	planar-flow casting
OHW	oxyhydrogen welding	PGAA	prompt gamma-ray activation analysis
OM	optical micrograph	PGM	platinum-group metal
OMCVD	organometallic chemical vapor deposition	pH	negative logarithm of hydrogen-ion activity
OMS	orbital maneuvering system (space shuttle)	PH	precipitation-hardenable

(continued)

Table 19.1 (continued)

PI	polyimide	PVA	polyvinyl alcohol
PIB	polyisobutylene	PVAC	polyvinyl acetate
PIC	pressure-impregnation-carbonization	PVAL	polyvinyl alcohol
PIXE	proton-induced x-ray emission; particle-induced x-ray emission	PVB	polyvinyl butyral
		PVC	polyvinyl chloride
pixel	picture element	PVD	physical vapor deposition
PLAP	pulsed-laser atom probe	PVDF	polyvinylidene fluoride
PLASTEC	Plastics Technical Evaluation Center	PVF	polyvinyl formal
PLC	programmable logic controller	PVP	polyvinylpyrrolidone
PLD	pulsed laser deposition	PVRC	Pressure Vessel Research Committee
PLZT	lead lanthanum zirconate titanate	PWB	printed wiring board
pm	picometer	PWHT	postweld heat treatment
PM	precious metal; paramagnet; permanent mold	PWR	pressurized water reactor(s)
		PZN	lead zinc niobate
P/M	powder metallurgy	PZT	lead zirconium titanate
PMMA	polymethyl methacrylate	q	fatigue notch sensitivity factor
PMN	lead magnesium niobate	Q	quench factor; heat removal rate
PMS	Process Management System; lead-molybdenum-sulfide ($PbMo_6S_8$)	Q-BOP	quick-quiet basic oxygen process
		QT	quenched and tempered
PMT	photomultiplier tube	r	radius
PNN	lead nickel niobate	R	roentgen; Rankine; rare earth
POC	metal powder cutting; products of combustion	R	radius; universal gas constant; ratio of the minimum stress to the maximum stress; bulk resistance; reluctance (reciprocal of permeance); rolling reduction ratio; resultant force
POD	probability of detection		
POM	polyoxymethylene		
PP	polypropylene		
ppb	parts per billion	R_a	surface roughness in terms of arithmetic average
PPB	prior particle boundary		
ppba	parts per billion atomic	R_{max}	maximum peak-to-valley roughness height
PPE	polyphenylene ether	R_q, R_{rms}	root-mean-square roughness average
ppm	parts per million	R_t	total roughness peak-to-valley
PPO	polyphenylene oxide	R_y	maximum peak-to-valley roughness height
PPRIC	Pulp and Paper Research Institute of Canada	R_z	ten-point height (roughness average)
PPS	polyphenylene sulfide; property prediction system	RA	recrystallization annealed; reduction of/in area; rosin activated
ppt	parts per trillion	rad	radiation absorbed dose; radian
PPTA	poly-p-phenylene terephthalamide	RAD	ratio-analysis diagram
PQ	physical quality	RAM	random access memory
Pr	Prandtl number	RBS	Rutherford backscattering spectroscopy/spectrometry
PREP	plasma rotating electrode process		
PROM	programmable read-only memory	RBSC	reaction-bonded silicon carbide
PS	polystyrene	RBSN	reaction-bonded silicon nitride
PSA	pressure-sensitive adhesive	RCF	rolling contact fatigue
psi	pounds per square inch	RCR	recrystallization controlled rolling
psia	pounds per square inch (absolute)	RD	rolling direction (of a sheet)
psid	pounds per square inch (differential)	RDF	radial distribution function (analysis)
psig	gage pressure (pressure relative to ambient pressure) in pounds per square inch	Re	Reynolds number
		RE	rare earth
PSII	plasma-source ion implementation	Ref	reference
PSP	plasma spraying	rem	remainder or balance; roentgen equivalent man
PSU	polysulfone		
PSZ	partially stabilized zirconia	REP	rotating-electrode process
PTA	plasma transferred arc	rf, RF	radio frequency
PTFE	polytetrafluoroethylene	RFEC	remote-field eddy current
PTH	plated through-holes	RGA	residual gas analyzer
PUCB	phenolic urethane cold box	RGB	red, green, blue
PUN	phenolic urethane no-bake	RH	relative humidity; refrigeration hardened
PUR	polyurethane	RHEED	reflection high-energy electron diffraction

(continued)

Table 19.1 (continued)

RIBAD	reactive ion-beam-assisted deposition	SCFH	standard cubic feet per hour
RIM	reaction injection molding	scfm	standard cubic foot per minute
RIP	reactive ion plating	SCFM	subcritical fracture mechanics
rms	root mean square	SCHAZ	subcritical heat-affected zone
ROC	rapid omnidirectional compaction/consolidation	SCR	silicon controlled rectifier
		SCRATA	Steel Castings Research and Trade Association
rpm	revolutions per minute		
RRIM	reinforced reaction injection molding	SDI	strategic defense initiative
RS	Raman spectroscopy; rapid solidification	SE	secondary electron
RSD	relative standard deviation	SEM	scanning electron microscope/microscopy
RSEW	resistance seam welding	SEN	single-edge notched (specimen)
RSPD	rapid-solidification plasma-deposition	SENB	single-edge notch beam
RSR	rapid solidification rate	SERS	surface-enhanced Raman spectroscopy
RSW	resistance spot welding	SF	slow-fast (wave form)
RT	room temperature	SFC	supercritical fluid chromatography
RTE	reversible temper embrittlement	sfm	surface feet per minute
RTM	resin transfer molding	SFSA	Steel Founders' Society of America
RTP	rapid thermal processing	SG	standard grade; spheroidal graphite; spin glass
RTR	real-time radiography		
RTV	room-temperature vulcanizing (rubber)	SGA	soluble-gas atomization
RWMA	Resistance Welder Manufacturers Association	Sh	Sherwood number
		SHC	synthetic hydrocarbon
s	second	SHE	standard hydrogen electrode
s	sample standard deviation	SHT	Sumitomo high toughness
S	siemens	SI	Système International d'Unités; silicone; spark ignition
S	nominal engineering stress or normal engineering stress		
		SIC	standard industry codes
S_a	alternating stress amplitude	SIMA	strain induced melt activated
S_C	carbon saturation	SIMS	secondary ion mass spectroscopy
S_m	mean stress	sin	sine
S_{max}	maximum stress	SIS	superconductor/insulating/superconductor
S_{min}	minimum stress	SLAM	scanning laser acoustic microscope/microscopy
S_r	range of stress		
SACD	spheroidized annealed cold drawn	SLEEM	scanning low-energy electron microscope/microscopy
SACP	selected-area channeling pattern		
SAD	selected-area diffraction	SLR	single-lens reflex (camera)
SADP	selected-area diffraction pattern	SMA	shape memory alloy
SAE	Society of Automotive Engineers	SMAW	shielded metal arc welding
SAM	scanning Auger microscopy; scanning acoustic microscopy	SMC	sheet molding compound
		SME	Society of Manufacturing Engineers; shape memory effect; solid-metal embrittlement
SAMPE	Society for the Advancement of Materials and Processing Engineering		
		SMIE	solid metal induced embrittlement
SAN	styrene-acrylonitrile	SMS	tin-molybdenum-sulfide ($SnMo_6S_8$)
SANS	small-angle neutron scattering	SMYS	specified minimum yield stress
SAP	sintered aluminum powder	S-N	stress-number of cycles (fatigue)
SATT	shear-area transition temperature	S/N	signal-to-noise (ratio)
SAW	submerged arc welding	SNECMA	Societe Nationale d'Etude et de Construction de Moteurs
SAXS	small-angle x-ray scattering		
SBC	steel-bonded (titanium) carbide	SNR	signal-to-noise ratio
SBR	styrene-butadiene rubber	SPA-LEED	spot profile analysis—low-energy electron diffraction
SBS	styrene-butadiene-styrene		
Sc	Schmidt number	SPC	statistical process control
SC	single-crystal	SPE	Society of Plastics Engineers
SCaM	scanning capacitance microscope/microscopy	spf	seconds per foot
		SPF	superplastic forming
SCC	stress-corrosion cracking	SPF/DB	superplastic forming/diffusion bonding
SCE	saturated calomel electrode	sp gr	specific gravity
SCEA	side cutting edge angle	SPI	Society of the Plastics Industry

(continued)

Table 19.1 (continued)

SPS	service propulsion system (space shuttle)	TAPPI	Technical Association of the Pulp and Paper Industry
SQ	structural quality		
SQC	statistical quality control	TB	torch brazing
SQUID	superconducting quantum interference device	TBC	thermal barrier coating
		TBCCO	Tl-Ba-Ca-Cu-O
SRB	sulfate reducing bacteria; solid rocket booster (space shuttle)	TC	total carbon
		TCA	trichloroethane
SRGHAZ	subcritically reheated grain-coarsened heat-affected zone	TCAB	twin carbon arc brazing
		TCE	trichloroethylene
SRIM	structural reaction injection molding	tcp	topologically close-packed
SRM	Standard Reference Material(s)	TCP	thermochemical processing
SRO	short-range order	TCT	thermochemical treatment
SRX	static recrystallization	TD	thoria dispersed; thorium dioxide dispersion strengthened; transverse direction (of a sheet); Toyota diffusion (process)
SS	slow-slow (wave form); Swedish Standard		
SSC	sulfide stress cracking; superconducting supercollider		
		TDA	Titanium Development Association
SSI	small-scale integration	TDI	toluene diisocyanate
SSIUS	Specialty Steel Industry of the United States	TEA	triethanolamine
SSMS	spark source mass spectrometry	TEG	triethylene glycol
SSPC	Steel Structures Painting Council	TEM	transmission electron microscope/microscopy; thermal energy method
SSRT	slow strain rate testing		
SSVOD	strong stirred vacuum oxygen decarburization	TF	thermal fatigue
		TFE	tetrafluoroethylene
SSW	solid-state welding	TFS	tin-free steel
St	stokes	TG	thermogravimetry
ST	short transverse (direction)	TGA	thermogravimetric analysis
STA	solution treated and aged	THSP	thermal spraying
STB	Sumitomo top and bottom blowing (process)	TIG	tungsten inert gas (welding)
		TIR	total indicator reading
std	standard	TMA	thermomechanical analysis
STEL	short-term exposure limit	TMCP	thermomechanical controlled processing
STEM	scanning transmission electron microscope/microscopy	TME	tempered martensite embrittlement
		TMF	thermomechanical fatigue
STLE	Society of Tribologists and Lubrication Engineers (formerly ASLE)	TMP	thermomechanical processing
		TMS	The Minerals, Metals, and Materials Society
STM	scanning tunneling microscope/microscopy	TNAA	transmission electron microscopy
STOL	short-time overload	TNT	2,4,6-trinitrotoluene
STP	standard temperature and pressure	TOF	time of flight
STQ	solution treated and quenched	tonf	tons of force
SUS	Saybolt universal second (measure of viscosity)	TP	thermoplastic
		TPI	thermoplastic polyimide; turns per inch
Sv	sievert	TPS	thermal protection system
SW	stud arc welding	TPUR	thermoplastic polyurethane
SWG	steel wire gage	TRIP	transformation induced plasticity
t	metric tonne	TRS	transverse rupture strength
t	thickness; time	tsi	tons per square inch
T	tesla	TSM	tandem scanning microscope/microscopy
T	temperature	TTS	tearing topography surface
T_b	boiling temperature	TTT	time-temperature-transformation
T_c	transition temperature from normal to superconducting state; critical ordering temperature; critical transition temperature; Curie temperature	TTU	through-transmission ultrasonics
		TTZ	transformation-toughened zirconia
		UBC	used beverage can/container
		UCL	upper control limit
T_g	glass transition temperature	UHC	ultrahigh carbon
T_L	liquidus temperature	UHF	ultrahigh frequency
T_m, T_M	melt/melting temperature	UHMWPE	ultrahigh molecular weight polyethylene
T_s	solidus temperature	UHV	ultrahigh vacuum
tan	tangent	UKAEA	United Kingdom Atomic Energy Authority

(continued)

Table 19.1 (continued)

UL	Underwriter's Laboratories	WRC	Welding Research Council
ULCB	ultralow-carbon bainitic	WS	wave soldering
UNI	Ente Nazionale Italiano di Unificazione	wt%	weight percent
UNS	Unified Numbering System (ASTM-SAE)	WXRFS	wavelength-dispersion x-ray fluorescence spectroscopy
UPS	ultraviolet photoelectron spectroscopy		
USAF	United States Air Force	XES	x-ray emission spectroscopy
USBM	United States Bureau of Mines	XLPE	cross-linked polyethylene
USDA	United States Department of Agriculture	XPS	x-ray photoelectron spectroscopy
USM	ultrasonic machining	XRD	x-ray powder diffraction
USSWG	United States steel wire gage	XRF	x-ray fluorescence
UST	underground storage tank	XRFS	x-ray fluorescence spectroscopy
USW	ultrasonic welding	XRPD	x-ray powder diffraction
UT	ultrasonic testing	XRS	x-ray spectroscopy
UTS	ultimate tensile strength	XTEM	cross-sectional transmission electron microscope/microscopy
UV	ultraviolet		
UV/VIS	ultraviolet/visible (absorption spectroscopy)	YAG	yttrium-aluminum-garnet
v	velocity	YBCO	Y-Ba-Cu-O
V	volt	YIG	yttrium iron garnet
V	volume; velocity	yr	year
VAC-ESR	electroslag remelting under reduced pressure	YS	yield strength
VAD	vacuum arc degassing	z	ion change
VADER	vacuum arc double-electrode remelting	Z	impedance; atomic number; standard normal distribution
VAR	vacuum arc remelted/remelting		
V-D	vacuum degassing	ZGS	zirconia grain stabilized
VHF	very high frequency	ZOLZ	zero-order Laue zone
VHP	vacuum hot pressing	ZR	zone refined
VHS	very high speed	ZTA	zirconia-toughened-alumina
VHSIC	very high speed integrated circuit	°	angular measure; degree
VI	viscosity index	°C	degree Celsius (centigrade)
VID	vacuum induction degassing	°F	degree Fahrenheit
VIDP	vacuum induction degassing and pouring	⇌	direction of reaction
VIM	vacuum induction melting	%	percent
VIM/VID	vacuum induction melting and degassing	α	angle of incidence; coefficient of thermal expansion
VIS	visible		
VLSI	very large scale integration	γ	surface energy; surface tension; shear strain
VM	vacuum melted	Δ	change in quantity; an increment; a range
VOD	vacuum oxygen decarburization (ladle metallurgy)	ΔG^0	standard free energy of formation change
		ΔH	change in enthalpy
VODC	vacuum oxygen decarburization (converter metallurgy)	ΔT	temperature difference
		ε	strain
VOID	vacuum oxygen induction decarburization	ε_{el}	elastic strain
vol	volume	ε_{in}	inelastic strain
vol%	volume percent	$\varepsilon_p, \varepsilon_{pl}$	plastic strain
voxel	volume element	$\dot{\varepsilon}$	strain rate
VPE	vapor-phase epitaxy	η	viscosity
w	whisker	θ	angle
W	watt	λ	wavelength
W	width; weight	μ	friction coefficient; magnetic permeability; x-ray absorption coefficient
Wb	weber		
WDS	wavelength dispersive spectroscopy	μB	Bohr magneton
WJM	waterjet machining	μin.	microinch
wk	week	μm	micrometer (micron)
WOL	wedge-opening load	μs	microsecond
WORM	write once read many	ν	Poisson's ratio
WPAFB	Wright-Patterson Air Force Base	ρ	density
WQ	water quench	σ	stress
WQT	water quenched and tempered	σ_y	yield stress

(continued)

Table 19.1 (continued)

σ_{uts}	ultimate tensile strength	Φ	angle of refraction
σ_{ys}	yield strength	₵	centerline
τ	shear stress	₽	parting line
Ω	ohm	√	surface roughness

Table 19.2 Mathematical signs and symbols

Symbol	Definition	Symbol	Definition
+	plus (sign of addition); positive	→≐	approaches
−	minus (sign of subtraction); negative	∝	varies as; is proportional to
± (∓)	plus or minus (minus or plus)	∞	infinity
×	times, by (multiplication sign); diameters (magnification)	√	square root of
		∥	parallel to
·	multiplied by	( ) [ ] { }	parentheses, brackets, and braces; quantities enclosed by them to be taken together in multiplying, dividing, etc.
÷	sign of division		
/	divided by; per		
:	ratio sign; divided by; is to	π	(pi) = 3.14159+
::	equals; as (proportion)	°	degrees
<	less than	′	minutes
>	greater than	″	seconds
<<	much less than	∠	angle
>>	much greater than	dx	differential of x
=	equals	Δ	(delta) difference
≡	identical with	∂	partial derivative
~	similar to; approximately	∫	integral of
≈	approximately equals	∮	line integral around a closed path
≅	approximately equals; congruent	Σ	(sigma) summation of
≤	equal to or less than	∇	del or nabla; vector differential operator
≥	equal to or greater than	|x|	absolute value of x
≠	not equal to		

Table 19.3 Greek alphabet

Upper and lower cases	Name	Upper and lower cases	Name	Upper and lower cases	Name	Upper and lower cases	Name
Α α	Alpha	Η η	Eta	Ν ν	Nu	Τ τ	Tau
Β β	Beta	Θ θ ϑ	Theta	Ξ ξ	Xi	Υ υ	Upsilon
Γ γ	Gamma	Ι ι	Iota	Ο ο	Omicron	Φ ϕ φ	Phi
Δ δ ∂	Delta	Κ κ	Kappa	Π π	Pi	Χ χ	Chi
Ε ε	Epsilon	Λ λ	Lambda	Ρ ρ	Rho	Ψ ψ	Psi
Ζ ζ	Zeta	Μ μ	Mu	Σ σ ς	Sigma	Ω ω	Omega

20 Directory of Standards Organizations and Technical Associations

Table 20.1 Technical associations and standards organizations located in North America

Aluminum Association, Inc.
900 19th Street, NW
Suite 300
Washington, DC 20006
(202) 862-5100

Aluminum Extruders Council (AEC)
1000 North Rand Road
Suite 214
Wauconda, IL 60084
(847) 526-2010

American Bearing Manufacturers Association (ABMA)
1200 19th Street, NW
Suite 300
Washington, DC 20036-2401
(202) 429-5155

American Bureau of Metal Statistics (ABMS)
400 Plaza Drive, Harmon Meadow
P.O. Box 1405
Secaucus, NJ 07094-0405
(201) 863-6900

American Electroplaters and Surface Finishers Society (AESF)
12644 Research Parkway
Orlando, FL 32826
(407) 281-6441

American Foundrymen's Society, Inc. (AFS)
505 State Street
Des Plaines, IL 60016-8399
(847) 824-0181

American Galvanizers Association (AGA)
12200 E. Iliff Ave.
Suite 204
Aurora, CO 80014
(303) 750-2900

American Gear Manufacturers Association (AGMA)
1500 King Street
Alexandria, VA 22314-2730
(703) 684-0211

American Institute of Mining, Metallurgical, and Petroleum Engineers (AIME)
345 E. 47th St., 14th Floor
New York, NY 10017
(212) 705-7695

American Institute of Steel Construction (AISC)
1 E. Wacker Drive
Suite 3100
Chicago, IL 60601-2001
(312) 670-2400

American Iron and Steel Institute (AISI)
1101 17th Street, NW
Suite 1300
Washington, DC 20036-4700
(202) 452-7100

American National Standards Institute (ANSI)
11 W. 42nd St., 13th Floor
New York, NY 10036
(212) 642-4900

(continued)

Table 20.1 (continued)

American Petroleum Institute (API)
1220 L Street, NW
Washington, DC 20005
(202) 962-4776

American Society for Nondestructive Testing (ASNT)
1711 Arlingate Lane
P.O. Box 28518
Columbus, OH 43228-0518
(614) 274-6003

American Society for Quality Control (ASQC)
611 E. Wisconsin Ave.
P.O. Box 3005
Milwaukee, WI 53201-3005
(414) 272-8575

American Society of Mechanical Engineers (ASME)
345 E. 47th Street
New York, NY 10017
(800) 843-2763

American Welding Institute (AWI)
10628 Dutchtown Road
Knoxville, TN 37932
(615) 675-2150

American Welding Society (AWS)
550 N.W. LeJeune Road
Miami, FL 33126
(800) 443-WELD (9353)

American Wire Producers Association (AWPA)
515 King Street
Suite 420
Alexandria, VA 22314
(703) 549-6003

American Zinc Association (AZA)
1112 16th Street, NW
Suite 240
Washington, DC 20036
(202) 835-0164

APMI International
105 College Road East
Princeton, NJ 08540-6692
(609) 452-7700

ASM International
9639 Kinsman Road
Materials Park, OH 44073-0002
(216) 338-5151

Association of Industrial Metallizers, Coaters,
and Laminators (AIMCAL)
211 N. Union Street
Suite 100
Alexandria, VA 22314
(703) 684-4868

Association of Iron and Steel Engineers (AISE)
Three Gateway Center
Suite 2350
Pittsburgh, PA 15222
(412) 281-6323

ASTM
1916 Race Street
Philadelphia, PA 19103-1187
(215) 299-5585

Canadian Institute of Mining, Metallurgy,
and Petroleum (CIM)
Xerox Tower
Suite 2110
3400 de Maisonneuve Blvd., W.
Montreal, Quebec
Canada, H3Z 3B8
(514) 939-2710

Canadian Standards Association (CSA)
178 Rexdale Blvd.
Rexdale, Ontario
Canada M9W 1R3
(416) 747-4000

Center for Information and Numerical Data Analysis
and Synthesis (CINDAS)
Purdue University
2595 Yeager Road
West Lafayette, IN 47906-1398
(800) 224-6327

Copper Development Association (CDA)
260 Madison Ave.
New York, NY 10016
(212) 251-7200

Ductile Iron Society (DIS)
28938 Lorain Road
Suite 202
North Olmstead, OH 44070
(216) 737-8040

Edison Welding Institute (EWI)
1100 Kinnear Road
Columbus, OH 43212
(614) 486-9400

Federation of Societies for Coating Technology (FSCT)
492 Norristown Road
Blue Bell, PA 19422
(215) 940-0777

Forging Industry Association (FIA)
25 Prospect Ave. West
Suite 300
Cleveland, OH 44115
(216) 781-6260

Industrial Fasteners Institute (IFI)
1717 E. 9th Street
Suite 1105
Cleveland, OH 44114-2879
(216) 241-1482

International Cadmium Association
12110 Sunset Hills Road
Suite 110
Reston, VA 22090
(703) 709-1400

(continued)

Table 20.1 (continued)

International Copper Association, Ltd.
260 Madison Ave.
New York, NY 10016
(212) 251-7240

International Lead Zinc Research Organization (ILZRO)
2525 Meridian Parkway
P.O. Box 12036
Research Triangle Park, NC 27709
(919) 361-4647

International Magnesium Association (IMA)
1303 Vincent Place
Suite 1
McLean, VA 22101
(703) 442-8888

International Precious Metals Institute (IPMI)
4905 Tilgman Street
Suite 160
Allentown, PA 18104
(610) 395-9700

International Titanium Association (ITA)
1781 Folsom Street
Suite 100
Boulder, CO 80302-5714
(303) 443-7515

Investment Casting Institute
8350 N. Central Expressway
Suite M-110
Dallas, TX 75206-1602
(214) 368-8896

Iron and Steel Society (ISS)
410 Commonwealth Drive
Warrendale, PA 15086-7512
(412) 776-1535

Iron Casting Research Institute (ICRI)
2938 Fischer Road, Unit B
Columbus, OH 43204
(614) 275-4203

Lead Industries Association
295 Madison Ave.
New York, NY 10017
(212) 578-4750

Materials Properties Council (MPC)
345 E. 47th Street
New York, NY 10017
(212) 705-7693

Materials Research Society (MRS)
9800 McKnight Road
Pittsburgh, PA 15237-6006
(412) 367-3012

Metal Fabricating Institute (MFI)
P.O. Box 1178
Rockford, IL 61105
(815) 965-4031

Metal Powder Industries Federation (MPIF)
105 College Road East
Princeton, NJ 08540-6692
(609) 452-7700

Metals Information Analysis Center (MIAC)
Purdue University
2595 Yeager Road
West Lafayette, IN 47906-1398
(3170 494-9393

Metal Treating Institute (MTI)
1550 Robert Drive
Jacksonville Beach, FL 32250
(904) 249-0448

NACE International
P.O. Box 218340
Huston, TX 77218-8340
(713) 492-0535

National Coil Coaters Association (NCCA)
401 N. Michigan Ave.
Chicago, IL 60611-4267
(312) 321-6894

National Institute of Standards and Technology (NIST)
Gaithersburg, MD 20899
(301) 975-2000

National Paint and Coatings Association (NPCA)
1500 Rhode Island Ave., NW
Washington, DC 20005
(202) 462-6272

National Technical Information Service (NTIS)
U.S. Department of Commerce
Technology Administration
Springfield, VA 22161
(703) 487-4650

Nickel Development Institute (NiDI)
214 King Street West
Suite 510
Toronto, Ontario
Canada M5H 3S6
(416) 591-7999

Non-Ferrous Founders Society (NFFS)
455 State Street
Suite 100
Des Plaines, IL 60016
(708) 299 0950

North American Die Casting Association (NADCA)
9701 W. Higgins Road
Suite 880
Rosemont, IL 60018
(708) 292-3600

Powder Coating Institute (PCI)
2121 Eisenhower Ave.
Suite 401
Alexandria, VA 22314
(703) 684-1770

(continued)

Table 20.1 (continued)

Rare-Earth Information Center
Iowa State University
Institute for Physical Research and Technology
Ames Laboratory
Ames, IA 50011-3020
(515) 294-2272

SAE International
400 Commonwealth Drive
Warrendale, PA 15096-0001
(412) 776-4970

Silver Institute (SI)
1112 16th Street, NW
Washington, DC 20036
(202) 835-0185

Society for the Advancement of Materials and
  Processing Engineering (SAMPE)
P.O. Box 2459
Covina, CA 91722
(818) 331-0616

Society of Manufacturing Engineers (SME)
1 SME Drive
P.O. Box 930
Dearborn, MI 48121-0930
(313) 271-1500

Society of Tribologists and Lubrication Engineers (STLE)
840 Busse Highway
Park Ridge, IL 60068-2376
(708) 825-5536

Society of Vacuum Coaters (SVC)
440 Live Oak Loop
Albuquerque, NM 87122
(505) 856-7188

Specialty Steel Industry of North America (SSINA)
3050 K Street, NW
Suite 400
Washington, DC 20007
(202) 342-8630

Spring Manufacturers Institute (SMI)
2001 Midwest Road
Suite 106
Oak Brook, IL 60521
(708) 495-8588

Steel Founders' Society of America (SFSA)
Cast Metals Federation Building
455 State Street
Des Plaines, IL 60016
(847) 299-9160

Steel Structures Painting Council (SSPC)
40 24th Street
Pittsburgh, PA 15222-4643
(412) 281-2331

The Metallurgical Society (TMS-AIME)
420 Commonwealth Drive
Warrendale, PA 15086-7514
(412) 776-9000

Welding Research Council (WRC)
345 East 47th St.
New York, NY 10017
(212) 705-7956

Wire Association International (WAI)
1570 Boston Post Road
P.O. Box H
Guilford, CT 06437
(203) 453-2777

Table 20.2 Selected international standards organizations arranged according to country/region of origin

See Table 20.1 for North American (U.S. and Canada) technical associations and standards organizations

ARGENTINA
Instituto Argentino de Racionalizacion
  de Materiales (IRAM)
Chile 1192
1098 Buenos Aires
Argentina
Tel: 54 1 383 37 51
Fax: 54 1 383 84 63

AUSTRALIA
Standards Australia (SAA)
P.O. Box 1055
Strathfield-NSW 2135
Australia
Tel: 61 2 746 4700
    61 2 746 4748
Fax: 61 2 746 8450
     61 2 746 4765

(continued)

Table 20.2 (continued)

AUSTRIA
Osterreichisches Normungsinstitut (ON)
Heinestrasse 38
Postfach 130
A-1021 Wien
Austria
Tel: 43 1 213 00
Fax: 43 1 21 30 06 50

BELARUS
Committee for Standardization, Metrology
  and Certification (BELST)
Starovilensky Trakt 93
Minsk 220053
Belarus
Tel: 375 172 37 52 13
Fax: 375 172 37 25 88

BELGIUM
Institut belge de normalisation (IBN)
Av. de la Brabanconne 29
B-1000 Bruxelles
Belgium
Tel: 32 2 738 01 11
Fax: 32 2 733 42 64

BRAZIL
Associacao Brasileira de Normas Tecnicas (ABNT)
Av. 13 de Maio, no 13, 27o andar
Caixa Postal 1680
20003-900 Rio de Janeiro-RJ
Brazil
Tel: 55 21 210 31 22
Fax: 55 21 532 21 43

BULGARIA
Committee for Standardization and Metrology
  at the Council of Ministers (BDS)
21, 6th September Str.
1000 Sofia
Bulgaria
Tel: 359 2 85 91
Fax: 359 2 80 14 02

CHILE
Instituto Nacional de Normalizacion (INN)
Matias Cousino 64—6o piso
Casilla 995—Correo Central
Santiago
Chile
Tel: 56 2 696 81 44
Fax: 56 2 696 02 47

CHINA
China State Bureau of Technical Supervision (CSBTS)
Dept. of Stand-4, Zhichun Road
Haidian District
P.O. Box 8010
Beijing 100088
Peoples Republic of China
Tel: 86 10 203 24 24
Fax: 86 10 203 10 10

CROATIA
State Office for Standardization and Metrology
  (DZNM)
Ulica grada Vukovara 78
10000 Zagreb
Croatia
Tel: 385 1 53 99 34

CZECH REPUBLIC
Czech Office for Standards, Metrology and Testing (COSMT)
Biskupsky dvur 5
113 47 Praha 1
Czech Republic
Tel: 42 2 232 44 30
Fax: 42 2 232 43 73

DENMARK
Dansk Standard (DS)
Baunegaardsvej 73
Hellerup DK-2900
Denmark
Tel: 45 39 77 01 01
Fax: 45 39 77 02 02

EGYPT
Egyptian Organization for Standardization
  and Quality Control (EOS)
2 Latin America Street
Garden City Cairo
Egypt
Tel: 20 2 354 97 20
Fax: 20 2 355 78 41

EUROPE
Association Europeenne des Constructeurs de Materiel
  Aerospatial (AECMA)
88 Bd Malesherbes
F-75008 Paris
France
Tel: 33 1 4563 82 85
Fax: 33 1 4225 15 48

Commission of the European Communities
  (CEC or CCE)
2100 M Street NW
7th Floor
Washington, DC 20037
USA
Tel: (202) 862 9500
Fax: (202) 429 1766

European Committee for Standardization (CEN)
36, Rue de Strassart
B-1050 Brussels
Belgium
Tel: 32 2 519 68 11
Fax: 32 2 519 68 19

FINLAND
Finnish Standard Association SFS (SFS)
P.O. Box 116
Fin-00241 Helsinki
Finland
Tel: 358 0 149 9331
Fax: 358 0 146 4925

(continued)

Table 20.2 (continued)

FRANCE
Delegation Generale pour L'Armement (AIR)
Centre de Documentation de l'Armement
26, Boulevard Victor
00460-Armees
France
Tel: 33 1 4552 45 24
Fax: 33 1 4552 45 74

Association Francaise de Normalisation (AFNOR)
Tour Europe-Cedex 7
F-92080 Paris la Defense
France
Tel: 33 1 42 91 55 55
    33 1 42 91 58 07
Fax: 33 1 42 91 56 56

GERMANY
DIN Deutsches Institut fur Normung e.V. (DIN)
Burggrafenstrasse 6
Postfach 1107
D-10787 Berlin
Germany
Tel: 49 30 2601 2344
Fax: 49 30 2601 1231

HUNGARY
Magyar Szabvanyugyi Hivatal (MSZH)
Postafiok 24
H1450 Budapest 9
Hungary
Tel: 36 1 218 30 11
Fax: 36 1 218 51 25

INDIA
Bureau of Indian Standards (BIS)
Manak Bhavan
9 Bahadur Shah Zafar Marg
New Delhi 110002
India
Tel: 91 11 323 79 91
Fax: 91 11 323 40 62

INDONESIA
Dewan Standardisasi Nasional-DSN
 (Standardization Council of Indonesia) (DSN)
Sasana Widya Sarwono Lantai 5
Jl. jend. Gatot Subroto No. 10
Jakarta 12710
Indonesia
Tel: 62 21 520 6574
Fax: 62 21 520 6574

INTERNATIONAL
International Organization for Standardization (ISO)
1, rue de Varembe
Case postale 56
CH-1211 Geneve 20
Switzerland
Tel: 41 22 749 0111
Fax: 41 22 733 3430

ITALY
Ente Nazionale Italiano di Unificazione (UNI)
Via Battistotti Sassi 11 b
I-20133 Milano
Italy
Tel: 39 2 70 02 41

JAPAN
Japanese Industrial Standards Committee (JISC)
c/o Standards Department
Ministry of International Trade and Industry
1-3-1, Kasumigaseki, Chiyoda-ku
Tokyo 100
Japan
Tel: 81 3 35 01 92 95
Fax: 81 3 35 80 14 18

Japanese Standards Association (JSA)
1-24-4, Akasaka
Minato-ku
Tokyo 107
Japan
Tel: 81 3 3583 8003
Fax: 81 3 3586 2029

KOREA
Korean National Institute of Technology
 and Quality (KNITQ)
1599 Kwanyang-dong
Dongan-ku, Anyang-city
Kyonggi-do 430-060
Republic of Korea
Tel: 82 3 43 84 18 61
Fax: 82 3 43 84 60 77

MALAYSIA
Standards and Industrial Research Institute of
 Malaysia (SIRIM)
Persiaran Dato Menteri, Section 2
P.O. Box 7035, 40911 Shah Alam
Selangor Darul Ehsan
Malaysia
Tel: 60 3 559 26 01
Fax: 60 3 550 80 95

MEXICO
Direccion General de Normas (DGN)
Calle Puente de Tecamachalco No 6
Lomas de Tecamachalco
Seccion Fuentes
Naucalpan de Juarez 53 950
Mexico
Tel: 52 5 729 93 00
Fax: 52 5 729 94 84

NETHERLANDS
Nederlands Normalisatie-instituut (NNI)
Kalfjeslaan 2
P.O. Box 5059
NL-2600 GB Delft
Netherlands
Tel: 31 15 2 69 03 90
Fax: 31 152 69 01 90

NEW ZEALAND
Standards New Zealand (SNZ)
Standards House
Private Bag 2439
Wellington 6020
New Zealand
Tel: 64 4 498 59 90
Fax: 64 4 498 59 94

(continued)

Table 20.2 (continued)

NORWAY
Norges Standardiseringsforbund (NSF)
P.O. Box 7020
Homansbyen
N-0306 Oslo
Norway
Tel: 47 22 46 60 94
Fax: 47 22 46 44 57

PAKISTAN
Pakistan Standards Institution (PSI)
39 Garden Road, Saddar
Karachi 74400
Pakistan
Tel: 92 21 772 95 27
Fax: 92 21 772 81 24

PAN AMERICA
Pan American Standards Commission (COPANT)
Avenida Andres Bello, Torre Fondo Comun
Piso 11
Caracas 1050
Venezuela
Tel: 58 2 5742941
Fax: 58 2 5742941

POLAND
Polish Committee for Standardization (PKN)
ul. Elektoralna 2
P.O. Box 411
PL-00-950 Warszawa
Poland
Tel: 48 22 620 54 34
Fax: 48 22 620 07 41

PORTUGAL
Instituto Portugues da Qualidade (IPQ)
Rua C a Avenida do Tres Vales
P-2825 Monte de Caparica
Portugal
Tel: 351 1 294 81 00
Fax: 351 1 294 81 01

ROMANIA
Institutul Roman de Standardizare (IRS)
Str. Jean-Louis Calderon Nr. 13
70201 Bucharest 2
Romania
Tel: 40 1 211 32 92
     40 1 615 58 70
Fax: 40 1 210 08 33
     40 1 312 47 44

RUSSIA
Committee of the Russian Federation for Standardization, Metrology and Certification (GOST R)
Leninsky Prospekt 9
Moskva 117049
Russian Federation
Tel: 7 095 236 40 44
Fax: 7 095 237 60 32

SAUDI ARABIA
Saudi Arabian Standards Organization (SASO)
Imam Saud Bin Abdul Aziz Bin Mohammed Road (West End)
P.O. Box 3437
Riyadh 11471
Saudia Arabia
Tel: 966 1 452 00 00
Fax: 966 1 452 00 86

SLOVAKIA
Slovak Office of Standards, Metrology and Testing (UNMS)
Stefanovicova 3
814 39 Bratislava
Slovakia
Tel: 42 7 49 10 85
Fax: 42 7 49 10 50

SLOVENIA
Standards and Metrology Institute (SMIS)
Ministry of Science and Technology
Kotnikova 6
SI-61000 Ljubljana
Slovenia
Tel: 386 61 1312 322
Fax: 386 61 314 882

SOUTH AFRICA
South African Bureau of Standards (SABS)
1 Dr Lategan Rd, Groenkloof
Private Bag X191
Pretoria 0001
South Africa
Tel: 27 12 428 79 11
Fax: 27 12 344 15 68

SPAIN
Asociacion Espanola de Normalizacion y Certificacion (AENOR)
Fernandez de las Hoz, 52
E-28010 Madrid
Spain
Tel: 34 1 432 60 00
Fax: 34 1 310 49 76

SWEDEN
SIS—Standardiseringen i Sverige (SIS)
St. Eriksgatan 115
Box 6455
S-113 82 Stockholm
Sweden
Tel: 46 8 610 30 00
Fax: 46 8 30 77 57

SWITZERLAND
Swiss Association for Standardization (SNV)
Muhlebachstrasse 54
CH-8008 Zurich
Switzerland
Tel: 41 1 254 54 54
Fax: 41 1 254 54 74

(continued)

Table 20.2 (continued)

THAILAND
Thai Industrial Standards Institute (TISI)
Ministry of Industry
Rama VI Street
Bangkok 10400
Thailand
Tel: 66 2 245 78 02
Fax: 66 2 247 87 41

TURKEY
Turk Standardlari Enstitusu (TSE)
Necatibey Cad. 112
Bakanliklar 06100 Ankara
Turkey
Tel: 90 312 417 83 30
Fax: 90 312 425 43 99

UKRAINE
State Committee of Ukraine for Standardization, Metrology and Certification (DSTU)
174 Gorky St.
GSP, Kiev-6, 252650
Ukraine
Tel: 380 44 226 29 71
Fax: 380 44 226 29 70

UNITED KINGDOM
British Standards Institution (BSI)
2 Park Street
London W1A 2BS
England
Tel: 44 71 629 9000
Fax: 44 71 629 0506

VENEZUELA
Comision Venezolana de Normas Industriales (COVENIN)
Avda. Andres Bello-Edf. Torre Fondo
Comun
Piso 12
Caracas 1050
Venezuela
Tel: 58 2 575 22 98
Fax: 58 2 574 13 12

YUGOSLAVIA (FORMER)
Savezni zavod za standardizaciju (SZS)
Kneza Milosa 20
Post Pregr. 933
Beograd
Serbia (former Federal Republic of Yugoslavia)
Tel: 381 11 64 35 57
Fax: 381 11 68 23 82

21 Bibliography of Selected References

Materials Properties and Selection

Ferrous Alloys

- *Properties and Selection: Irons, Steels, and High-Performance Alloys,* Vol 1, *ASM Handbook,* ASM International, 1990
- *Materials Selection and Design,* Vol 20, *ASM Handbook,* ASM International, 1997
- J.E. Bringas, Ed., *Ferrous Metal,* Vol 1, *The Metals Black Book,* Casti Publishing, 1995
- *Iron and Steel Products,* Section 1, *Annual Book of ASTM Standards,* ASTM, revision issued annually (contains 7 volumes)
- J.R. Davis, Ed., *ASM Specialty Handbook: Carbon and Alloy Steels,* ASM International, 1996
- P. Harvey, Ed., *Engineering Properties of Steels,* American Society for Metals, 1982
- J.R. Davis, Ed., *ASM Specialty Handbook: Stainless Steels,* ASM International, 1994
- P. Lacombe, B. Baroux, and G. Beranger, Ed., *Stainless Steels,* Les Éditions de Physique, 1993
- R. Lula, Ed., *Stainless Steel,* American Society for Metals, 1985
- D. Peckner and I.M. Bernstein, Ed., *Handbook of Stainless Steels,* McGraw-Hill, 1977
- G. Krauss, Ed., *Tool Steels,* 5th ed., ASM International, 1997
- J.R. Davis, Ed., *ASM Specialty Handbook: Tool Materials,* ASM International, 1995
- M. Blair, T.L. Stevens, and B. Linskey, Ed., *Steels Castings Handbook,* 6th ed., Steel Founders' Society of America and ASM International, 1995
- J.R. Davis, Ed., *ASM Specialty Handbook: Cast Irons,* ASM International, 1996
- C.F. Walton and T.J. Opar, Ed., *Iron Castings Handbook,* Iron Castings Society, 1981

Nonferrous Alloys and Superalloys

- *Properties and Selection: Nonferrous Alloys and Special-Purpose Materials,* Vol 2, *ASM Handbook,* ASM International, 1990
- *Properties and Selection: Irons, Steels, and High-Performance Alloys,* Vol 1, *ASM Handbook,* ASM International, 1990
- *Materials Selection and Design,* Vol 20, *ASM Handbook,* ASM International, 1997
- J.E. Bringas, Ed., *Nonferrous Metals,* Vol 2, *The Metals Red Book,* Casti Publishing, 1993
- *Nonferrous Metals Products,* Section 2, *Annual Book of ASTM Standards,* ASTM, revision issued annually (contains 5 volumes)
- J.R. Davis, Ed., *ASM Specialty Handbook: Aluminum and Aluminum Alloys,* ASM International, 1993
- J.E. Hatch, Ed., *Aluminum: Properties and Physical Metallurgy,* American Society for Metals, 1984

- K.R. Van Horn, Ed., *Aluminum,* Vol 1–3 American Society for Metals, 1967
- L.F. Mondolfo, *Aluminum Alloys: Structure and Properties,* Butterworths, 1976
- R. Boyer, T. Collings, and G. Welsch, Ed., *Materials Properties Handbook: Titanium Alloys,* ASM International, 1994
- E.W. Collings, *The Physical Metallurgy of Titanium Alloys,* American Society for Metals, 1984
- M.J. Donachie, Jr., Ed., *Titanium: A Technical Guide,* ASM International, 1988
- J.R. Davis, Ed., *ASM Specialty Handbook: Tool Materials,* ASM International, 1995
- J.R. Davis, Ed., *ASM Specialty Handbook: Heat-Resistant Materials,* ASM International, 1997
- E.F. Bradley, Ed., *Superalloys, Supercomposites, and Superceramics,* Academic Press, 1989
- J.K. Tien and T. Caulfield, Ed., *Superalloys, Supercomposites, and Superceramics,* Academic Press, 1989
- C.T. Sims, N.S. Stoloff, and W.C. Hagel, Ed., *Superalloys II,* John Wiley & Sons, 1987

Failure Analysis, Fatigue, and Fracture

Failure Analysis

- *Failure Analysis and Prevention,* Vol 11, *ASM Handbook,* ASM International, 1986
- D.J. Wulpi, *Understanding How Components Fail,* American Society for Metals, 1985
- K.A. Esaklul, Ed., *Handbook of Case Histories in Failure Analysis,* Vol 1 and 2, ASM International, 1992–1993
- C. Brooks and A. Choudary, *Metallurgical Failure Analysis,* McGraw-Hill, 1993
- J.A. Collins, *Failure of Materials in Mechanical Design: Analysis, Prediction,* Prevention, John Wiley & Sons, 1993
- D.R.H. Jones, *Engineering Materials 3: Materials Failure Analysis—Case Studies and Design Implications,* Pergamon Press, 1993
- S. Nishida, *Failure Analysis in Engineering Applications,* Butterworths-Heinemann, 1992
- P.F. Timmins, *Fracture Mechanics and Failure Control for Inspectors and Engineers,* ASM International, 1994
- R. Viswanathan, *Damage Mechanisms and Life Assessment of High-Temperature Components,* ASM International, 1989
- V.S. Goel, Ed., *Analyzing Failures,* American Society for Metals, 1986
- F.R. Hutchings and P.M. Unterweiser, Ed., *Failure Analysis: The British Engine Technical Reports,* American Society for Metals, 1981

Fatigue and Fracture

- *Fatigue and Fracture,* Vol 19, *ASM Handbook,* ASM International, 1996
- *Fractography,* Vol 12, *ASM Handbook,* ASM International, 1987
- N.E. Dowling, *Mechanical Behavior of Materials: Engineering Methods for Deformation, Fracture, and Fatigue,* Prentice-Hall, 1993
- J.M. Barsom and S.T. Rolfe, *Fracture and Fatigue Control in Structures,* 2nd ed., Prentice-Hall, 1987
- R.C. Rice, Ed., *Fatigue Design Handbook,* Society of Automotive Engineers, 1988
- *Fatigue Data Book: Light Structural Alloys,* ASM International, 1995
- H.E. Boyer, Ed., *Atlas of Fatigue Curves,* American Society for Metals, 1986
- O.F. Deveraux, A.J. McEvily, Jr., and R.W. Staehle, Ed., *Corrosion Fatigue: Chemistry, Mechanics, and Microstructure,* National Association of Corrosion Engineers, 1971

- J.A. McEvily, Jr., Ed., *Atlas of Stress-Corrosion and Corrosion Fatigue Curves,* ASM International, 1990
- J.E. Campbell, W.W. Gerberich, and J.H. Underwood, Ed., *Application of Fracture Mechanics for Selection of Metallic Structural Materials,* American Society for Metals, 1982
- D. Broeck, *The Practical Use of Fracture Mechanics,* Kluwer Academic Publishers, 1989
- D. Broeck, *Elementary Fracture Mechanics,* 4th ed., Martinus Nijhoff, 1986
- R.W. Hertzberg, *Deformation and Fracture Mechanics of Engineering Materials,* 3rd ed., John Wiley & Sons, 1989
- T.L. Anderson, *Fracture Mechanics, Fundamentals, and Applications,* 2nd ed., CRC Press, 1995
- J.F. Knott, *Fundamentals of Fracture Mechanics,* Butterworths, 1973

Corrosion and Wear

Corrosion

- *Corrosion,* Vol 13, *ASM Handbook,* ASM International, 1987
- B. Craig and D. Anderson, Ed., *Handbook of Corrosion Data,* 2nd ed., ASM International, 1995
- G.Y. Lai, *High Temperature Corrosion of Engineering Alloys,* ASM International, 1990
- R.H. Jones, Ed., *Stress-Corrosion Cracking: Materials Performance and Evaluation,* ASM International, 1992
- R. Gibala and R.F. Hehemann, Ed., *Hydrogen Embrittlement and Stress-Corrosion Cracking,* American Society for Metals, 1984
- I.M. Bernstein and A.W. Thompson, Ed., *Hydrogen Effects in Metals,* TMS-AIME, 1981
- M.G. Fontana, *Corrosion Engineering,* 3rd ed., McGraw-Hill, 1986
- H.H. Uhlig and R.W. Revie, *Corrosion and Corrosion Control,* 3rd ed., John Wiley & Sons, 1985
- R. Baboian, Ed., *Corrosion Tests and Standards: Application and Interpretation,* ASTM, 1995
- W.H. Ailor, Ed., *Handbook on Corrosion Testing and Evaluation,* John Wiley & Sons, 1971
- C.G. Munger, *Corrosion Prevention by Protective Coatings,* National Association of Corrosion Engineers, 1985

Wear

- *Friction, Lubrication, and Wear Technology,* Vol 18, ASM Handbook, ASM International, 1992
- M.B. Peterson and W.O. Winer, Ed., *Wear Control Handbook,* American Society of Mechanical Engineers, 1980
- A.Z. Szeri, *Tribology: Friction, Lubrication, and Wear,* McGraw-Hill, 1980
- J.A. Schey, *Tribology in Metalworking: Friction, Lubrication, and Wear,* American Society for Metals, 1983
- D.A. Rigney, Ed., *Fundamentals of Friction and Wear of Metals,* American Society for Metals, 1981

Steelmaking

- W.T. Lankford, N.L. Samways, R.F. Craven, and H.E. McGannon, Ed., *The Making, Shaping and Treating of Steel,* Association of Iron and Steel Engineers, 1985
- *Continuous Casting,* Vol 1–4, The Iron and Steel Society, 1983–1988

- W.L. Roberts, *Hot Rolling of Steel,* Marcel Dekker, 1983
- W.L. Roberts, *Cold Rolling of Steel,* Marcel Dekker, 1979
- W.L. Roberts, *Flat Processing of Steel,* Marcel Dekker, 1988

Forming Processes

General Overviews

- C. Wick, J.T. Benedict, and R.F. Veilleux, Ed., *Forming,* Vol 2, 4th ed., *Tool and Manufacturing Engineers Handbook,* Society of Manufacturing Engineers, 1984
- S. Kalpakjian, *Manufacturing Processes for Engineering Materials,* Addison-Wesley, 1984
- L.E. Doyle, C.A. Keyser, J.L. Leach, G.F. Schrader, and M.B. Singer, *Manufacturing Processes and Materials for Engineers,* Prentice-Hall, 1985
- *Machinery's Handbook,* 25th ed., Industrial Press, 1996

Forming and Forging

- *Forming and Forging,* Vol 14, *ASM Handbook,* ASM International, 1988
- T. Altan, S.I. Oh, and H.L. Gegel, *Metal Forming Fundamentals and Applications,* American Society for Metals, 1983
- G.D. Lascoe, *Handbook of Fabrication Processes,* ASM International, 1988
- K. Lange, Ed., *Handbook of Metal Forming,* McGraw-Hill, 1985
- T.G. Byrer, Ed., *Forging Handbook,* Forging Industry Association and American Society for Metals, 1984
- C. Wick, J.T. Benedict, and R.F. Veilleux, Ed., *Forming,* Vol 2, 4th ed., *Tool and Manufacturing Engineers Handbook,* Society of Manufacturing Engineers, 1984

Machining

- *Machining,* Vol 16, *ASM Handbook,* ASM International, 1989
- T.J. Drozda and C. Wick, Ed., *Machining,* Vol 1, 4th ed., *Tool and Manufacturing Engineers Handbook,* Society of Manufacturing Engineers, 1983
- *Machining Data Handbook,* Vol 1 and 2, 3rd ed., Metcut Research Associates, 1980
- M.C. Shaw, *Metal Cutting Principles,* Oxford University Press, 1984
- G.F. Benedict, *Nontraditional Manufacturing Processes,* Marcel Dekker, 1987
- R.I. King and R.S. Hahn, *Handbook of Modern Grinding Technology,* Chapman & Hall, 1986

Casting

- *Casting,* Vol 15, *ASM Handbook,* ASM International, 1988
- M. Blair, T.L. Stevens, and B. Linskey, Ed., *Steels Castings Handbook,* 6th ed., Steel Founders' Society of America and ASM International, 1995
- C.F. Walton and T.J. Opar, Ed., *Iron Castings Handbook,* Iron Castings Society, 1981
- E.L. Kotzin, *Metalcasting and Molding Processes,* American Foundrymen's Society, 1981
- *Aluminum Casting Technology,* 2nd ed., American Foundrymen's Society, 1993
- J. Campbell, *Castings,* American Foundrymen's Society, 1992

Powder Metallurgy

- *Powder Metallurgy,* Vol 7, *ASM Handbook,* ASM International, 1984
- F.V. Lenel, *Powder Metallurgy,* Metal Powder Industries Federation, 1980

- R.M. German, *Powder Metallurgy Science,* Metal Powder Industries Federation, 1994
- F. Thummler and R. Oberacker, *Introduction to Powder Metallurgy,* The Institute of Materials, 1994
- H.H. Hausner and M.K. Mal, *Handbook of Powder Metallurgy,* 2nd ed., Chemical Publishing, 1982
- C.G. Goetzel, *Treatise of Powder Metallurgy,* Vol 1–4, Interscience, 1949

Heat Treating

Ferrous Alloys

- *Heat Treating,* Vol 4, *ASM Handbook,* ASM International, 1991
- G. Krauss, *Steels: Heat Treatment and Processing Principles,* ASM International, 1990 (updated version of M.A. Grossmann and E.C. Bain, Principles of Heat Treatment, 5th ed., American Society for Metals, 1964)
- H.E. Boyer, Ed., *Practical Heat Treating,* American Society for Metals, 1984
- G.E. Totten, C.E. Bates, and N.A. Clinton, *Handbook of Quenchants and Quenching Technology,* ASM International, 1992
- C.S. Siebert, D.V. Doane, and D.H. Breen, *The Hardenability of Steels—Concepts, Metallurgical Influences, and Industrial Applications,* American Society for Metals, 1977
- D.V. Doane and J.S. Kirkaldy, Ed., *Hardenability Concepts with Application to Steel,* 1978
- *Heat Treater's Guide: Practices and Procedures for Irons and Steels,* 2nd ed., ASM International, 1995
- G.F. Vander Voort, Ed., *Atlas of Time-Temperature Diagrams for Irons and Steels,* ASM International, 1991
- M. Atkins, *Atlas of Continuous Transformation Diagrams for Engineering Steels,* British Steel Corporation, Sheffield, 1977 (revised U.S. edition published in 1980 by American Society for Metals)
- *Atlas of Isothermal Transformation and Cooling Transformation Diagrams,* American Society for Metals, 1977
- *Atlas of Isothermal Transformation Diagrams,* 3rd ed., U.S. Steel, 1963

Nonferrous Alloys

- *Heat Treating,* Vol 4, *ASM Handbook,* ASM International, 1991
- C.R. Brooks, *Heat Treatment, Structure and Properties of Nonferrous Alloys,* American Society for Metals, 1982
- *Heat Treater's Guide: Practices and Procedures for Nonferrous Alloys,* ASM International, 1996
- G.F. Vander Voort, Ed., *Atlas of Time-Temperature Diagrams for Nonferrous Alloys,* ASM International, 1991

Joining

Welding

- *Welding, Brazing, and Soldering,* Vol 6, *ASM Handbook,* ASM International, 1993
- *Welding Technology,* Vol 1, 8th ed., *Welding Handbook,* American Welding Society, 1987 (updated version of Volumes 1 and 5 from the 7th ed.)

- *Welding Processes,* Vol 2, 8th ed., *Welding Handbook,* American Welding Society, 1991 (updated version of Volumes 2 and 3 from the 7th ed.)
- *Metals and Their Weldability,* Vol 4, *Welding Handbook,* American Welding Society, 1982
- H.B. Cary, *Modern Welding Technology,* 2nd ed., Prentice Hall, 1989
- R.D. Stout, *Weldability of Steels,* 4th ed., Welding Research Council, 1987

Brazing

- *Welding, Brazing, and Soldering,* Vol 6, *ASM Handbook,* ASM International, 1993
- *Ceramics and Glasses,* Vol 4, *Engineered Materials Handbook,* ASM International, 1991 (see Section 7, "Joining," which includes articles on ceramic/metal and glass/metal seals)
- M.M. Schwartz, *Brazing,* ASM International, 1987
- M.M. Schwartz, *Ceramic Joining,* ASM International, 1990
- *Brazing Handbook,* American Welding Society, 1991
- G. Humpston and D.M. Jacobson, *Principles of Soldering and Brazing,* ASM International, 1993

Soldering

- *Welding, Brazing, and Soldering,* Vol 6, *ASM Handbook,* ASM International, 1993
- *Packaging,* Vol 1, *Electronic Materials Handbook,* ASM International, 1989 (see Section 6, "Soldering and Mounting Technology," and Section 9, "Failure Analysis," both of which contain articles on soldering materials and processing and solder failure mechanisms)
- H.H. Manko, *Solders and Soldering,* 3rd ed., McGraw-Hill, 1992
- H.H. Manko, *Soldering Handbook for Mounted Circuits and Surface Mounting,* Van Nostrand Reinhold, 1986
- R.J. Klein Wassink, *Soldering in Electronics,* Electrochemical Publications, 1989
- G. Humpston and D.M. Jacobson, *Principles of Soldering and Brazing,* ASM International, 1993

Adhesive Bonding

- *Adhesives and Sealants,* Vol 3, *Engineered Materials Handbook,* ASM International, 1990, (see Section 4, "Surface Considerations," and Section 10, "Bonded Structure Repair," both of which include articles on adhesive of bonding of metals, particularly aluminum)
- E.W. Thrall and R.W. Shannon, Ed., *Adhesive Bonding of Aluminum Alloys,* Marcel Dekker, 1985
- S. Semerdjiev, *Metal-to-Metal Adhesive Bonding,* Business Books Ltd., London, 1970
- I. Skeist, Ed., *Handbook of Adhesives,* 3rd ed., Van Nostrand Reinhold, 1990

Surface Engineering

- *Surface Engineering,* Vol 5, *ASM Handbook,* ASM International, 1994
- *Friction, Lubrication, and Wear Technology,* Vol 18, *ASM Handbook,* ASM International, 1992 (see the Section "Surface Treatments and Coatings for Friction and Wear Control")
- *Corrosion,* Vol 13, *ASM Handbook,* ASM International, 1987 (see the Section "Corrosion Protection Methods")
- C. Wick and R.F. Villeux, Ed., *Materials, Finishing and Coating,* Vol 3, 4th ed., *Tool and Manufacturing Engineers Handbook,* Society for Manufacturing Engineers, 1985
- P. Swaraj, *Surface Coatings,* John Wiley & Sons, 1985
- F.A. Lowenheim, *Electroplating,* McGraw-Hill, 1978

- W.A. Safranek, *The Properties of Electrodeposited Metals and Alloys: A Handbook,* 2nd ed., American Electroplaters and Surface Finishers Society, 1986
- L.J. Durney, Ed., *Electroplating Engineering Handbook,* 4th ed., Van Nostrand Reinhold, 1984
- G.O. Mallory and J.B. Hajdu, Ed., *Electroless Plating: Fundamentals and Applications,* American Electroplaters Society, 1990
- J.D. Keane, Ed., *Steel Structures Painting Manual,* Steel Structures Painting Council, 1989
- C.G. Munger, *Corrosion Prevention by Protective Coatings,* National Association of Corrosion Engineers, 1985
- K.H. Stern, Ed., *Metallurgical and Ceramic Protective Coatings,* Chapman and Hall, 1996
- S.M. Rossnagel, J.J. Cuomo, and W. Westwood, Ed., *Handbook of Plasma Processing Technology: Fundamentals, Etching, Deposition, and Surface Interactions,* Noyes Publications, 1990
- K.K. Shuegrat, Ed., *Handbook of Thin Film Deposition Processes and Techniques,* Noyes Publications, 1988

Testing

Mechanical Testing

- *Mechanical Testing,* Vol 8, *ASM Handbook,* ASM International, 1985
- *Welding, Brazing, and Soldering,* Vol 6, *ASM Handbook,* ASM International, 1993 (see the Section "Weldability Testing")
- *Forming and Forging,* Vol 14, *ASM Handbook,* ASM International, 1988 (see the Sections "Evaluation of Workability" and "Evaluation of Formability for Secondary [Sheet] Forming")
- *Friction, Lubrication, and Wear Technology,* Vol 18, *ASM Handbook,* ASM International, 1992 (see the Section "Micromechanical Properties Techniques")
- *Fatigue and Fracture,* Vol 19, *ASM Handbook,* ASM International, 1996 (see the Sections "Fatigue Mechanisms, Crack Growth, and Testing" and "Fracture Mechanics, Damage Tolerance, and Life Assessment")
- *Metals—Mechanical Testing; Elevated and Low-Temperature Tests; Metallography,* Vol 03.01, *Metals Test Methods and Analytical Procedures,* Section 3, *Annual Book of ASTM Standards,* ASTM, revision issued annually
- P. Han, Ed., *Tensile Testing,* ASM International, 1992
- *Hardness Testing,* ASM International, 1987
- H.E. Davis, G.E. Troxell, and G.F.W. Hauck, *The Testing of Engineering Materials,* 4th ed., McGraw-Hill, 1982

Corrosion Testing

- *Corrosion,* Vol 13, *ASM Handbook,* ASM International, 1987 (see the Section "Corrosion Testing and Evaluation")
- *Mechanical Testing,* Vol 8, *ASM Handbook,* ASM International, 1985 (see the Sections "Fatigue Testing" and "Corrosion Testing")
- *Wear and Corrosion; Metal Corrosion,* Vol 03.02, *Metals Test Methods and Analytical Procedures,* Section 3, *Annual Book of ASTM Standards,* ASTM, revision issued annually
- R. Baboian, Ed., *Corrosion Tests and Standards: Application and Interpretation,* ASTM, 1995
- W.H. Ailor, Ed., *Handbook on Corrosion Testing and Evaluation,* John Wiley & Sons, 1971

Nondestructive Testing

- *Nondestructive Evaluation and Quality Control,* Vol 17, *ASM Handbook,* ASM International, 1989

- *Nondestructive Testing,* Vol 03.03, *Metals Test Methods and Analytical Procedures,* Section 3, *Annual Book of ASTM Standards,* ASTM, revision issued annually
- L. Cartz, *Nondestructive Testing,* ASM International, 1995
- *Nondestructive Testing Handbook,* Vol 1–7, American Society for Nondestructive Testing, 1982–1990
- R. Halmshaw, *Nondestructive Testing,* Edward Arnold, 1987
- D.E. Bray and R.K. Stanley, *Nondestructive Evaluation: A Tool for Design, Manufacturing, and Service,* McGraw-Hill, 1989
- J.J. Burke and V. Weiss, Ed., *Nondestructive Evaluation of Materials,* Plenum Press, 1979

Metallography and Materials Characterization

Metallography

- *Metallography and Microstructures,* Vol 9, *ASM Handbook,* ASM International, 1985
- G.F. Vander Voort, *Metallography: Principles and Practice,* McGraw-Hill, 1984
- *Metals—Mechanical Testing; Elevated and Low-Temperature Tests; Metallography,* Vol 03.01, *Metals Test Methods and Analytical Procedures,* Section 3, *Annual Book of ASTM Standards,* ASTM, revision issued annually
- L.E. Samuels, *Metallographic Polishing by Mechanical Methods,* 3rd ed., American Society for Metals, 1982
- L.E. Samuels, *Optical Microscopy of Carbon Steels,* American Society for Metals, 1980
- G. Petzow, *Metallographic Etching,* American Society for Metals, 1978
- E. Beraha and B. Shpigler, *Color Metallography,* American Society for Metals, 1977
- J.L. McCall and W.M. Mueller, Ed., *Metallographic Specimen Preparation,* Plenum Press, 1974
- V.A. Philips, *Modern Metallographic Techniques and Their Applications,* Interscience Publishers, a division of John Wiley & Sons, 1971

Materials Characterization

- *Materials Characterization,* Vol 10, *ASM Handbook,* ASM International, 1986
- *Surface Engineering,* Vol 5, *ASM Handbook,* ASM International, 1994 (see the Section" Testing and Characterization of Coatings and Thin Films")
- *Friction, Lubrication, and Wear Technology,* Vol 18, *ASM Handbook,* ASM International, 1992 (see the Section "Laboratory Characterization Techniques")
- *Analytical Chemistry for Metals, Ores and Related Materials (I),* Vol 03.05, *Metals Test Methods and Analytical Procedures,* Section 3, *Annual Book of ASTM Standards,* ASTM, revision issued annually
- *Analytical Chemistry for Metals, Ores and Related Materials (II),* Vol 03.06, *Metals Test Methods and Analytical Procedures,* Section 3, *Annual Book of ASTM Standards,* ASTM, revision issued annually
- B.D. Cullity, *Elements of X-Ray Diffraction,* 2nd ed., Addison Wesley, 1979
- D.C. Joy, A.D. Romig, Jr., and J.I. Goldstein, Ed., *Analytical Electron Microscopy,* Plenum Press, 1986
- L.E. Murr, *Electron Optical Applications in Materials Science,* McGraw-Hill, 1970
- B.L. Gabriel, *SEM: A User's Manual for Materials Science,* American Society for Metals, 1985
- J.I. Goldstein, D.E. Newbury, P. Echlin, D.C. Joy, C. Fiori, and E. Lifshin, *Scanning Electron Microscopy and X-Ray Microanalysis,* Plenum Press, 1981
- A.W. Czanderna, Ed., *Methods of Surface Analysis,* Elsevier, 1975

- V. Thomsen, *Modern Spectrochemical Analysis of Metals: An Introduction for Users of Arc/Spark Instrumentation,* ASM International, 1996

General Engineering Reference

- H.E. Boyer and T.L. Gall, Ed., *Metals Handbook Desk Edition,* American Society for Metals, 1984 (revised Volume to be published in 1998)
- M.M. Gauthier, Ed., *Engineered Materials Handbook Desk Edition,* ASM International, 1995
- H. Baker, Ed., *Alloy Phase Diagrams,* Vol 3, *ASM Handbook,* ASM International, 1992
- J.R. Davis, Ed., *ASM Materials Engineering Dictionary,* ASM International, 1992
- S.P. Parker, Ed., *McGraw-Hill Dictionary of Scientific and Technical Terms,* 4th ed., McGraw-Hill, 1989
- J. Frick, Ed., *Woldman's Engineering Alloys,* 8th ed., ASM International, 1994
- M.L. Bauccio, Ed., *ASM Metals Reference Book,* 3rd ed., ASM International, 1993
- M.L. Bauccio, Ed., *ASM Engineered Materials Reference Book,* 2nd ed., ASM International, 1994
- *Worldwide Guide to Equivalent Irons and Steels,* 3rd ed., ASM International, 1992
- W. Mack, Ed., *Worldwide Guide to Equivalent Nonferrous Metals and Alloys,* ASM International, 1995
- A.S. Melilli, Ed., *Comparative World Steel Standards,* ASTM, 1996
- C.W. Wegst, Ed., *Stahlschlüssel* (Key to Steel), 17th ed., Verlag Stahlschlüssel, 1995
- D.L. Potts and J.G. Gensure, *International Metallic Materials Cross-Reference,* Genium Publishing, 1989
- *Metals and Alloys in the Unified Numbering System,* 7th ed., SAE International, 1996